Sergio Petrozzi

Practical Instrumental Analysis

Related Titles

Prichard, E., Barwick, V.

Quality Assurance in Analytical Chemistry

2007
Hardcover
ISBN: 978-0-470-01203-1

Funk, W., Dammann, V., Donnevert, G.

Quality Assurance in Analytical Chemistry

Applications in Environmental, Food and Materials Analysis, Biotechnology and Medical Engineering

2007
Hardcover
ISBN: 978-3-527-31114-9

Ratliff, T. A.

The Laboratory Quality Assurance System

A Manual of Quality Procedures and Forms

2005
E-Book
ISBN: 978-0-471-72166-6

Kellner, R., Mermet, J.-M., Otto, M., Valcarcel, M., Widmer, H. M. (eds.)

Analytical Chemistry

A Modern Approach to Analytical Science

2004
Hardcover
ISBN: 978-3-527-30590-2

Miller, J. M., Crowther, J. B. (eds.)

Analytical Chemistry in a GMP Environment

A Practical Guide

2000
Hardcover
ISBN 978-0-471-31431-8

Strobel, H. A., Heineman, W. R.

Chemical Instrumentation

A Systematic Approach

1989
Hardcover
ISBN: 978-0-471-61223-0

Sergio Petrozzi

Practical Instrumental Analysis

Methods, Quality Assurance and Laboratory Management

WILEY-VCH Verlag GmbH & Co. KGaA

The Author

Sergio Petrozzi
Zurich University of Applied Sciences
Institute of Chemistry and Biological Chemistry
Einsiedlerstrasse 31
8820 Waedenswil
Switzerland

Library of Congress Card No.: applied for

British Library Cataloguing-in-Publication Data
A catalogue record for this book is available from the British Library.

Bibliographic information published by the Deutsche Nationalbibliothek
The Deutsche Nationalbibliothek lists this publication in the Deutsche Nationalbibliografie; detailed bibliographic data are available on the Internet at <http://dnb.d-nb.de>.

Cover Design Adam-Design, Weinheim
Typesetting Toppan Best-set Premedia Limited, Hong Kong
Printing and Binding Markono Print Media Pte Ltd, Singapore

Print ISBN: 978-3-527-32951-9
ePDF ISBN: 978-3-527-66592-1
ePub ISBN: 978-3-527-66591-4
mobi ISBN: 978-3-527-66590-7
oBook ISBN: 978-3-527-66589-1

Printed in Singapore
Printed on acid-free paper

Contents

Preface to the German Edition

An integral part of the technical–chemical training of a student is the teaching laboratory in analytical chemistry. Analytical chemistry is a partial discipline that develops and applies methods, technologies and strategies in order to gain information on chemical compounds and processes. Principles of organic, inorganic and biological chemistry are supported by analytics, and the ever-increasing demand of society to make important decisions based on watertight data and validated methods requires the reinforcement of analytical education and research.

Following several years of being a teaching adviser in a university chemistry laboratory, I concerned myself with the validation of methods in environmental analytics in my capacity as an application chemist for a manufacturer of analytical equipment. In so doing I became aware of the different priorities placed on analytical processes in these various organizations.

In an academic environment the analytical chemistry laboratory serves, by way of experiments, to deepen what has been learned, to consolidate knowledge and to help apply it in a practical manner. The teaching laboratory complements the theoretical knowledge and understanding of analytical procedures learned during lectures. It serves to convey the knowledge and facilitate working techniques and practical skills through exercises. The lab courses are meant to consolidate the careful and independent execution of one's own experiments, to train the ability to observe, and to encourage professional scientific work. Nowadays, the reports written after each experiment are evaluated with regard to the formulation, validation and agreement of hypotheses. Often the basis for this is measurement results achieved in an academic environment. Perhaps even aspects of quality assurance are taken into consideration in the evaluation process. This all happens at educational institutions on a more or less voluntary basis and seldom is there the necessity to apply international guidelines or norms.

In analytical laboratories in chemical industry the quality management system consists of aspects of corporate policy, ergonomics, measurement technology, product assessment and electronic data processing, and is certified by quality audits. Such extensive quality assurance measures are conducted for ecological and economic reasons. The maintenance of product quality, and thus of competitiveness, obliges the companies to comply with extensive legal requirements. The work to be carried out within the framework of quality assurance is not only necessary to guarantee correct and accurate analytical values, but can be absolutely necessary for Good Laboratory Practice (GLP)-compliant laboratories or laboratories that want to be accredited, and are, at the same time, critical for the survival of corporations.

After destiny led me back to the university, I decided to write this book to sensitize students to the issue of quality assurance. Analytical validation is a topic that has become increasingly important for practice over the past years as a result of increased efforts in quality assurance, and which will also continue to play an important role in the future. Validations are an integral part of analytical processes and are a part of the repertoire of chemists who work in analytics, such as study of the literature, experimental work and reporting. Reports of experiments run during lab classes are intended to illustrate theoretical explanations and have to contain experimental–methodological information about quality assurance measures. There is a growing need to take account of the statistical evaluation of the results. All validation steps, from the analytical question to the evaluation of the results, are involved in the analytical process of a lab experiment and described together with plausibility assessments. What is to be conveyed to the student is the fact that analysts, in their capacity as problem solvers, perform a service for certain groups of customers. Based on information supported by the strategy, structure and culture of the customer, analysts apply their techniques and broad knowledge. The results must be precise and exact to be used by the customers for their purposes. Students must be made aware of aspects such as personnel costs, equipment, methods and improvements of the analytical process achieved by rationalization, increased competition and customer satisfaction. The solution to the problem should in any case be processed in such a way as to be “fit for purpose”. Moreover, analysts must be able to communicate clearly complex workflows and connections as well as results to customers in management, marketing or legal departments who are not versed in the subject.

There are several good books on analytical chemistry, but most of them go into such detail on the theory that the relationship to the practical implementation is diluted. The standard procedure for finding solutions to analytical problems outlined here should be a model for contemporary and practical work in analytical laboratories. It has been consciously kept general to be applicable to a wide diversity of problems and customer specifications. In the modern-day laboratory practice there is quite a series of variations and modifications of the procedural model presented in this book. However, everyone is familiar with the phased approach, from the order entry to the end of the project, which is also an essential part of the workflow presented in this book.

In view of the fact that approximately eighty percent of analysts will work in industry in future, I regard it as crucial to illustrate clearly to students the role of analytical chemists in the industrial environment. In this environment analysts are engaged as problem solvers and will be confronted with corporate requirements, such as time, costs and precision. Students will learn that there are no general values for the acceptance of measurement insecurity, selectivity or quality in an analytical process, but rather that the values must be defined prior to starting the project. The results achieved must suffice with regard to the purpose and depend on the legal goals and limitations of the customer. The selection and degree of optimization of the methods, as well as the scope of validation and documentation, are reflected in the order and its execution.

This book is intended to serve as a practical support for students and as an aid to internship supervisors. It is meant as a guideline and attempts to be an understandable course book on the present-day status of practice. The trials run during laboratory classes and described here are formulated as projects and are meant to be carried out by the students in teamwork. In so doing, not only will specialist knowledge be deepened, but one will also learn to make decisions in a broader context. Analytical problems vary from trial to trial. However, the solution process is retained in the examples given during the internship and consists of the following:

- understanding of the chemistry and physics on which the analytical method is based
- knowledge of the analytical processes from the order entry to reporting
- awareness that analytical work is service

Wädenswil, Switzerland, October 2009 *Sergio Petrozzi*

Preface to the English Edition

The German edition of this book aroused lively interest, and this has incited me to write an extended version for translation into English.

The range of analytical methods introduced has been substantially enlarged and the number of the laboratory experiments has also been extended. The description of the new experiments also follows the proven standard and each starts with a small introduction (real presentation of problems involving different analyses) to the background and a short introduction to the procedure chosen by the analyst.

Translated from the originally published edition in German under the title *Instrumentelle Analytik* by Mrs. Maria Schmitz.

In her capacity as a freelance lecturer for the English Department of the University of Applied Sciences in Mainz, Economic Faculty, Mrs. Schmitz lectures on Intercultural Competence – German and American Business Styles. In addition, she lectures in the Communication Skills Faculty of the private academy accadis Bildung.

Zürich, Switzerland, May 2012 *Sergio Petrozzi*

Dedication

My sincere thanks go to Prof. Dr. Georg Schwedt for the foreword to this book and for giving me the opportunity to share his rich experience.

I owe a special thanks to Dr. Huldrych Egli (Büchi Labortechnik AG, Flawil/CH) for working out the project "*N*-Protein Determination according to Kjeldahl".

For his contribution to the project Field-Flow Fractionation (FFF) I thank Dr. Tino Otte (Postnova Analytics GmbH, Landsberg/DE).

I am indebted to Dr. Oliver Mandal (Büchi Labortechnik AG, Flawil/CH) for his NIR project contribution.

For his contribution to the project GC-MS I would like to thank Lucio D'Ambrosio.

A hearty thanks to Dr. Martin Brändle (Chemistry Biology Pharmacy Information Center, ETHZ) for the section "Literature and Database Research" and Appendix A.

I would like to convey my appreciation to Dr. Markus Juza (DSM Nutritional Products, CH), Dr. Sebastian Opitz and Dr. Marc Bornand (Zurich University of Applied Sciences, ZHAW), and Dr. Thomas Schmid (Laboratory of Organic Chemistry, ETHZ) for their proofreading of selected chapters.

I wish to thank Dr. Markus Zingg (ZHAW) for his critical scrutiny of the manuscript, Nick Bell (Language Services, ZHAW) for linguistic proofreading and Roland Kaulbars (Shimadzu GmbH, CH) for his support with the MS projects.

Thanks to Dr. Nestor Pfammatter (Safety, Security, Health and Environment, ETHZ) for his permission to implement the "Safety Manual" worked out for the ETH Zürich.

For their permission to use passages from the scripts of their lecture or laboratory course documentation my thanks go to

- Prof. Dr. Gustav Peter (TWI Ingenieurschule Winterthur, CH)
- Prof. Dr. Eduard Gamp (ZHAW)
- Prof. Dr. Renato Zenobi and Dr. Thomas Schmid (Laboratory of Organic Chemistry-ETHZ, Analytical Chemistry Laboratory)
- Prof. Thomas M. Lüthi (ZHAW)

The following projects were worked out thanks to valuable support and the willingness to provide company documents such as training material, manuals, operating instructions and monographs:

- High Performance Liquid Chromatography Coupled with Mass Selective Detection – LC/MS (Agilent Technologies Sales & Service GmbH & CoKG, Waldbronn/DE and Shimadzu Schweiz GmbH/CH)
- High-Performance Thin Layer Chromatography – HPTLC (CAMAG Laboratory, Muttenz/CH)
- Near-Infrared – NIR-Spectrometry (Büchi Labortechnik AG, Flawil/CH)
- Atomic Absorption Spectrometry – AAS (PerkinElmer AG, Schwerzenbach/CH)
- Karl Fischer Water Determination and Ion Chromatography (Metrohm Schweiz AG, Herisau/CH)
- *N*-Protein Determination according to Kjeldahl (Büchi Labortechnik AG, Flawil/CH)
- Field-Flow Fractionation – FFF (Postnova Analytics GmbH, Landsberg/DE)

Finally, I would like to thank everyone whom I have forgotten. Rest assured that this was not intentional.

Foreword

Results from the application of instrumental analytical methods today often form the basis for political, legal and medical decisions that pertain not only to the recycling and maintenance of the quality of air, water, the earth or food, but also to the overall "quality of life". In the medical field, particularly biochemical analytics and pharmaceutical research depend on instrumental analytics. In the field of material sciences and in technical subjects, knowledge of the content of trace elements is, for example, an important prerequisite for ascertaining physical properties. Even such diverse sciences as geology, archeology and the restoration of works of art and antique books make use of analytical methods to solve the problems of their respective fields. In fact, there is hardly an area of experimental natural sciences that does not make use of chemical analytics in some form or another.

This makes it all the more important for students to become acquainted with the practice and problem orientation of instrumental analytics. That means they must learn to select a method suitable to a given task. Analytical procedure, in general, consists of several phases, ranging from sampling and sample preparation to the evaluation and assessment of the results of the analysis. In so doing, planning and assessing analytical data with regard to their reliability and accuracy (in other words the validation of an analytical procedure) plays a very important role. For students it is therefore necessary to learn not only the theoretical aspect of chosen analytical methods and applications, but also to practise how to perform the necessary steps to solve analytical tasks efficiently. This laboratory course book can be an essential contribution to such a procedure.

Bonn, Germany, June 2009 *Prof. Dr. Georg Schwedt*

1 Introduction

1.1 Analytical Chemistry – The History

Already in ancient Egypt, knowledge of chemistry existed and was used when embalming pharaohs and dignitaries. Greek philosophers such as Plato (428–347 BC), Aristotle (384–322 BC) or Empedocles began to look for rules to explain natural phenomena. Plato believed that diverse atoms could be differentiated by their constitution. According to this theory, the atoms of an element could be changed into those of another element by simply modifying the form. *Aristotle* postulated that all elements and the substances formed by them were composed of a kind of original substance. This original substance, however, could, in the course of time, take on many forms, such as shape and color. Empedocles, who lived sometime between 490 and 430 B.C., explained that all substances were composed of four elements, namely fire, earth, water and air.

Ever since the beginning of alchemy in the Middle Ages, men have sought for a material with which to best convert metal fastest and most simply and that could be exploited best. That was the stone of wisdom. The ability to illustrate this was generally regarded as an Act of God's mercy, even if someone owned a functioning specification, it would have been useless without God's intervention. The Phlogiston Theory was introduced by the German physician and chemist Georg Ernst Stahl (1659–1734) in 1697. According to this theory, all flammable substances contained Phlogiston (gr. phlox, the flame). When it was burned and/or oxidized the flame escaped as a gas-shaped something. Phlogiston was a hypothetical substance that could be used without having necessarily to be proven. This assertion was applied to all appearances of fire in nature and ruled the thoughts of chemists for almost one hundred years. A substance would burn more readily, the more Phlogiston it had.

The spiritual aspect went hand in hand with the scientific aspect. Aristotle added a fifth, supernatural element to these four. The quintessence as the most inner core of all substances and to which a sustaining and healing power was attributable. Quintessences were gained by extraction, that is, by separating all ineffective or unpurified ingredient. These were material essences which were the sum of a body's own effective powers and/or qualities. The idea of a microcosm and a

Practical Instrumental Analysis: Methods, Quality Assurance and Laboratory Management, First Edition.
Sergio Petrozzi.

macrocosm stated that everything that occurred in the universe (macrocosm) had its correspondence and effect on earth (microcosm). As early as in Babylonian astronomy the planets were linked to certain materials (e.g., moon–silver, sun–gold). The constellation of the planets was important for chemical reactions to be successful.

The Renaissance witnessed the birth of attempts to renew chemistry. People wanted to get rid of everything, one by one, which was not rationally explicable. The making of gold and the associated magic, astrology and magic methods could not be reconciled with chemistry which was grounded in insights and based on reason. More and more chemists turned their backs on alchemy and finally began to fight against it. The chemists upheld research and the critical ability to think, reason, as the highest judge of the truth of a theory. In parallel to chemistry, analytical chemistry developed its own experimental skills. The first quantitative determination in chemistry was conducted by A.L. Lavoisier (1743–1794) and as early as the nineteenth century, analytical chemistry had become an established branch of chemistry. In a book published in 1894, W. Ostwald described "The Scientific Foundations of Analytical Chemistry". In this book he introduced dissociation constants, solubility products, ion products, water ion products and indicator equilibrium into analytical chemistry.

Analytics today determines the success of science, technology and medicine and is an interdisciplinary field. At the beginning of this development, analytical investigations were limited to the composition of substances and/or of mixed substances with regard to their main components. At the same time the need for generalization of analytical methods, not only on the basis of the theory of chemical reactions, but also on the basis of the physical theory of the structure of atoms and molecules arose. Later, procedures were developed to analyze trace amounts of an element or a chemical compound in a mixture. Determination of the structure of molecules and investigation of the structure of solids also became important fields for the analyst.

1.2 Analytical Chemistry and Its Role in Today's Society

Analytics is an interdisciplinary, scientific discipline also termed "analytical chemistry". The terms quantity and quality owe their existence to the results of analytics. Analytical issues are all-pervasive, and by no means only a part of scientific discipline. Rather, analytics often has a predominant role in the industrial value chain. Increasingly, more quality characteristics are being allotted to products and processes, which increasingly correspond to the need for analytics in all areas of life. Our society is demanding analytically secured data and judgments instead of empirical or traditional foundations for general or industrial decision-making. In this manner medical diagnostics, for example, is being shaped more and more by methods of analytical and bioanalytical chemistry. Buzzwords such as food security or water contamination, greenhouse gases or doping tests, gene analysis or certification of genuineness are visibly tied to the performance of analytical chemistry, visible for every citizen. Good analytics create trust and thus are a prerequisite for production and marketing.

Responsible political and economic decisions have long been based on ecological insights, that is, findings based on environmental analytics. The concept of sustainability will be even more important in future, and this draws on analytical competence even more than any concept of human action has ever done. In a nutshell: More and more ideological, medical, legal and economic decision-making rests on analytical data. This applies both to the governmental control of such areas as health, the environment, security and resources, and to the control of trade and economic processes. Similarly, the decisive spurt in the development of high technology (microchips, high-tensile materials, medical diagnostics) is always based on highly developed analytics. Increasing globalization and the progressive growing together of the countries of Europe have reduced trade barriers at their borders. The goal is to facilitate a free and unprohibited exchange of products and services. With this development, it is becoming more necessary than ever, however, to make the quality of goods transparent, because only supplementary information on the composition, purity or reliability of goods makes them salable commodities.

In order to guarantee uniform procedures across national borders when collecting analytical information, international guidelines, such as Good Laboratory Practices (GLP), Good Manufacturing Practices (GMP) or standards for good analytical work (e.g., International Organization of Standardization – ISO 17025) have been introduced.

These aspects make it clear that analytics is of fundamental significance and that this trend will continue. Analytics is the common task of several partners including universities, industry, analytics laboratories, the equipment industry and the authorities. Thus Europe will in future increasingly need the respective analysts, laboratories, educational and research institutions, more than ever before.

Only 40% of German universities have specialist analytical departments in the faculty of chemistry. This is the result of a study by the Society of German Chemists (SGCh). In some 50% of these it is tied to the subject "inorganic chemistry" since, traditionally, beginners in chemistry were introduced to the subject by way of simple analytical laboratory tasks. A cross-discipline like analytical chemistry, with increasing research roles in the whole area of material sciences, food science and medicine suffers from such a wrong allotment or subordination.

The concept of sustainability, which also takes a central position in the code of conduct of the GDCh (https://www.gdch.de/home.html, accessed May 2012), requires an extended concept of education: Beyond the difficult issue of the sciences the education must take into consideration the consequences of insights and their material implementation. Herein lies one of the most demanding tasks of the universities that have to convey high chemical analytical understanding. The specialist areas/faculties of chemistry and the university management are called upon to reinforce analytical chemistry in the further and new development of the curricula and to ensure and make use of their interdisciplinary function. Research promotion should clearly set priorities. Chemical analytical research is dependent on taking a leading position in methodology development and in giving preference to the application of expressly demanding issues. Only a quality-oriented promotion of research in stable interdisciplinary research structures of analytics can create the prerequisites that pave the way for industry (equipment manufacturers and users) in increasingly more areas to successfully offer automated or reliable practicable methods, and to market and use these globally.

It is therefore in the long-term interest of universities, politicians and the economy to firmly establish strong structures of education and research in analytical chemistry.

> Excerpt from: Society of German Chemists (GDCh) – Memorandum Analytics 2003.
>
> The Gesellschaft Deutscher Chemiker (GDCh) is the largest chemical society in continental Europe with members from academe, education, industry and other areas. The GDCh supports chemistry in teaching, research and application and promotes the understanding of chemistry in the public.

I'm not afraid of storms, for I'm learning to sail my ship.
Aeschylus

2
Introduction to Quality Management

Quality. There is hardly another term mentioned more often in connection with products and services. Entire branches thrive from "selling" quality or from supporting corporations in their quality efforts. However, what is actually the essence of quality? Where does quality come from and how has the term changed over the centuries?

General definition:
Quality is understood to be synonymous with high value. It is not measurable, but rather it can merely be grasped (subjective term) by experience. Unsuited to corporate practice.

Product-related definition:
Quality is interpreted as being measurable. It becomes an objective characteristic, whereby subjective criteria are eliminated. (Example: The larger the cucumbers the more valuable, the better the quality)

User-oriented definition:
Quality only exists from the point of view of the user, that is, the customer.

Process-related definition:
Quality is equated with compliance with specifications. There exists a zero-defect policy (do it right the first time). (Example: Punctuality of a means of transport, a zero-defect product)

Value-related definition:
Taking into consideration the cost and/or price of a service, quality means a favorable price/performance ratio.

The consequence for corporate practice is that the various functional areas in a corporation develop varying opinions of just what quality is.

2.1
Historical Background

The pre-industrial technical production (as opposed to agrarian production) was handicraft production, whereby the so-called system of guilds played a decisive

Practical Instrumental Analysis: Methods, Quality Assurance and Laboratory Management, First Edition.
Sergio Petrozzi.

role from an organizational standpoint. The guilds determined the manufacturing procedures, tool types, tool use and even production quantities. On the one hand, they gave the craftsman social stability and ensured his company. On the other hand, their rigidity impeded innovation, technical progress and the expansion of production.

The link between the production of the craftsman according to the guild system and modern industrial production according to the factory system was the already more strongly centralized manufacturing system. They then continued to process the raw material on their own, or under their own control in centralized production sites.

With the industrial revolution in Great Britain, that reached its zenith between 1780 and 1820, technology changed drastically, as new machines were provided (for example, Watt's steam engine).

The transition of the production method from individual producers to the factory first took place in the English wool-producing textile industry from about 1800 to 1820. Technically the era was characterized by the rapid transition from hydroelectricity to the steam engine, as the driving force for the new textile machines which led the mechanization process. What was also characteristic was the system by which work was divided, that is, the division of production into individual work steps and the distribution to hundreds and thousands of workers. Industrial work was now finally concentrated in one place, the factory and synchronized with the running machines. The economic consequences of the factory system were mass production, the fall in the price of industrial products and a distribution and marketing system that catered to mass-produced products.

The corresponding development and spread of precision tool machines, as existed in the USA, were indispensable prerequisites. In the 1890s the automation of work processes already existed with the use of these machines.

What was also of great significance was the rationalization of the workflow under the influence of the concept of scientific management developed in 1895 by the American production engineer Frederick Winslow Taylor (http://www.skymark.com/resources/leaders/taylor.asp, accessed May 2012).

This involved, for example, breaking down the production process into calculable elements and recording and eliminating redundant movements and hidden breaks.

Mass production of goods also induced a different understanding of quality. It was largely unclear why products did not have the required quality, although their production required only a few twists. The production processes were not predictable and people worked feverishly to control them. Walter Shewhart played a key role here (http://www.skymark.com/resources/leaders/shewart.asp, accessed May 2012).

2.2 Variability

All systems and processes demonstrate a certain variability. This variability is typical of each process. No two things are exactly identical. Even if they appear to be identical at a first glance, they turn out to have differences upon closer inspection that had been hidden from the observer. When we become interested in quality, we must understand where this variability comes from and how it influences the process or the work results, respectively. Shewhart, one of the early pioneers of quality management, divides variability into two categories according to its origin:

- systematic influences (*common causes*), and
- random influences (*special causes*)

For example: Whenever you write your signature ten times, you realize that no two signatures are exactly the same. Not even when you make the greatest effort. The deviations are predictable and move within certain limits (common cause). Should someone bump into you while you are signing a document, the signature seems totally different. You experience an extraordinary dispersion (special cause). This leads us to deduce that when extraordinary dispersions occur, the causes of the deviations must be immediately investigated, in order to re-establish the controllability of the process.

Accidental influences are numerous and mostly minimal in their impact. They are always present and have random (unpredictable) effects on the process and the work results.

Shewhart noticed that these random influences obeyed simply statistical laws.

As a result, he developed a concept that found its way into the literature as statistical process control (SPC) and that has been developed further by several people in quality management. According to Shewhart: *"A phenomenon is regarded as controlled, when future behaviour can be forecast by experience at least in a limited area. A prediction within certain limits means that the likelihood that the phenomenon will move within certain limits in future can be estimated"* [1].

A quality control chart (Figure 2.1a) is a form on which the test results of sample tests are graphically illustrated in a time sequence. Test values determined from the continuous production process indicate the current quality level.

To have a record of quality, control chart samples are taken at prescribed intervals. The indicators are calculated from the sample and recorded on the quality control chart.

Previously calculated warning (upper limits/lower limits) and intervention limits (upper intervention limits/lower intervention limits) represent limit values which signal an impermissible worsening of quality whenever the limits exceed/fall short of these values.

If the intervention limits are exceeded, the causes must be analyzed and suitable corrective action taken (Figure 2.1b).

A process is regarded as controlled when the changes caused by such a process are effected exclusively within the control limits (Figure 2.1c).

The consistent, relative changes in the variables of interest are defined as a trend. In order to recognize trends by virtue of their measurement values, trend rules are defined, for example.

(a)

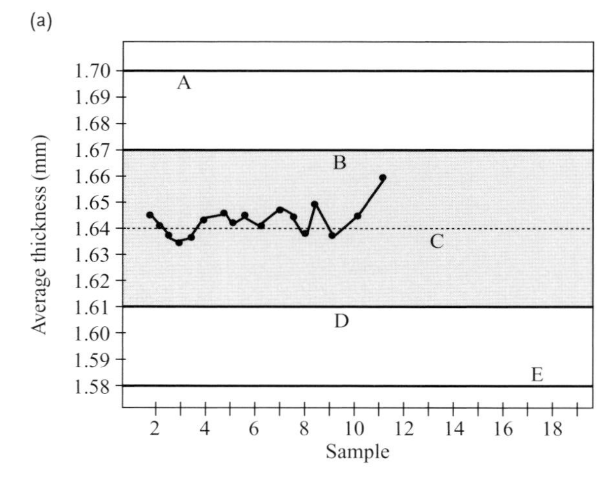

(b)

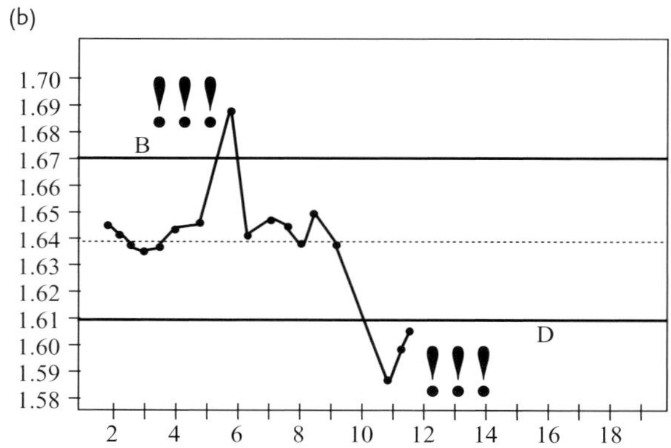

(c)

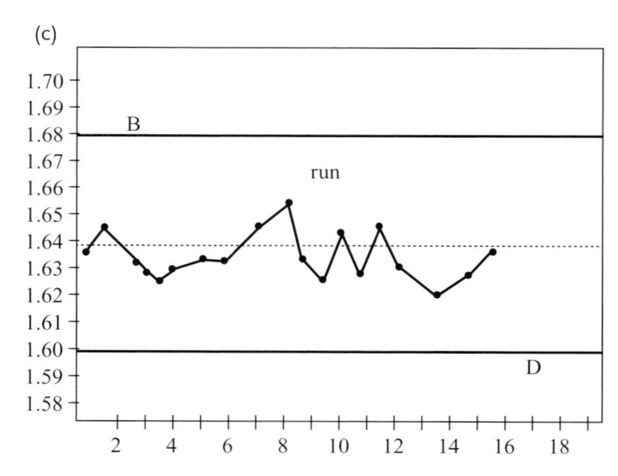

Figure 2.1 (a) Example of a quality control chart (**A** upper limit value, **B** upper intervention limit, **C** process mean, **D** lower intervention limit, **E** lower limit value), (b) intervention limits exceeded, (c) controlled process, (d) trend behavior.

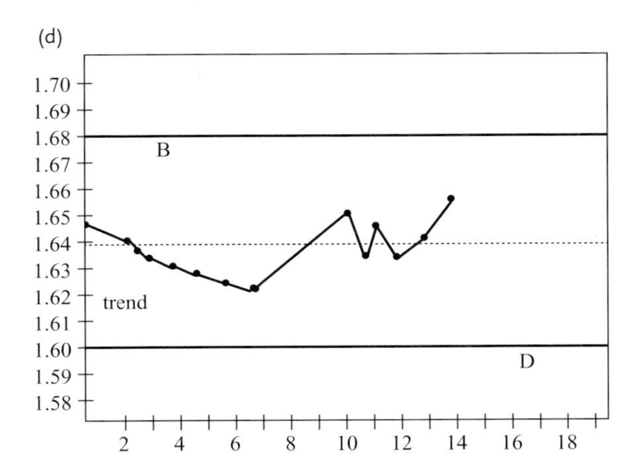

Figure 2.1 (Continued)

- 7 consecutive values are above the mean
- 7 consecutive values are below the mean
- 7 consecutive values demonstrate a falling tendency
- 7 consecutive values demonstrate a rising tendency

One of the first steps towards improving the process is to find out whether dispersion is systematic or accidental because only when all systematically conditioned dispersions have been eliminated can attempts be made to reduce the accidental dispersion.

In the case of a systematic dispersion, the process is said to be out of control. In the case of accidental dispersion, the process is said to be stable. Only in this case can it be expected that the points will always lie once on one side and once on the other side of the process mean. As long as this is the case, the process remains predictable with regard to consistent quality, productivity and costs. As a consequence, statistical process monitoring becomes an integral part of corporate management.

A further merit of Shewhart was the development of the so-called Plan–Do–Check–Act (PDCA)-control circuit, a problem-solving technique that allows systematic proceedings in uncontrolled processes.

This circuit consists of four consecutive steps:

With consistent use, this control circuit leads to continuous improvement as a basic model. As a result, this circuit has been used and propagated by several authors. In the literature it is variously called the Deming circuit, Shewhart circuit or Ishikawa circuit (http://www.skymark.com/resources/leaders/deming.asp, http://www.skymark.com/resources/leaders/ishikawa.asp, accessed May 2012).

2.3
The Four Pillars of Wisdom (from Shewhart to Deming)

William Edwards Deming was a scholar of Shewhart. Through close contact with Shewart, Deming had already learnt about and understood the principle of variation in manufacturing products during the early years of his career. After the Second World War, Deming was invited to Japan, which had a ruined economy, to demonstrate his basic model of continuous improvement, based on Shewart's PDCA circuit.

Deming was the first to understand the organization as a process and depict it accordingly. He understands the system as a quantity of elements, parts of an organization, function, duties, which together work towards achieving the goal of the entire system. There must be a goal. The aim of the organization must be known, because without an essential aim, there is no system. The goal of a system brings profit for everyone, for shareholders, employees, suppliers, customers, the society, the environment and the economy. The goal of knowledge management and quality management is the customer orientation, whereby the term "customer" refers to both external and internal customers. System optimization is the common goal.

Deming's teachings are presented in the literature as *"The System of Comprehensive Knowledge"*. The system is composed of four main components:

- Understanding the system
- Understanding dispersion (in the meaning of the work published by Shewhart)
- Understanding the theory of knowledge (How do we learn? How do we improve ourselves?)
- Understanding psychology and human nature.

2.4
Zero-Defect Tolerance

The American Philip Crosby (http://www.skymark.com/resources/leaders/crosby.asp, accessed May 2012) developed the Zero Defects Concept, which was aimed at eliminating deviations in production without scrap and rework. His insight that in corporations several practices exist that run counter to this goal, such as, for example, determining deviation quotas, accepting rework as being unavoidable, and the attitude that manufacturing quality produces costs, led him to demand a corporation-wide transition based on four principles:

- Quality does not mean excellence but rather meeting the specifications (conformity).
- Quality does not arise by discovering deviations, but by avoiding them.
- The aspired quality standards are not acceptable quality levels, but zero defects.
- Quality is not measured by indices, but by the price of deviation.

2.5
Why Standards?

In order to guarantee that milestones and delivery parts, respectively, comply with the finished project and finished product, respectively, a common language had to be invented to ensure that misunderstandings, false interpretations of regulations and so on were not the cause of failures. Many things had to be standardized. Thus the first "norms" originated.

Norms ensure that one thing fits the other, be it as a market ordering instrument for the elimination of trade barriers, for alleviating routine tasks or guaranteeing safety and health by defining the status of technology for products and services.

Norms are not passed "top down", but are made precisely by those who need them: the economy, consumers, administration and science. Their representatives invest time and know-how in creating norms – in their own interest, and also in the interest of the public at large.

Norms are – put simply – rules of technology. The demand rationalization, facilitates quality assurance, serves workplace safety, and standardizes test methods, for example, in environmental protection, and facilitates in general the communication between the economy, technology, science, administration and public life, to name just a few examples.

2.6
The Controlled Process

Basically, a process describes a workflow and pursues a superior goal. Characteristically, a process is distinguished by the fact that it supersedes functional, organizational or personal limits.

Someone is made responsible for each process – the "process owner". One person can be responsible for several process steps.

The person responsible for the individual process step must be able to assume that the input received complies with the quality specifications. This also applies to the next step. That means, therefore, that the owner's responsibility lies in letting the small process run as required and guaranteeing the next owner the required input.

Clear instructions must be given for each process step. Only in this manner is the responsible person in a position to fulfill their duties. The instructions must, as far as possible, be given in such a way as to enable the online monitoring of the process step.

Should the process get out of control, that is, should certain limits be exceeded, the person in charge must react and bring the process back into line using continuous improvement methods (for example, the PDCA control circuit, Figure 2.2).

There are, therefore, two essential measures to point out:

- Limits which, when exceeded, establish that the process or the process step is out of control and no longer guarantees product quality.

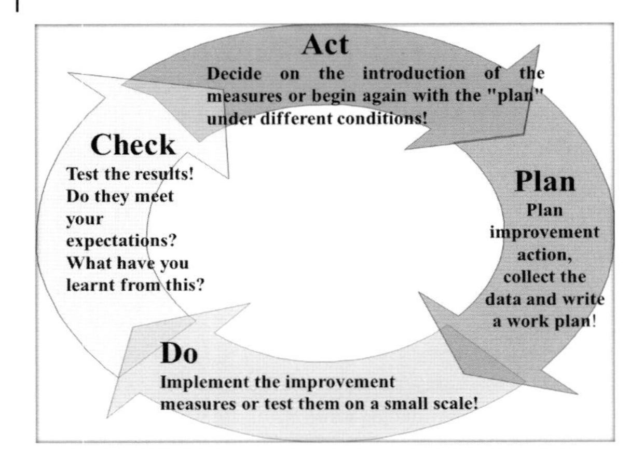

Figure 2.2 PDCA-control circuit.

- Specifications that tell the person in charge how to react in the case of exceeding the limits.

The measures to be taken can vary widely:

- interrupt the process
- call the superior
- change machine fittings
- and so on.

Disturbances in the process or process step are all those factors that can cause the process to get out of control. Knowledge of these disturbances is of vital significance for controlling the process. Awareness of these factors permits monitoring, the specification of limits and the definition of measures. Knowledge of the disturbances and their possible effects is processed in cause and effect, and danger and risk analyses.

Important instruments for recording and evaluating deviations can be any of the following:

- The Ishikawa or fishbone diagram (Figure 2.3)
- Hazard analysis and critical control point (HACCP)
- Failure mode and effect analysis (FMEA)

2.7 ISO Guidelines 9004

ISO 9004:2009 provides guidance to organizations to support the achievement of sustained success by a quality management approach. It is applicable to any organization, regardless of size, type and activity. The standard is based on quality management (QM) principles, which promote understanding of QM and its use

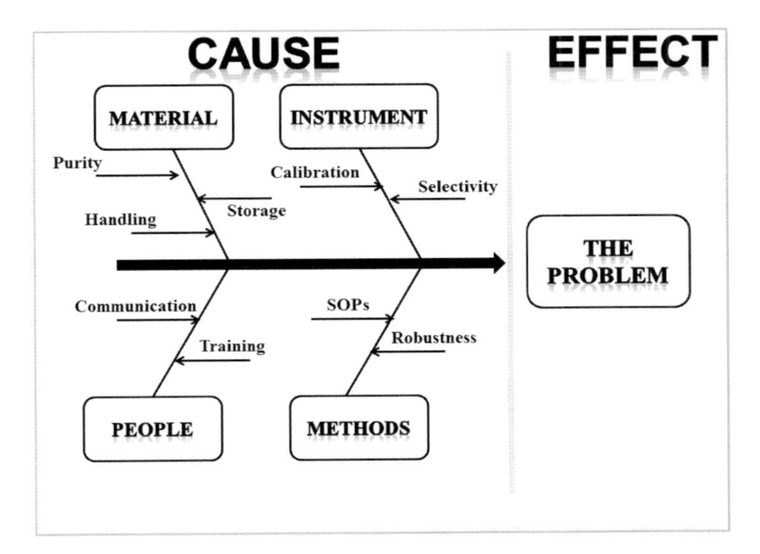

Figure 2.3 Ishikawa diagram.

to improve the management of an organization (company). The standard is not a guideline for implementing the ISO 9001 norm. Nor is it meant for certification. The primary element is to attain customer satisfaction based on an organization that not only seeks to be effective, but also to be efficient. The entire standard observes the quality of the whole organization from an entrepreneurial standpoint. Apart from the classical customers who buy the products, other interested parties are the investors, employees, suppliers and society.

ISO 9004:2000 contains a section dedicated to self-evaluation of a company that has implemented a quality management system. Even if this is still just a guide, some companies may find that guide useful, especially when limited experience exists. The questions set in the guide are mainly general, but may be detailed according to the specifics of the company. The performance process consists of activities performed to get a better understanding and control of application performance. A maturity level is a well-defined evolutionary plateau towards achieving a mature process. Five levels of maturity are identified:

- Level 1

  No formal approach. Companies with quality management systems which have this level have unpredictable results, and are characterized also by lack of evidence regarding quality. They run a business in a very unpredictable way, rather than managing the business.

- Level 2

  Reactive approach. Usually, companies that fall into this level have a system based on corrections for solving problems. From time to time, minimum data regarding the results of improvements are taken into consideration. For them, the rule "make and correct" is a usual style of life, neglecting the understanding of what they are doing or how capable they are. Production costs in such companies are high and also, in many situations, there is no or poor vertical communication. Employees are seen as simple machines, and the superiority of managers over them is similar to that of the human race over monkeys. Unless changes occur in such companies, in order to improve things, most will end up closing their business.

- Level 3

  Stable approach of a formal system. This level characterizes companies which have a systematic process approach. Also, there are visible signs of systematic improvement (continual improvements, most of the time, but only simple or very simple). There is available data regarding the conformance with stated objectives. Managers of such companies take into consideration the trends of processes and have a suitable understanding of what their business means. However, elements of a level 2 maturity still exist, such as the reactive approach. Part of the middle management of these companies has had or attends management training regard. Often, their achievements are not used within the company as they should be. These companies have lows and highs in their activity, and are not capable all the time of fully understanding the real causes of this. They will be made happy by highs, and upset by lows. Recovering from lows is a painful process, not always successful. Most companies have systems which fall into this level.

- Level 4

  Sustained continual improvement. This level characterizes companies that have learned the useful lessons of their level 3 life. They understood that middle management has a key role in organization and employees are part of it. They are not simple machines, they have needs and expectations, which the company struggle to meet. A very good level of communication exists in these companies, no matter the direction. Plenty of data are recorded and analyzed and good decisions are made based on facts, not suppositions. Constant high quality of product is the focus in these companies and there exist signs that they are fighting also to exceed expectations, beyond fulfillment of requirements. Such companies have visions for the future and their mission is most of the time adequate. Improvements exist most of the time; they can be small or large and the company finds the necessary resources for them. Planning is an important part of the company's life.

- Level 5

  Best performance in class. At this level, companies are doing detailed benchmarking to prove their competitiveness against competitors. Their processes are strongly integrated. The future of such companies is sure, the security of jobs is provided. Customers have a high degree of confidence in the products or services of the company. Every person in the company is involved in processes, not because of constricts, but because of awareness. A high level of training is provided to personnel, according to their specific needs. Fulfillment of customer requirements is like "another day at the office" and anticipation of customer expectation is a constant focus. New products are made after extensive research of that initial expectation of the clients, to become new requirements for the company. Companies having this level of maturity are the best examples of how to manage a business rather than running the business.

2.8 Quality Management System (QMS) Requirements

The standards of the ISO 9000 series distinguish between QMS requirements and product requirements. The QMS requirements are laid down in ISO 9001. QMS requirements are of a general nature and apply to organizations in any industrial or economic sector, irrespective of the products offered. The ISO 9000 series does not determine any product requirements. Product requirements can be determined either by the customers (= specifications) or by the authorities (e.g., legal specifications).

Development of a QMS according to ISO 9000

A QMS according to ISO 9000 can be developed as follows:

- by ascertaining the requirements and expectations of the customer and other interested partners and parties
- by determining the organization's quality policy and quality goals
- by determining the necessary processes and responsibilities, in order to achieve the defined quality goals
- by determining and providing the necessary resources, to achieve the quality goals
- by introducing methods to measure the effectiveness and efficiency of each process
- by applying this measurement to ascertain the current effectiveness and efficiency of each process
- by determining the means of preventing deviations and eliminating the causes of such deviations.
- by introducing and applying a process for the continuous improvement of the QMS.

> Quality ensues when satisfaction is achieved. The needs of the user are defined by measurable criteria, which in turn become the guiding measure for product development.
>
> *W.A. Shewhart*

The dependence on controls to improve quality must stop. In particular, comprehensive monitoring is superfluous when quality is built into the products by controlled processes.

E.W. Deming

The Japanese understanding of quality is based on the following pillars: “Quality first”–quality goals as a management priority, “conformance to consumer requirements”–the consumer defined quality, involvement of all corporate departments and all levels, continuous improvement and a social system that forms the basis of meaning and wellness for the employee.

K. Ishikawa

Quality means to keep the promise.

P.B. Crosby

3
Fundamentals of Statistics

3.1
Basic Concepts

An analysis result void of statistical evaluation is essentially worthless. Besides the actual numerical value, what is therefore important is a statement on its accuracy and on its uncertainty.

Books on mathematics and statistics are, with regard to their practical use and interpretation, essentially written in a complex and abstract manner. The students are penalized all the more with formulas, derivations and definitions which often prove to be somehow more confusing than they are helpful, at least in the case of a practice-related application (Figure 3.1). In addition, specialist literature describes procedures and explanatory approaches to determining some statistical parameters, for example, the limit of detection, in different ways. For the most part, statistics is taught separately from the accompanying concrete and practical issue, which makes the subject appear rather dry. In its practical application, a statistical method gains meaning for the overall result and delivers statements on the quality of the results.

Statistics offers methods to summarize a collection of data. These methods may be graphical or numerical, both of which have their own advantages and disadvantages. Graphical methods are better suited for the recognition of patterns in the data, whereas numerical methods give well-defined measures of some properties. In general, it is recommended to use both approaches for the description of data. For the correct evaluation of analytical results only a basic knowledge of statistics is required in laboratory classes of analytical chemistry, which, in this section, is conveyed with a focus on the purely practical application. Only what is absolutely necessary is taught in an exemplary manner. Wherever it was acceptable, the author worked with estimated values, approximations and rules of thumb.

An important element of statistics is concerned with measurement deviations or accurate measurement uncertainty and was formerly called error calculation. This allows statements on the "true" value of the measured dimension and an estimation of the accuracy of the measurement. (Please note: the definition of true value according to the International Union of Pure and Applied Chemistry (IUPAC) Compendium of Chemical Terminology 2nd Edition, 1997, is "The value that

Practical Instrumental Analysis: Methods, Quality Assurance and Laboratory Management, First Edition.
Sergio Petrozzi.

Figure 3.1 Let's keep it as simple as necessary.
(Source of picture: Richard R. Rediske, Ph.D., Grand Valley State University).

characterizes a quantity perfectly in the conditions that exist when that quantity is considered. It is an ideal value, which could be arrived at only if all causes of measurement deviation were eliminated, and the entire population was sampled".) Despite the increasing acceptance of understanding measurement uncertainty determinations as a characteristic of the quality and reliability of a measured value, many university laboratories still find it difficult to follow this path.

This book is not a substitute for a course book on statistics. What is important is that students of chemistry are in a position to recognize and apply basic correlations in statistics paying attention to the different definitions of measurement terms. Application of what is learned can then follow, either manually or computer-aided. Simple table calculation programs or even pocket calculators that are excellent value for money contain statistical data that facilitate the calculation of statistical key figures. In addition, numerous software programs are available on the market with which statistical operations for validating findings and methods are possible, as, for example, required by the German Institute for Standardization (DIN) norms (see Section 4.5.2).

The following terms must be understood:

- Population (universe) and sample
- Distribution of values, in particular the normal distribution
- Means, especially the arithmetic mean
- Deviation from the mean, in particular the standard deviation and confidence interval
- Accuracy, absolute deviation, relative deviation
- Precision, variance
- Significant figures
- Measurement uncertainty
- Regression, linear, nonlinear

Thus in statistics there are, apart from the distribution curve, other methods of distributing values (log-normal distribution, Poisson distribution), other means than the arithmetic means (the mode and the median, geometric means, harmonious means), other dispersion masses than the standard deviation (percentiles) and other regressions than linear regression. They are not considered here, because they do not play a great role in analytical quality control.

Statistics as methodology is the scientific discipline of responsible data analysis. It serves, on the one hand, to illustrate large quantities of data as transparently and tangibly as possible and, on the other hand, to substitute simple mass numbers such as means and dispersion, thus leading to a data reduction and better comparability of various methods.

What does statistics have to do with an analysis?

- Averaging for quantitative determinations
- Standard deviation of analysis values (characterization by dispersion mass)
- Regression analysis for calibration lines for quantitative determinations

3.1.1 Population and Sample

In applied statistics there is a strict difference between *population* and *sample*. Just imagine a box with 5000 marbles *(statistical units)* in three different colors *(characteristic of the unit)*. This quantity of marbles is called the *population (test object)*. It is the collection of all values of interest. The mass of descriptive statistics, such as the mean or the standard deviation or, like in our example, the number of black, green and blue marbles, are called *parameters*, if they are calculated from the population. When 50 marbles are chosen from the box a subset of the *population* is created, this amount of 50 marbles is called a *sample (test sample)*. A descriptive mass, here the number of black marbles from the sample of 50 marbles, is called the *estimate (analyte)* (Figure 3.2).

As it is, for the most part, impossible to ascertain the population practically, in practice samples are mainly selected. The laws of statistics apply exactly only to basic wholes (infinite measurement values). However, if the sample of measured data is representative (the whole must be exactly determined by rational, spatial and time-based criteria, whereby the criteria are based on the investigation goal) of the population, then likelihood can be expressed with regard to the accuracy of the measurement. An experimenter should always ask himself whether the selected samples are representative of the population he is interested in.

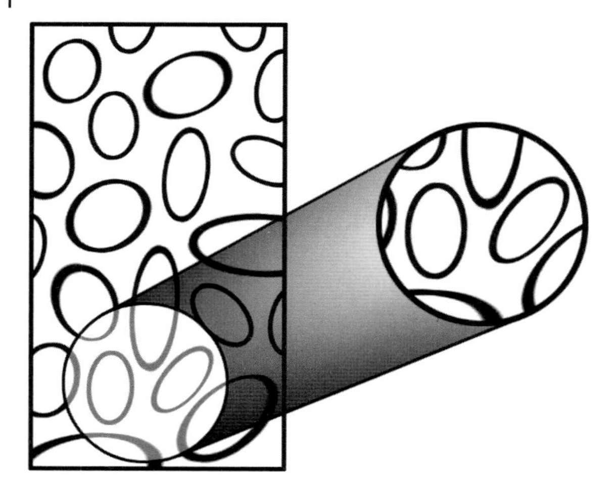

Figure 3.2 Population and sample.

3.1.2
Distribution of Values

When the same analysis is repeated several times, the result is dispersed measurement values. An important aid is to inspect the frequency distribution in a predefined interval.

The normal distribution curve of a set of infinite data can be described as bell-shaped, also called a Gaussian curve (Figure 3.3), after the nineteenth-century German mathematician Karl Friedrich Gauss. It is symmetrical with a central point of maximum frequency (arithmetic mean). *The probability of deviations from the mean is comparable in either direction.*

The graph of the normal distribution depends on two factors – the mean and the standard deviation (SD). The mean of the distribution determines the location of the center of the graph, and the standard deviation determines the spread of the graph. The more precise and/or the more reproducible it is, the narrower the Gaussian curve.

There is a relationship between normal distribution (or the values that lie under the curve of normal distribution) and the standard deviation:

68.3% of all values lie within $\bar{x} \pm 1$ standard deviation $P = 68\%$
95.4% of all values lies within $\bar{x} \pm 2$ standard deviations $P = 95\%$
99.7% of all values lies within $\bar{x} \pm 3$ standard deviations $P = 99\%$
where P is the level of confidence

The bell-shaped (Gaussian) curve decorated the 10 DM banknote of the Federal Republic of Germany beside the portrait of Carl Friedrich Gauss until 2001 (Figure 3.4).

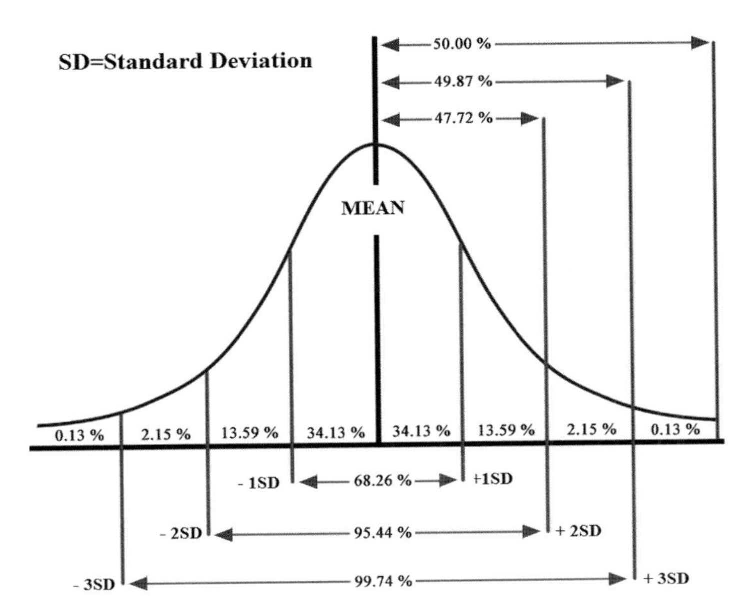

Figure 3.3 Normal distribution (Gaussian distribution).

Figure 3.4 10 DM banknote of the Federal Republic of Germany.

Theoretically it would be possible to analyze the population in the form of infinite determinations. Of course, this is not done in practice: The standard deviation and the mean are determined from samples. Parameters such as standard deviation and mean values of the population are designated by *Greek* letters, estimates by *Roman* letters (Table 3.1).

Table 3.1 The customary way of abbreviating means and standard deviations.

Value determined from	Mean	Standard deviation
Sample ($n \leq 10$)	$\bar{x}$	s
Population ($n = \infty$)	μ	σ

3.2 Important Terms

3.2.1 Mean, Arithmetic Mean, Average ($\bar{x}$)

Calculating the arithmetic mean $\bar{x}$ for n determinations

- All values are added together.
- The result is divided by the number of values (n).
- The arithmetic mean ($\bar{x}$) has the same unit as the individual values, for example, $g\,l^{-1}$ or $mmol\,l^{-1}$ and so on.

$$\bar{x} = \frac{\sum_{i=1}^{n} x_i}{n} \tag{3.1}$$

where x_i is a single value, n the number of measurements, i the running number of measurements ($i = 1,2,3 \ldots n$).

If there are no systematic deviations, the maximum of the Gaussian curve indicates the best possible approximation to the "true" content of a component μ, which is the closest in value to the arithmetic mean from a sufficiently large number of analysis data (most likely value):

$$\bar{x} \approx \mu \text{ (for } n \leq 10)$$

$$\bar{x} \approx \mu \text{ (for } n = \infty)$$

3.2.2 Standard Deviation (σ, s)

Precision is described by the standard deviation. The standard deviation is a so-called dispersion mass and is a measure for dispersing measurement values around the mean and applies to measurements proceeding from the population.

$$\sigma = \sqrt{\frac{\sum_{i=1}^{n}(x_i - \mu)^2}{n}} \tag{3.2}$$

where σ is the population standard deviation
and in individual measurements (samples):

$$s = \sqrt{\frac{\sum_{i=1}^{n}(x_i - \bar{x})^2}{n-1}} \tag{3.3}$$

where s is the empirical standard deviation.

As mostly the population is not measured, this formula is, as a rule, used for calculating the standard deviation; for example, in Excel it is defined as follows:

The calculated standard deviation (function STABW) is a mutually compatible estimate of the standard deviation of the population. This means that it is divided by $n - 1$ rather than by n.

The following applies to large samples, that is, to large values of n:

$s \approx \sigma$ (for $n \leq 10$)

$s = \sigma$ (for $n = \infty$)

Why dividing by $n - 1$ is more suitable than dividing by n is explained in the specialist literature with the help of the term "mutual compatibility" and can be proven with the help of estimation theory [2].

The definition of standard deviation according to the IUPAC Compendium of Chemical Terminology 2nd Edition, 1997, is "The positive square root of the sum of the squares of the deviations between the observations and the mean of the series, divided by one less than the total number in the series. The standard deviation is the positive square root of the variance, a more fundamental statistical quantity".

3.2.3
Variance (Var,V)

The spread of a set of data points around their mean value can be described by various parameters of which the standard deviation is the most common for people doing scientific work. Another term to report the precision of data is the variance.

The variance is the square of the standard deviation. In order not to calculate with negative figures, the standard deviation is squared. Variance (Var) is given by the equation:

$$\mathrm{Var} = s^2 \tag{3.4}$$

The definition of variance according to the IUPAC Compendium of Chemical Terminology 2nd Edition, 1997, is "The sum of the squares of the standard deviations, expressing the contributions to the overall precision from various sources of uncertainty."

3.2.4
Standard Deviation of Mean Values ($\overline{S}$)

If the standard deviation is calculated by means gained from grouped analyses of size n of the same population, the standard deviation is given by the equation:

$$\overline{S} = \frac{s}{\sqrt{n}} \tag{3.5}$$

($\overline{S}$) Standard deviation of mean values.

3.2.5
Relative Standard Deviation (RSD) and Coefficient of Variation (CV)

The relative standard deviation can be calculated from the standard deviation and the mean. It gives the ratio of the standard deviation s and the mean expressed as a percentage:

$$\mathrm{RSD} = \frac{s}{\overline{x}} \tag{3.6}$$

where RSD is the relative standard deviation, $\overline{x}$ the mean and s the standard deviation.

Specifying the standard deviation is more or less useless without the additional specification of the mean value (and of course the type of distribution). It makes a big difference if $s = 5$ with a mean of $\overline{x} = 200$ or with a mean of $\overline{x} = 8$. Relating the standard deviation to the mean resolves this problem.

The relative standard deviation multiplied by 100 gives the coefficient of variation (CV)

$$CV = \frac{s}{\overline{x}} \cdot 100\% \tag{3.7}$$

The advantage of the coefficient of variation is that it is possible to compare dispersion measures having values of completely different orders of magnitude.

3.2.6
Confidence Interval (CI), Confidence Limits

The confidence interval is the range around the mean, within which the "true" value can be expected to lie with a certain confidence level. The boundaries, the lower or upper limit of the confidence interval are known as *confidence limits.*

$$\Delta\bar{x} = t(P, f)\cdot\frac{s}{\sqrt{n}} \tag{3.8}$$

where t is the Student factor (see Appendix B, Table B.1), P the confidence level, f the degrees of freedom, and n the number of measurements.

The *student factor* is dependent on the number of degrees of freedom (f) and the confidence level (P), showing a dependence of the confidence interval on the number of measured samples and the selected confidence level.

The number of degrees of freedom shows the number of independent results that enter into the calculation of the standard deviation.

The confidence level is defined as $1 - \alpha$ (1 minus the significance level). Thus, when we have a 95% confidence interval, we are stating that there is a 95% probability that the true population mean is covered by the interval.

Usually, the significance level is chosen to be 0.05 (or equivalently, 5%).

The t-factor was developed by William Gosset. The work was first published in 1908, while Gosset was employed at the Guinness Brewery to statistically analyze the results of alcohol determinations in beer. As his employer did not agree to publish the company's confidential data, Gosset published it under the pseudonym Student.

For mean values $f = n - 1$
For data derived from linear regression $f = n - 2$

3.3
Quality of Results (Accuracy and Precision)

Measurement uncertainty that must be attributed to a measurement value is an inevitable byproduct of each measurement. Measurement uncertainty is a parameter that is gained from repeated measurements and which, together with the measurement results, serves to characterize a value range for the "true" value of the measurement parameter.

In connection with measurement processes, the term “error”, however, used in older literature (including measurement error, error calculation, error propagation, etc.,) poses a problem, as the general linguistic meaning of the term “error” implies that something is wrong.

The deviation from a measurement result of the true value of a measurement parameter does not, however, mean that the measurement was wrong. Rather, it is caused by the fact that each measurement procedure bears the risk of certain uncertainty.

A quantitative analysis result without information on its uncertainty is essentially worthless, as the result did not assess (e.g., threshold compliance, Figure 3.5) nor can it be compared to other values. A well-reasoned uncertainty analysis is a characteristic of professionalism in metrology.

Figure 3.5 shows a test report where four measurements are compared with a threshold limit value (TLV). The results A + B are clearly below or above the threshold.

In cases C and D the test report points out that the threshold is within the uncertainty limit of the result. In order to obtain a closer confidence range and thus create a better basis on which to make decisions, further samples must be analyzed.

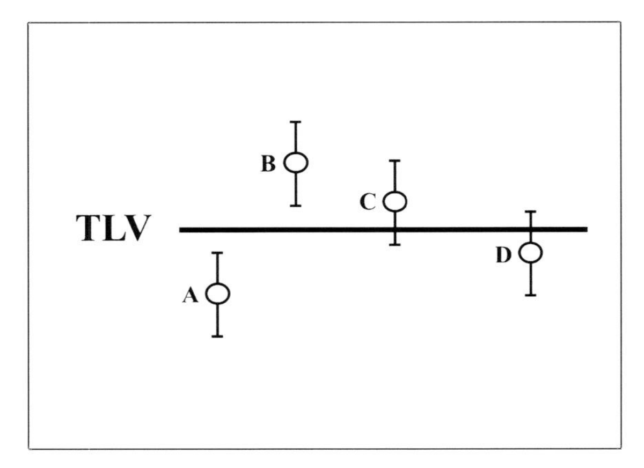

Figure 3.5 Evaluation of a possible threshold is exceeded (○ = points, I = deviation bars).

Wednesday, June 29, 1994
San Francisco Chronicle

Simpson to Give Sample Of His Hair for Testing – Defense asks to do its own DNA comparison

Author: *Michael Janofsky: New York Times*

The lawyer for O.J. Simpson told prosecutors yesterday that they could have a sample of his client's hair for scientific analysis. But throughout the short hearing in Los Angeles Municipal Court, the lawyer, Robert Shapiro, wrangled with prosecutors over virtually every assertion they made. It was an aggressive tactic that is likely to continue – and prolong – every phase of the case against Simpson, who has been charged with murdering his former wife and a male friend of hers. The only major point the two sides agreed on was that they are ready to proceed tomorrow with a preliminary hearing to determine whether enough evidence exists to put Simpson on trial.

With his client seated quietly at his side, Shapiro said that in response to the prosecutors' request Monday, he will provide a hair sample from Simpson. But then, he pressed the prosecutors for separate samples of blood, hair and other evidence so the defense team could have its own forensic experts analyze them. Shapiro said an "over-the- shoulder" examination by defense experts would be inadequate. "It does us no good to simply watch what somebody else does," he said. "We want to conduct our own tests at our own laboratory."

He also questioned the integrity of the prosecutors' testing laboratory and procedures that are being used in testing the evidence collected.

As the case has moved along, prosecutors have said they have collected more than 60 pieces of evidence – although not the murder weapon – which they contend would be more than enough to link Simpson to the crimes in a preliminary hearing that is expected to last at least one week.

The evidence found at the scene of the killings includes various blood samples and a dark blue knit cap that contained hair. The prosecutor mentioned other evidence without saying where it was found, including a stained glove.

About 3 p.m. yesterday, police officers with a search warrant went to Simpson's home and searched for 3½ hours in bushes, vehicles and other areas for further evidence. They towed a black Rolls-Royce and left with brown grocery bags and police evidence bags. They would not say what the bags held. In pressing his case to examine every shred of evidence, Shapiro asked not only for samples but also written notes from the medical examiner's office, the methods its scientists used for analysis and *the laboratory's rate of error.*

When Marcia Clark, the lead prosecutor, said she had been told that the testing lab had no percentage of error because "they have not committed any errors," Shapiro raised an eyebrow and asked how long the records went back. Neither Clark nor her co-counsel, William Hodgman, responded.

In other attempts to demonstrate that prosecutors are being as responsive as possible, Clark said that the defense has been given all available evidence but that she "would be delighted" to provide any additional items as they become available. She said that the police crime laboratory is still conducting some tests and that the results will not be available until tomorrow.

Clark argued that "hairs cannot be split," a choice of words that caused laughter in the courtroom. Her point was that any hair sample still attached to a follicle could be tested only once, in which case she invited Shapiro's forensic experts to observe the test.

Without a murder weapon or any witness, yet known, to the killings, forensic evidence, particularly results of genetic testing of blood and hair, could play a major role in the prosecution's case against Simpson, 46, the former football star and actor. But genetic testing is not always conclusive, and Shapiro hinted after his appearance in court that he might oppose the introduction of such testing in a trial. Meanwhile, Louis Brown, the father of Simpson's slain ex-wife, disputed a coroner's document that said his daughter was on the phone with her mother at 11 p.m. the night she was killed. Brown said the call was closer to 10 p.m. Simpson lawyer F. Lee Bailey had cited the document in an attempt to bolster Simpson's alibi. Simpson says he was home waiting for a limousine to take him to the airport at 11 p.m.

3.3.1
Measurement Deviations

The determination of measurement deviations (old term: measurement error) should, on the one hand, serve to approximate as closely as possible the "true" value, and, on the other hand, to estimate (determine) the reliability of a measurement method.

A differentiation must be made between

- random deviations
- systematic deviations and
- gross error.

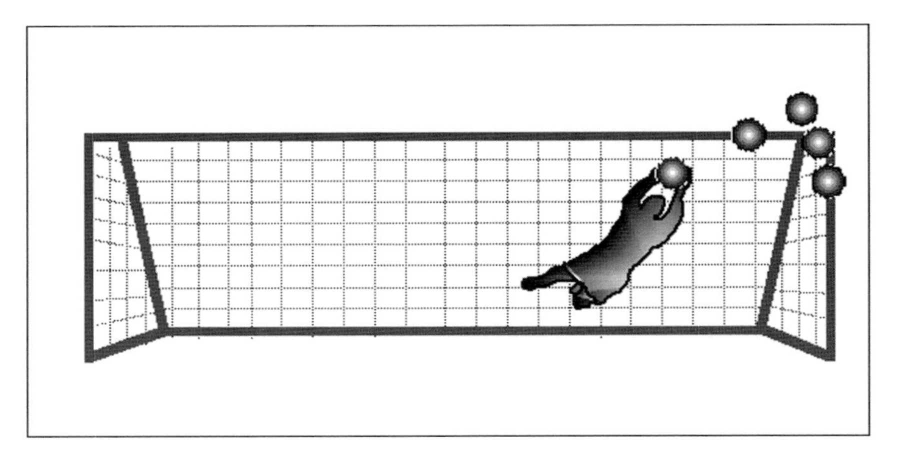

Figure 3.6 Goalkeeping example: accurate but imprecise – random deviations – gusts of wind, unevenness of the grass.

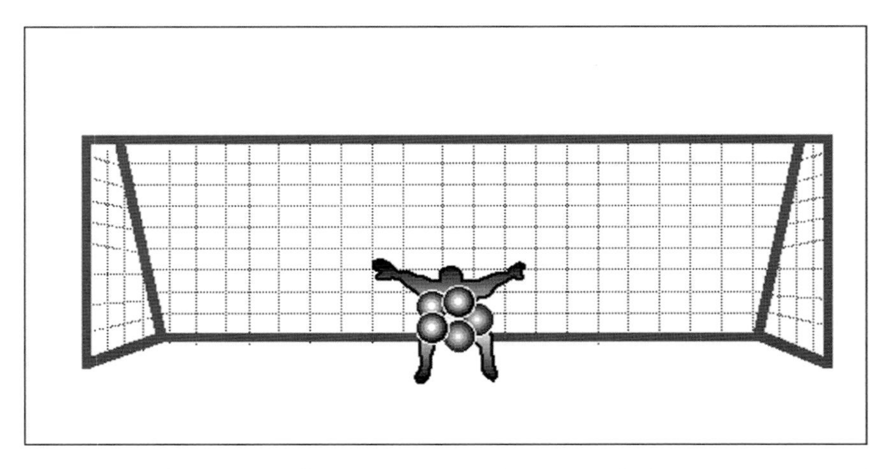

Figure 3.7 Goalkeeping example: inaccurate but precise – systematic deviations – it is assumed that the goalkeeper always jumps into the right-hand corner.

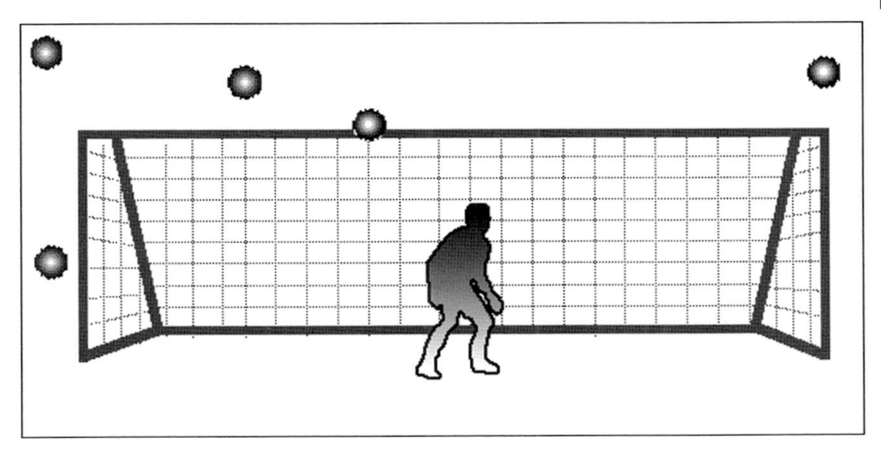

Figure 3.8 Goalkeeping example: inaccurate and imprecise–gross errors–incapability or incompetence.

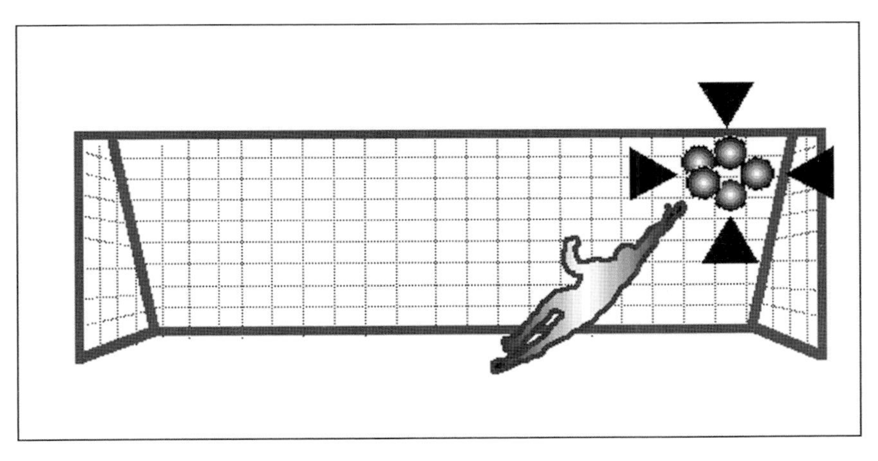

Figure 3.9 Goalkeeping example: accurate and precise–"true" value–perfect penalty shoot-out into the corner.

With a goal size of 7.32×2.44 m the goalkeeper had to cover 18 m^2. According to measurements, the speed of the kicked soccer ball is 90–100 km h^{-1}. The ball only takes 0.4 s to travel the 11 m to the goal middle, and 0.42 s to travel the 11.59 m to the goal end. Mathematicians from the University of Erlangen have calculated that the goalkeeper would have to fly into the corner with the speed of a 100 m runner in order to reach the ball. That is why a perfect penalty shot into the corner is (almost) impossible to stop (Figures 3.6–3.9).

3.3.2
Random Deviations – Influence on Precision

Intermediate or random deviations (also known as experimental errors) exist in every measurement and cannot be influenced. They are unavoidable, but can, however, be minimized by repeated measurements (system of averaging) and the use of suitable evaluation methods.

3.3.2.1 Precision

Precision is a measure of the spread of analytical values and is mostly recorded as a standard deviation or relative standard deviation from the results of statistical evaluation. The distinction is made between measurement precision (system precision) and methodological precision.

DIN-ISO: 3534-1 [3] points out that investigation procedures are all the more precise the smaller the resulting random deviations of the procedure. Random deviations are the sum of several small causes of deviations. Once they have been identified, they become systematic and can be eliminated or are considered in calculations.

> ***Precision**, The closeness of agreement between independent test results obtained by applying the experimental procedure under stipulated conditions. The smaller the random part of the experimental errors, which affect the results, the more precise the procedure. A measure of precision (or imprecision) is the standard deviation.*

3.3.2.2 Determination of Random Deviations

Measurement precision: A measure of the fluctuations caused by the analyzer itself. The determination of these fluctuations is performed by multiple analysis of a standard.

Method precision: A measure of the dispersion of analytical results.

(Laboratory) precision (intermediate precision)

Precision, which results from repetitions of the determination in a laboratory on the same sample under different conditions: a sample on different days in a laboratory under various conditions (people, lots of reagents, solvents, equipment, environmental conditions, etc.), but with the same procedures to determine.

(Comparative) precision (reproducibility)

Precision under comparable conditions: Conditions during the attainment of independent investigation results, consisting in applying the same method to an identical test object, in different laboratories by different operators, with different equipment devices.

(Repeatability) precision (within-run)

Precision under repeatability conditions: Conditions during the attainment of independent investigation results, consisting in applying the same method to an identical test object in the same laboratory by the same operator using the same device equipment at short time intervals.

In-house Laboratory Precision

(Repeatability conditions: a laboratory, an analyst, a device)

3.3.2.3 Causes of Random Deviations

- Instrumental errors, for example, electronically induced noise
- External influences, for example, light irradiation

3.3.3 Systematic Deviations – Influence on Accuracy

Basic deviations from the true value, that is, values being either too high or too low, are described as systematic deviations. Accuracy is the measure of systematic deviation. Systematic deviations are avoidable. They are unidirectional, that is, the magnitude and the sign are the same. Systematic deviations are reproducible, but not accurate (precise – inaccurate).

Constant Systematic Deviation

The magnitude of the systematic deviation is independent of the magnitude of the analysis results (for example, always $10\,mg\,l^{-1}$, too small to be measured).

Proportional Systematic Deviation

The magnitude of the systematic deviation increases and falls with the magnitude of the analysis result (for example, always 10%, too small to be measured)

3.3.3.1 Accuracy/Trueness

Accuracy is defined as the degree of closeness of agreement between an individual value obtained by measurement and the "true" or accepted reference value.

Trueness is defined as the degree of closeness of agreement between the mean of a large series of investigation results and the "true" or accepted reference value.

3.3.3.2 Bias

In analytical chemistry bias describes the total systematic deviation. One or more systematic deviation components can contribute to the bias. A larger systematic deviation from the accepted reference value is reflected by a larger bias value. A frequently used representation of the expected value of a measured result is the mean. In this case the bias represents the deviation between the mean of a number of results and the accepted reference value.

***bias**, Definition according to IUPAC Compendium of Chemical Terminology, 1985.*

Bias characterizes the systematic error in a given analytical procedure and is the (positive or negative) deviation of the mean analytical result from the (known or assumed) true value.

3.3.3.3 **Causes of Systematic Deviations**

- Instrumental errors, for example, change of settings on the devices, contamination, aging
- Method, for example, false measurement instruments, false concentrations
- Use, for example, reaction time not observed, indicator not optimal
- Environmental influences, for example, refraction, vibrations of the background, and so on
- Reference standards, for example, incorrect value for purity

3.3.3.4 **Effects on the Measurement**

Constant: for example, wrong bringing up to mark of a volumetric flask.

The smaller the measurement value, the bigger the constant deviation.

Proportional: proportional to the measurement parameters, for example, contaminated standard; the more standard used, the bigger the deviation. Thereby, the size of the measured value itself is irrelevant.

3.3.3.5 **Determination of Systematic Deviations**

Only indirect methods are possible:

- *Analysis of reference material*

  Certified reference materials (CRM), the content of which is well-known and correct, are, for example, tested by some institutes such as the National Institute of Standards and Technology (NIST) or the German counterpart Physikalisch-Technische Bundesanstalt (PTB) with methods that display slight systematic deviations.

  The huge range of different sample matrices, requiring many different analytical determinations to be carried out, results in the need for that number and range of reference materials. Information on the availability of reference materials and CRMs can be obtained from a number of sources [4].

  Target performance comparison can also be executed with materials in your own laboratory. If the confidence range of the results lies within a defined dispersion, the method is said to be correct.

  Apart from the set value-t-Test [5] other tests are applied, such as the *Doerffel*-Test [5] and the *Wilcoxon*-Test [5].

- *Independent Analyses Using Other Methods*

  Collaborative tests: Several samples are distributed to different participants and analyzed. The results are compared and evaluated.

- Accuracy and precision can be regarded as relative terms: They depend very much on the requirements of the client (target values), the purpose of the investigations and of the specifications/limits of the investigated analytical method.
- Trueness can be regarded as the umbrella term for the contribution to uncertainties of precision (random deviation) and accuracy (systematic deviation).

3.3.3.6 Recovery Experiments

If there are no reference samples available, accuracy can be evaluated using the recovery rates (RR). By increasing a sample with a known quantity of analyte (*spike*), the recovery rate can be calculated.

The entire method can be evaluated with the aid of the recovery rate.

1. Determination of the recovery rate with analyte-free samples and a matrix

$$\mathrm{RR}(\%) = \frac{(c_{\text{spike/sample}} - c_{\text{matrix}})}{c_{\text{spike}}} 100\% \tag{3.9}$$

2. Ascertaining the recovery rate with analyte and a matrix (for an example, see Table 5.1, B1.1)

Three solutions are made and analyzed:

Solution 1: Add V1 ml standard analyte solution to V2 ml sample solution.

The result is signal 1

Solution 2: Add V1 ml solvent to V2 ml sample solution.

The result is signal 2

Solution 3: Add V1 ml standard analyte solution to V2 ml solvent.

The result is signal 3

$$RR(\%) = \frac{(S_1 - S_2)}{S_3} 100\% \tag{3.10}$$

3.3.4 Gross Errors

Gross errors are "inadmissible" and deviations should be avoided through education, training, self-discipline, critical care and checking the results.

3.3.4.1 Causes of Gross Errors

- A lack of knowledge
- Laziness
- Negligence
- Bad luck

Gross errors cannot be detected with the help of theoretical models. The results thus gained are incorrect and must be repeated.

3.3.5 Uncertainty of Measurement Results

Definition (Quantifying Uncertainty in Analytical Measurements, Eurachem/Citag Guide, ISBN 0-948926-08-2, 2000)

Measurement uncertainty is a parameter that is allotted to the measurement results and characterizes the dispersion of those values, which can reasonably be attributed to measurement parameters. Measurement uncertainty can, e.g. be expressed as a standard deviation (or a multiple thereof) or as a confidence interval.

The uncertainty of a measurement result is determined by the uncertainty of all experimental parameters which in any way have an impact on the result, for example, because they appear in the calculation formula, or because they in turn influence one of the calculation parameters.

Measurement uncertainty that must be attributed to a measurement value is an inevitable byproduct of each measurement. Measurement uncertainty is a parameter that is gained from repeated measurements and which, together with the measurement results, serves to characterize a value range for the "true" value of the measurement parameter.

The uncertainty in the result of a measurement generally consists of several components which may be grouped into two categories according to the way in which their numerical value is estimated.

Type 1. Those which are evaluated by statistical methods
Type 2. Those which are evaluated by other means

3.3.5.1 Standard Uncertainty of Single Measurements

The term standard uncertainty is used when the uncertainty is expressed as *one* standard deviation ($\pm 1\ s$).

The measurement is repeated several times. From several single measurements, the sample size n is defined. All subordinate influencing parameters must be able to vary within the framework of their own uncertainty. If this is the case, one speaks of statistically independent single measurements.

The uncertainty of the result (u) is expressed as the empirical standard deviation (s)

$$\pm u = s \tag{3.11}$$

The measurement uncertainty can also be expressed as a confidence interval ($\Delta\overline{x}$)

$$\pm u = \Delta\overline{x} \tag{3.12}$$

How good the estimate of the measurement uncertainty is depends on the precision of the method and on the number of samples measured.

3.3.5.2 Combined Uncertainty

This value, together with the measurement result, characterizes a value range that contains the "true" value of the measurand with high probability.

It is not always possible to perform several independent measurements to determine the uncertainty of an analytical result, for reasons such as cost, lack of time or because the sample quantity is not sufficient.

Sometimes the single measurements are also not really independent, for example, when the same calibration data have been used for all assessments. The uncertainty of this calibration then has equal effect on all measurements thus becoming a systematic deviation that no longer appears in the dispersion of the individual results.

These problems can be largely avoided, if, using the methods of the "deviation propagation calculation" the uncertainty u_x of an analysis result from the uncertainties u_i (i = 1, 2, . . . , n) of all faulty individual steps of the analytical process is estimated.

$$u_x = f(u_1, u_2, u_3 \ldots u_n) \tag{3.13}$$

where u_s is the combined uncertainty of the measurement result x.

Table 3.2 shows an example of the estimation of measurement uncertainties.

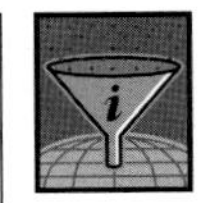

> When calculating combined uncertainty, all individual contributions must be of the same type, for example, standard deviations. If the uncertainty of a measurement parameter is expressed differently (e.g., as a confidence interval or as tolerance), there must be computational reconstruction to the standard deviation (s).
>
> If the uncertainties are expressed in different units, the relative standard deviations are added.

3.3.5.3 Procedure for Determining the Combined Uncertainty

1) Formulate the complete formula for the final result:

   In the interests of clarity it is advisable to subdivide the analytical process into several partial steps. This can be achieved by determining the measurement insecurity in individual steps.

Table 3.2 Estimation of measurement uncertainties (usual practice).

Experimental procedure stage	Source of uncertainty	Estimation (%) min.	max.
Sampling	Various steps	Difficult to assess	Difficult to assess
Sample preparation Gravimetrics	Weighing and transfer of the sample in a volumetric flask	0.01	0.5
	Dosages of 150 ml pentane	0.1	0.5
	Pipetting 50 ml pentane	0.05	0.1
	Weighing the empty calibration flask	0.01	0.5
	Arriving at a constant weight	0.1	1.5
	Dilution to the mark	0.05	0.1
	Weighing the filled calibration flask	0.01	0.5
Preparation of an internal standard (IStd)-solution	Uncertainty contribution of the standard (purity)	0.1	1.0
	Weighing	0.01	0.5
	Dilution	0.01	0.1
Preparation of a GC sample solution	Weighing	0.01	0.5
	Pipetting the IStd-solution	0.1	1.5
	Operator skill	0.1	1.0
	Dilution	0.01	0.1
	Uncertainty contribution of the GC system	0.1	0.5
Calculation	Integration	0.1	1.0

2) Identify all parameters characterized by uncertainty:

For this the fishbone–illustration according to Ishikawa (Figure 2.3) is very useful.

The cause and effect diagram is a simple aid in the form of a fishbone for the systematic identification of the causes of a problem. In the process, the possible causes of a certain effect are broken down into main and incidental causes. This is followed by structuring the causes in a diagram to have a clear overview. In this way, all problem causes should be identified and their dependences illustrated with the help of the diagram.

3) Ascertain or estimate the standard deviation for each parameter:

Calculating the combined measurement uncertainty is simplified when all uncertainties of an individual step in weighing are added together to form one new uncertainty (3.13) as an interim result. The same applies to the dilution step, and so on.

4) Ascertain the combined measurement insecurity from the individual contributions:

The standard deviations of the sums and differences are preferably calculated first (Eq. 3.14). This applies to identical units, otherwise relative standard deviations are used for calculation.

By comparing the individual relative variances it can also be conjectured at this stage which measurement parameters' contributions to the final results are negligible and/or wherein the major sources of insecurity in the procedure lie (Insecurity discussion = integral part of the validation of a method).

As a rule, parameters whose relative uncertainty is less than 10% of those with the most insecure parameters need not be considered.

5) Calculating measurable influences on the precision of the process:

   Perhaps the analytical process contains steps which do not feature as parameters in the calculation formula, but which still exert an influence on the precision of the analysis, for example, homogeneity of the problem material, work techniques and loss of analytes. These uncertainties are taken into consideration by adding still relative variances (possible realistic estimates) to the relative variance of the final results.

The formula used here only applies to uncertainty propagation in equations which contain sums, differences, multiplication or division. For other functional types such as exponential or logarithmic correlations, the method of partial differentiation is used (this book does not cover this topic).

3.3.5.4 Rules for Uncertainty Propagation

Sum or Difference The absolute variance of a sum or difference is equal to the sum of the absolute variances of the summands:

Measurement result x = partial step A + partial step B

Measurement result x = partial step A − partial step B

$$u_x^2 = u_A^2 + u_B^2 \tag{3.14a}$$

$$u_x = \sqrt{u_A^2 + u_B^2} \tag{3.14b}$$

Product or Quotient The relative variance of a product or quotient is equal to the sum of the relative variances of the factors:

Measurement result x = partial step A × partial step B

Measurement result x = partial step A / partial step B

$$\left(\frac{u_x}{x}\right)^2 = \left(\frac{u_A}{A}\right)^2 + \left(\frac{u_B}{B}\right)^2 \tag{3.15a}$$

$$u_x = x \cdot \sqrt{\left(\frac{u_A}{A}\right)^2 + \left(\frac{u_B}{B}\right)^2} \quad (3.15b)$$

3.3.5.5 Extended Uncertainty

By multiplying the measurement uncertainty (u) by a factor k, the extended uncertainty is calculated. The extended uncertainty defines an interval in which the value of the measurement parameter lies with high probability. In most cases, a factor of $k = 2$ is considered appropriate. For more demanding applications $k = 3$ is recommended.

This coverage factor or "safety factor" is applied to aim at higher statistical security ($k = 2$ for 95% confidence level, $k = 3$ for 99% confidence level)

$$\pm U = k \cdot u \quad (3.16)$$

k Coverage factor

3.3.6 Non-statistical Methods of Estimation

The standard deviation of influential parameters or behavioral steps cannot always be ascertained by means of several repetitive steps but it must be estimated. The basis for such an estimation is provided by

- Older measuring data
- Experience gained
- Biographical references
- Manufacturing references (for example, accuracy of a balance, injector precision, etc.)

3.3.6.1 Tolerance

If the respective standard deviations are not known, the tolerances must be used for their construction. To get the tolerance (T) to the standard deviation (s) we must make assumptions about the distribution. The most important are:

If the respective standard deviations are unknown, the tolerances arising from production information are to be applied. To reach the standard deviation (s) from the tolerance (T) assumptions must be made on the distribution. The most important of these are:

normal distribution

$$s = \frac{T}{3}* \quad (3.17)$$

rectangular distribution

$$s = \frac{T}{\sqrt{3}}* \quad (3.18)$$

* Quality level 3s

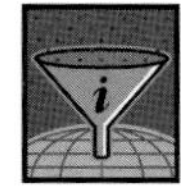

The uncertainty of the volume of volumetric flasks or pipettes is mentioned by several manufacturers as a so-called tolerance that is listed in the documentation or also directly etched on the volumetric flasks, for example, as (500.0 ± 0.1) ml. This means that the manufacturer guarantees that the volume certainly lies between 499.9 and 500.1 ml. Smaller or larger values can be excluded. Within the volume, however, all values are probably the same. This behavior correlates with a so-called rectangular distribution. These estimated values that are not based on statistical methods can be just as good as a standard deviation ascertained through experiments, or even better if the latter are based only on a few repeated measurements.

3.3.7 Expressing Analytical Results

A complete result will always contain the quantitative information of the measured variable (result) which has been experimentally determined and the measurement uncertainty.

Depending on which norm is applied, the expression of the result can vary:

Widely accepted in practice is

$$x_i = \overline{x} \pm U(\text{corresponding unit}), (n = 10, P\ 95\%) \tag{3.19}$$

where x_i are the measurement results, $\overline{x}$ the best estimate of the "true" value, U the extended measurement uncertainty, P the confidence level, and n the number of measurements.

3.3.7.1 Expressing the Measurement Uncertainty in the Value of a Quantity

The uncertainty that is associated with the estimated value of a quantity should be evaluated and expressed in accordance with the Guide to the Expression of Uncertainty in Measurement [ISO, 1995]. The standard uncertainty (i.e., estimated standard deviation, coverage factor $k = 1$) associated with a quantity x is denoted by $u(x)$. A convenient way to represent the uncertainty is given in the following example:

$$m_\text{n} = 1.674\ 927\ 28\ (29) \times 10^{-27}\ \text{kg}.$$

where m_n is the symbol for the quantity (in this case the mass of a neutron), and the number in parenthesis is the numerical value of the combined standard uncertainty of the estimated value of m_n referred to the last two digits of the quoted value; in this case

$$u(m_\text{n}) = 0.000\ 000\ 29 \times 10^{-27}\ \text{kg}.$$

If any coverage factor, k, different from one, is used, this factor must be stated.

3.3.7.2 Accordance with the National Institute of Standards and Technology, U.S. Department of Commerce (NIST)

Example (http://physics.nist.gov/cuu/Uncertainty/examples.html, accessed May 2012):

The following is an example of uncertainty statements as would be used in publication or correspondence. The quantity whose value is being reported is assumed to be a nominal 100 g standard of mass x_i.

x_i = (100.021 47 ± 0.000 70) g, where the number following the symbol ± is the numerical value of an expanded uncertainty $U = k.u$, with U determined from a combined standard uncertainty (i.e., estimated standard deviation) $u = 0.35$ mg and a coverage factor $k = 2$. Since it can be assumed that the possible estimated values of the standard are approximately normally distributed with approximate standard deviation u_c, the unknown value of the standard is believed to lie in the interval defined by U with a confidence level of approximately 95%.

3.3.8
Significant Figures – "Box-and-Dot" Method

In calculations the result can never be more exact than the most inaccurate number which goes into the computing (even if the pocket calculator displays 10 digit places). The significant figure reflects the exactitude of a measurement result and describes all figure places that are definitely known plus one digit place that is estimated.

It is not correct to add zeros; on the other hand, zeros should not be omitted, when they are significant.

Example: If a measuring flask permits measurement accuracy of 0.0001 l and if one measures 3 l of a solution with this accuracy, then the volume is to be recorded as 3.0000 l (or 3000.0 ml) – that is, with five significant figures.

In general, one should use only as many decimal places as are compatible with the precision of the measurements.

Stephenson [6] introduces a simple strategy to determine significant figures in numerical computations. He calls this method *box-and-dot*. This method is based on three steps:

Step 1

All numbers that are *not* zeros are provided with a frame (*box*)

for example, B.0.00243100   0.00[2431]00

Step 2

If there is a decimal point in the number (*dot*), a box is also drawn around the zeros and placed on the *right* (*terminally*) of the framed numbers that do not correspond with zero.

0.00 [2431] 00   0.00 [2431] [00]

Step 3

All numbers framed are added.

Answer: The number 0.00243100 has six significant figures

Please note:

- Exponents (scientific notation) are not taken into consideration.
- What is significant are only zeros in the middle and at the end.
- The position of the decimal point is not essential for determining the significant digit places
- Significant digit places have *nothing* to do with the number of decimal places.

Further examples

- number 2431.02460
  - 2431.02460 all the numbers that are *not* zero are *"boxed"*
  - a decimal point *is* present
  - 2431.0246 0 *end* zeros are *"boxed"*

  Nine figures are boxed = nine significant figures available
- Number 243102460
  - 24310246 0 all numbers that are not *zero* are *boxed*
  - decimal points are *not* available
  - 24310246 0 *end* zero is *not "boxed"*

  Eight figures are boxed = eight significant figures are available
- Number 10000.0
  - 1 0000.0 all numbers that are *not* zero are *boxed*
  - decimal point *is* present
  - 1 0000.0 *end* zeros are *"boxed"*

  Six figures are boxed = six significant figures are available
- Number 0.00010
  - 0.000 1 0 all numbers that are *not* zero are *boxed*
  - decimal point *is* present
  - 0.000 1 0 *end* zero is *"boxed"*

  Two figures are boxed = two significant figures are available
- Number 1.60×10^{-19} *scientific notation*

  In scientific notation the exponent is omitted
  - 1.6 0 all numbers that are *not* zero are *boxed*
  - decimal point *is* present
  - 1.6 0 *end* zero is *boxed*

  Three figures are boxed = three significant figures are available
- Number written in various ways:

  a) 8000, b) 8.0×10^3, c) 8000×10^3

  a) 8000 decimal point *not* present

  One significant figure

  b) 8.0×10^3 → 8.0 → 8. 0. decimal point *is* present

  Two significant figures

  c) 8.000×10^3 → 8.000 → 8. 000 decimal point *is* present

  Four significant figures

In different experiments different measurement values are linked together using different mathematical operations (additive, multiplicative, exponential).

The rule of thumb is:

- In a sum or difference, the number of significant digit places corresponds to the smallest number of significant figures.
- In the case of multiplications, divisions and square roots one never gets more significant figures than any of the factors in the expression.

3.3.9 Outlier Tests

Data are described as outliers when compared to the predominant majority of all other data they assume extreme values. Outliers can be easily seen in a scatter-plot, as they are far separated from the remaining cloud diagram. Although it is tempting to remove these suspect points from the statistical processing no data should be discarded prior to outliner test assessment.

Statistical surveys on outliers can also be conducted mathematically by conducting outlier tests. Outliers can be a great impediment for regression analysis. First, it is necessary to examine potential outliers more closely. Why is the data point so far off? Could it be a measurement error or a typing error? What significance does the data point have with regard to content?

Individual deviating values can be ascertained by statistical test procedures. Multiple replications of analyses provides a value that obviously deviates far from the mean without a reason having to be given. When calculating the results of an analysis, it must be decided whether this value has to be considered or whether it can be rejected as an "outlier" (gross error). To make this decision objective, screening tests are used with agreed limits for accepting or rejecting values.

3.3.9.1 The 2.5 s Barrier

Assuming normal distribution following a simple and very rough test, a value – from at least 10 values (parallel or repetitive measurements of a sample) – that appears to have a significant difference from the rest of the data can be eliminated when it lies outside the range $\bar{x} \pm 2.5s$, $\bar{x}$ the arithmetic mean and s the standard deviation are then recalculated by omitting the value "suspected of being an outlier".

3.3.9.2 Test according to Grubbs

In a sample there can be individual comparative figures that deviate more from the mean than the remaining values. The reason for this can be that an error occurred during documentation.

The test described here ascertains these possible outliers. However, it does not do this strictly according to percentage deviation from the mean, but includes the individual dispersion of the samples in the search for outliers. In the case of the *Grubbs* test [7], the test value is calculated for a suspicious value (maximum or minimum comparative value), which describes the distance from the mean in relationship to the standard deviation. If this relationship is larger than the Grubbs comparative value, the corresponding comparative value is described as an outlier.

The outlier thus discovered is then removed from the list of comparative values and the new minimum and maximum values are used to calculate a new test. The value suspected of being an outlier is included in the calculation each time. For the value suspected of being an outlier, the test value is calculated using the following formula:

$$\mathrm{TV} = \frac{(x_1 - \overline{x})}{s} \tag{3.20}$$

where TV is the test value, x_i the value suspected of being an outlier, $\overline{x}$ the arithmetic mean, and s the standard deviation.

The comparative value CV for a chosen significant level, scope n of the samples, is taken from the table in Appendix B, B.2.

If TV is larger than CV, the comparative value is regarded as an outlier and not included in the continued assessment. It should, however, remain visible as an eliminated value. The outlier test is repeated when the next suspicious value arises. Thereby new mean values and standard deviations arise.

3.4 Regression

3.4.1 Regression Analysis

The regression analysis is suited to treatment of various kinds of issues.

- Recognition of correlations
  A functional correlation can be recognized between two characteristics that have hitherto not been examined.
- Proof of correlations
  A correlation between two characteristics is suspected. Regression can be used to prove such a correlation.
- Estimation of the Nature and Size of correlations
  The type and size of a known correlation can be estimated with the regression analysis.
- Prognosis of Missing or Future Values
  Regression can be used to forecast missing or future values (values that lie between the measuring points).

regression analysis*, Definition according to IUPAC Compendium of Chemical Terminology 2nd Edition, 1997. Regression analysis is the use of statistical methods for modelling a set of dependent variables, Y, in terms of combinations of predictions, X. It includes methods such as multiple linear regression* (MLR) *and partial least squares* (PLS).

3.4.2 Calibration Function

The calibration function is the functional correlation between the measurement, for example, the absorbance and the content, for example, of a mass concentration.

The calibrating function can be either linear or nonlinear. Under certain circumstances it can become necessary to subdivide the calibration range into several subparts. Both linear and nonlinear calibrating functions are suitable for determining the concentration, as long as there is a clear reproducibility of the functional determination. Frequently, there is a linear correlation.

The simplest regression model, *the model of linear correlation between two characteristics,* is used for simple derivation.

Basically, the simple regression analysis consists of three components.

- A value x that is also described as "the independent variable".
- A value y that is also described as "the dependent variable".
- The assumption that between both these values there is a functional correlation in the form of $y = f(x)$.

***calibration**, Definition according to IUPAC Compendium of Chemical Terminology 2nd Edition, 1997. The set of operations which establish, under specified conditions, the relationship between values indicated by the analytical instrument and the corresponding known values of an analyte.*

3.4.3
The "Optimal" Trend Line

We are looking for that trend line to which the following applies:

- The trend line is clearly calculable.
- It is not about whether the trend line goes exactly through all points, but about being as close to all points as possible.
- The line generated is the one that minimizes the sum of the squares of the residuals for all points.
- Scatter-plot and correlation must be taken into consideration.

For every data analysis, a scatter-plot is first drawn to visualize any possible patterns. With this pattern an attempt is made to conduct a meaningful regression using an appropriate mathematical model. The pattern can be linear, or deviate from linear behavior, that is, the trend line does not necessarily have to be linear.

It is not the best approach to fit a line to data showing a visibly curved pattern (Figure 3.10). Apart from linear regression, there are other types of trends or regression, for example, logarithmic, polynomial, exponential, and so on. In the case of most analytical methods, however, the analytical function is determined by means of linear regression (exceptions, e.g., ion-selective electrodes, immunoassays, HPTLC and the X-ray fluorescence method).

The central question of regression analysis is now how can a line be "best" adapted to the given data points?

(a)

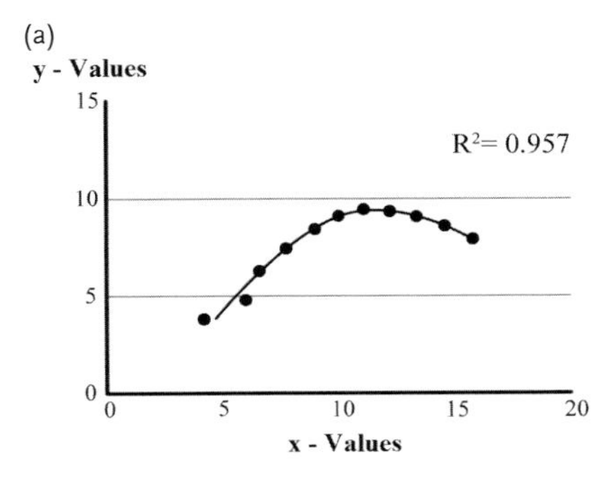

(b)

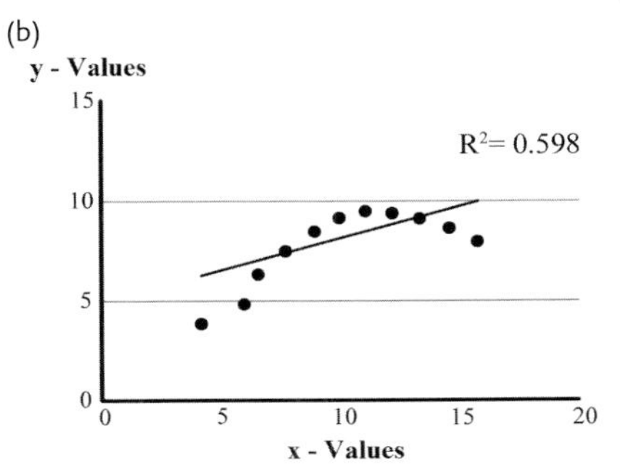

Figure 3.10 Optimal adaptations to regression.

Fulfillment to be regarded as the "best" line: The sum of the squared distances (residuals) of all points from the line (curve) must be as small as possible (least-squares method).

3.4.4 Linear Regression

The simplest kind of functional relation is the linear relation. This model is based on the idea that the correlation between the characteristics *X* and *Y* can be described by a straight line. A linear correlation can be recognized by the fact that all data points can be described by a line, at least approximately.

Without going into the individual calculation stages in detail, the following solution results for the regression curve:

$$y = ax + b \tag{3.21}$$

where y (signal) is the response variable, a (slope) the sensitivity, x (e.g., concentration) the explanatory variable and b (intercept) the blind signal (y value when $x = 0$).

3.4.4.1 Linearity

Linearity describes the propensity of a method to show results within a given range of concentration directly proportional to the concentration of the analytes. This works on the assumption that the correlation between signal and concentration is expected to be linear and can be described mathematically by the linear regression model.

Test of Linearity

- In a visual linearity test deviations from linearity are easily recognized.
- Conduct regression using suitable mathematical models. The correlation coefficient is (still) a widely accepted measure of linearity when using a linear calibration function of the first degree.

3.4.4.2 Statistical Information from Linear Regression

In quantitative analysis the regression calculation, especially in the case of the calibration, can offer very good service by deriving the measurement functions and their parameters (sensitivity, blind signal) from a number of calibration measurements. The calculation method of the linear regression may, however, only be used when there is really a linear correlation between measurement values and quantitative content, and when there is a statistically comparable dispersion or repeatedly measured values at the upper and lower end of the calibration function. Then we talk about variance homogeneity.

The parameters determined from a regression calculation are logically all the more insecure the worse the experimental supporting points are illustrated by the mathematical model. If the experimental points are strongly dispersed around the curve, both the calculated sensitivity and the blind signal bear the risk of a high degree of insecurity. Most pocket calculators and also Excel offer firmly programmed functions to calculate these optimal parameters from a table of values. The statistical functions of Excel thus enable not only the calculation of the slope (= sensitivity E) and the intercept (A_B), but also calculate standard deviations at the same time.

3.4.4.3 Analytical Sensitivity

The gradient of the measurement function in linear regression, given by the slope value, corresponds to the analytical sensitivity. The steeper the line of the measurement function, the more sensitively the signal reacts to changes in the analyte content of the samples. If sensitivity remains constant, one is dealing with a linear measurement function, which is particularly easy to discuss mathematically and statistically. Nonlinear measurement functions can sometimes be made linear by logarithmizing one or both axes.

3.4.4.4 Correlation Coefficient (*R*)

There are several different correlation masses. In the following, reference is always made to the linear product-moment correlation (*Bravais–Pearson*) when not expressly said to be otherwise.

The quality of the regression (correlation) can be a maximum of ±1, namely when the dispersion of the estimates is equal to the dispersion of the data points. This is exactly the case when the data points themselves already lie exactly on a curve. Then data points and estimates are identical and the dispersions are, of course, also equal in size. The further the data points are from the curve, the larger the dispersion of the data points in relation to the dispersion of the estimates.

Mostly, the correlation coefficient (R) is used as a mass for the correlation. The correlation coefficient only gives information as to how strongly the supporting points are scattered around the mathematical function (linear correlation of curve "slenderness" of the swarm of points). The values of the correlation coefficient range from −1 to +1, whereby $R_{xy} = 0$ indicates that there is no correlation between the two variables. $R_{xy} = -1$ indicates a perfect negative and $R_{xy} = +1$ a perfect positive correlation.

The statistical significance of the correlation coefficient is dependent on the number of data points in the calibration.

3.4.4.5 Coefficient of Determination (R^2)

The square of the correlation coefficient is called the coefficient of determination (R^2). The advantage of the coefficient of determination as an index number is that R^2 is always positive.

However, no conclusions can be drawn from these index numbers as to whether a linear or square adaptation would be more favorable. As the correlation need not necessarily be linear, it would not make sense to adapt a curve in Figure 3.10b with the mathematical calibration function ($y = a.x + b$). On the contrary, the points can be well adapted using a polynomial function of the second degree ($y = a.x^2 + b.x + c$) (Figure 3.10a). Consequently, the correlation coefficient determines the quality of a correlation (when using suitable mathematical functions) both in the case of curves and of nonlinear functions.

> If there is a linear correlation between two variables, the correlation coefficient (R) shows this accordingly. A reversal does not necessarily apply. A high correlation coefficient should not necessarily be taken to mean a linear correlation.

***linear range**, Definition according to IUPAC Compendium of Chemical Terminology 2nd Edition, 1997. Concentration range over which the intensity of the signal obtained is directly proportional to the concentration of the species producing the signal.*

3.4.4.6 Regression Equation

The regression equation helps to illustrate the correlation between a dependent and an independent variable. The regression equation supplies the value of the dependent variables, when the independent value is known. An important aspect of regression analysis poses the question what type of correlation exists between both characteristics x any y. Is it justified to adapt a curve? Should a curve be adapted or should a regression analysis perhaps be rejected in the first place?

After all, only when the basic model can adequately represent the reality can correct conclusions and forecasts be made for unmeasured values. For this reason it is vital to examine whether the curve really represents a good adaptation when executing a linear regression analysis. *Anscombe*'s classical example shows the possibilities and limits of linear regression analysis (Table 3.3).

At a first glance, these four records have nothing in common with the same number of elements, with some exceptions. Even when observed in the scatterplot, no commonalities become evident. Without taking a number of arguments into consideration, which speak against a linear regression analysis in three of the four cases, we would like to calculate a linear regression for all four records.

The solutions of the regression curves are as follows:

$$\textit{Regression function data quartet}\ \ 1\text{–}4\ (\text{Eq. 3.21}):\quad y = 0.5x + 3$$

Obviously, the "optimum curve" is exactly the same in all four cases. (Of course the examples have been chosen precisely so as to let the curve be always the same.)

What does this insight show? The example is meant to illustrate that in many cases it is not simply possible to describe the correlation between two characteristics by using one curve. In observing it must be carefully examined whether a linear correlation exists and whether the corresponding regression analysis can be justified. Only then do we get the desired results which can allow meaningful forecasts.

Let us again turn our attention to the four records and go through them individually.

A linear correlation can be recognized in the scatter-plot from data record 1 (Figure 3.11), as all data points could, at least approximately, be described by an even number.

The data points in Figure 3.12 clearly show that there is no linear correlation. To adapt an even number here would not lead to any meaningful results, as the

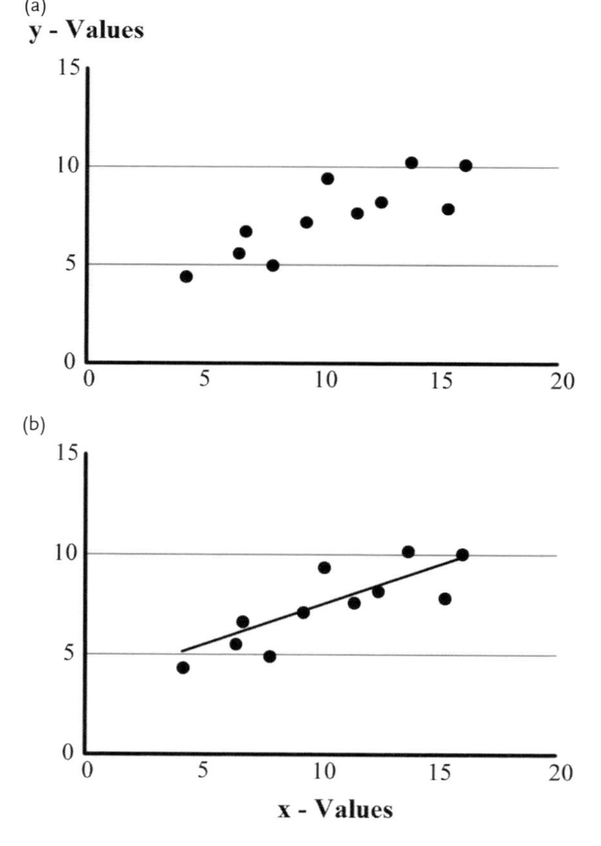

Figure 3.11 Scatter-plot from data record 1 (Table 3.3).

Table 3.3 Data quartet according to *Anscombe* test [8].

Record 1		Record 2		Record 3		Record 4	
x_1	y_1	x_2	y_2	x_3	y_3	x_4	y_4
8	6.95	8	8.55	8	6.77	8	5.76
13	7.58	13	8.74	15.6	13.95	8	7.71
8.63	9.29	9	9.2	9	7.11	8	8.72
11	8.33	11	9.26	11	7.81	8	8.47
14	9.96	14	8.1	14	8.84	8	7.04
6	7.24	6	6.13	6	6.08	8	6.26
4	4.26	4	3.1	4	6.39	19.2	12.79
12	10.25	12	9.13	12	8.15	8	5.56
7	4.82	7	7.26	7	6.42	8	7.91
6	5.68	6	4.74	10	7.46	8	6.89
10	8.04	10	9.33	6	6.73	8	6.58

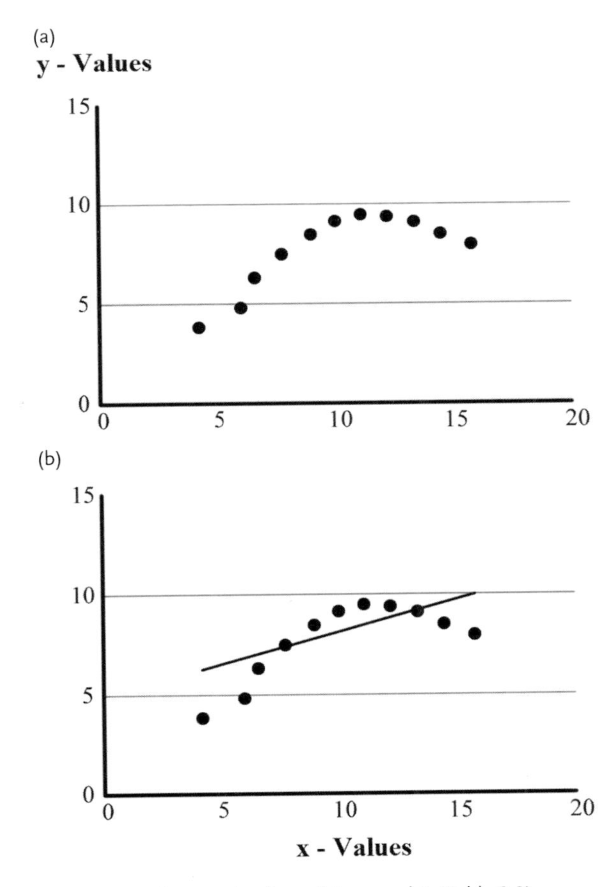

Figure 3.12 Scatter-plot from data record 2 (Table 3.3).

data points are not randomly dispersed around the even numbers. An adaptation here would have to be executed with a second-order polynomial.

The *Anscombe* data set is excellently suited to demonstrate the possibilities and the limits of (linear) regression analysis. It emphasizes the significance of diagram analysis, which immediately permits one to grasp visually whether or not there is a linear correlation between pairs of values.

Even in this case, (Figure 3.13), the thoughtless use of a linear regression analysis would not be advisable. The individual outlier impairs the analysis and has a strongly distorting influence on the even numbers. In practice it should be exactly verified whether in the case of an outlier a random deviation could not be the cause and/or whether it would be justifiable to exclude this one point from the analysis.

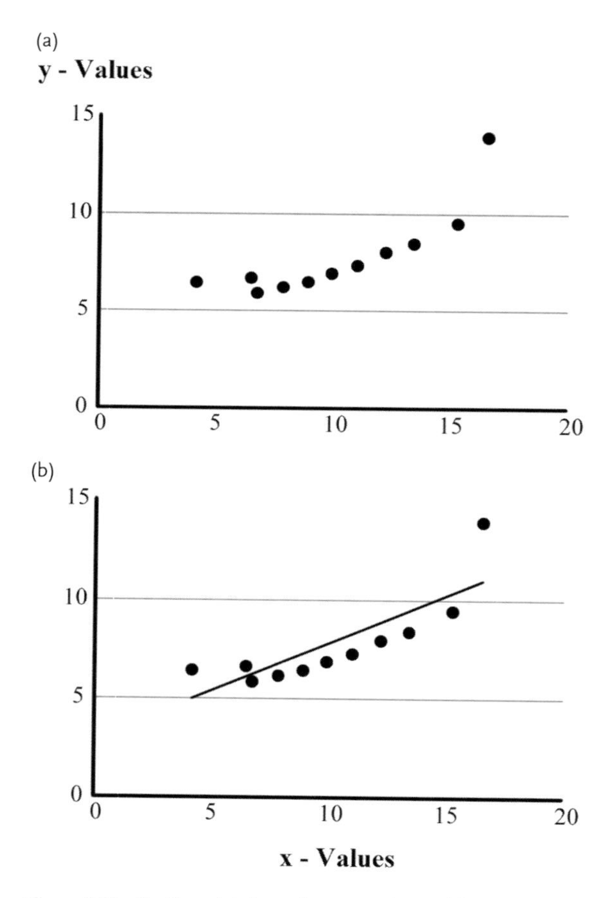

Figure 3.13 Scatter-plot from data record 3 (Table 3.3).

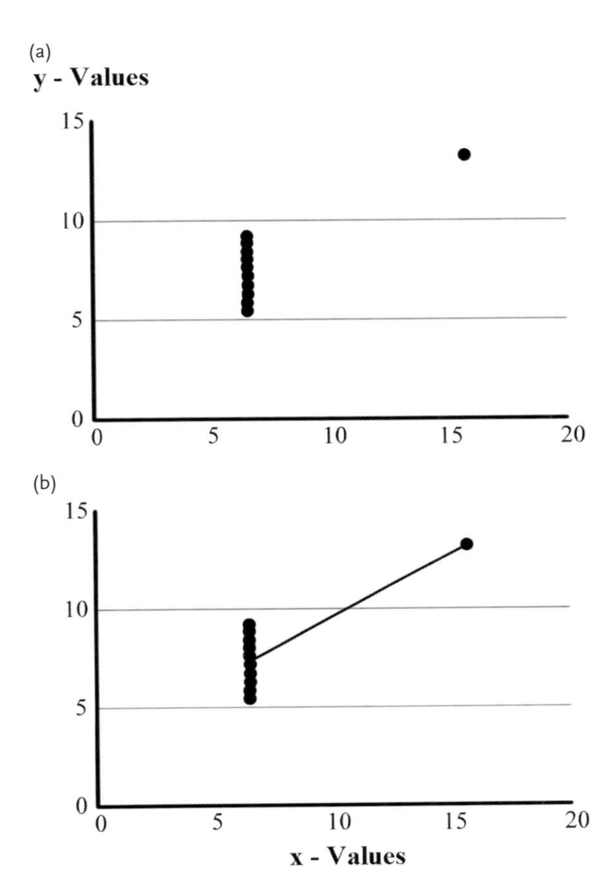

Figure 3.14 Scatter-plot from data record 4 (Table 3.3).

This distribution of data points in Figure 3.14 is not suitable for regression analysis, as there are only two different forms of the x-characteristic (whereby the second form is additionally based on an outlier).

> Did you hear about the statistician who put his head in the oven and his feet in the refrigerator? He said, "On average, I feel just fine."

4
The Analytical Process

4.1
The Analytical Process in the Overall Context

The analytical process is essentially all about obtaining qualitatively valuable chemical information taking into consideration the optimal use of resources (economic, material, personnel) and minimizing risks. This information deficit (that requires answers to questions such as how much, why, when, where?) is defined as an analytical "problem" and must be seen as a two-way interaction between the client and the analytical chemist [9]. The information required from the client is processed by the analyst taking into consideration social, scientific, legal, economic and ecological aspects. This assumes, however, that prior agreement has been reached with the client as to what results or goals are to be aimed at (*Doing the Right Things*, see Section 4.2), comprising the criteria to be used and the acceptability limits of the method.

The analyst uses his expertise and skills as well as intuition to solve the problem of the client/end user. As with every process, the analytical process is embedded in a larger context and is subject to internal and external factors (Figure 4.1).

What is "right" is decided based on the client's requirements (*fit for purpose*).

How the analytical process runs in analytical laboratories depends on how the strategy is embedded in the operational, structural and process organization of an enterprise or an academic institute.

For reasons of simplicity, the entire analytical process is divided into clear, manageable individual tasks (phases), as shown in Figure 4.2.

The workflow can, in the simplest case, be limited to simply examining a sample without sample processing. On the other hand, it can involve the planning of sampling, sampling, sample conservation, sample processing, measurement, calibration, through to the evaluation of the analysis results or storage of retained samples and include statistical validation and accreditation.

Practical Instrumental Analysis: Methods, Quality Assurance and Laboratory Management, First Edition.
Sergio Petrozzi.

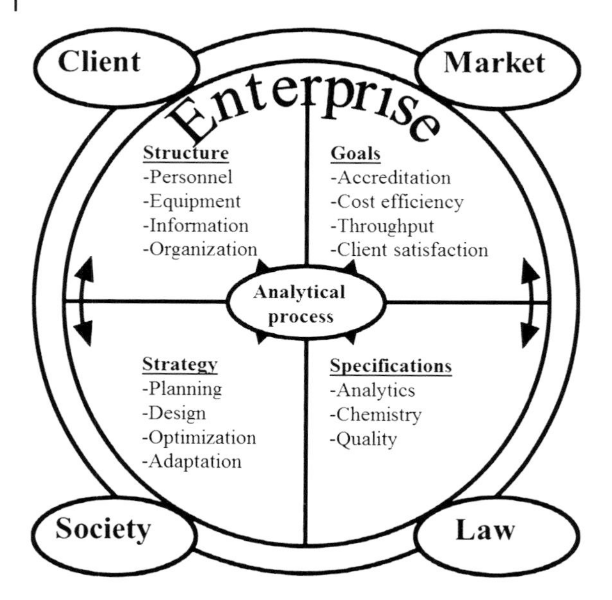

Figure 4.1 The analytical process in the overall context.

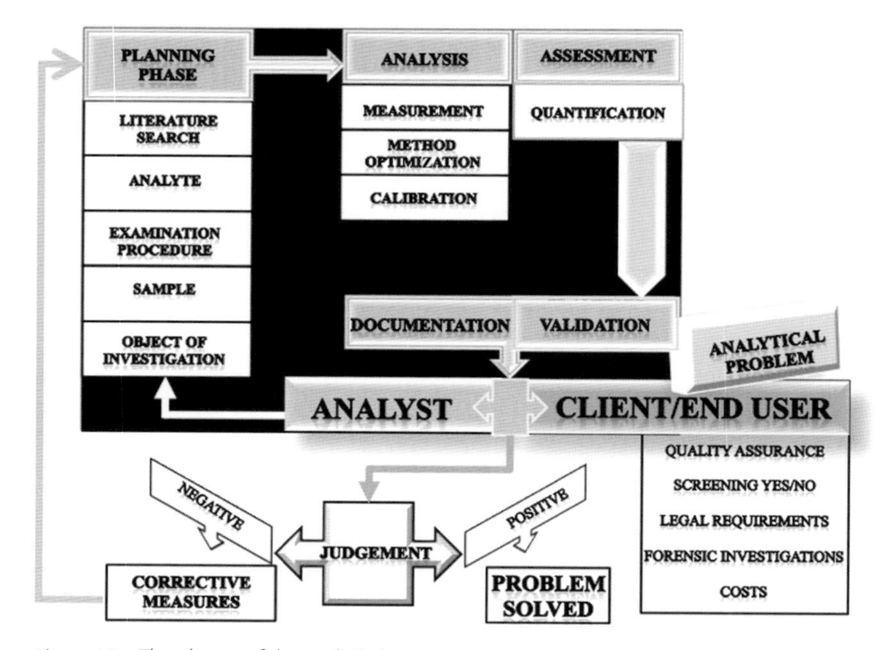

Figure 4.2 The phases of the analytical process.

At the beginning of each analysis there is always an overall question, the *analytical problem*, thus a lack of knowledge that can be eliminated by applying targeted analytical methods.

In the case of most analytical tasks it is about getting necessary information on an *object of investigation* or *investigation system* (system from which chemical information is needed). For this one takes an examination *sample* (representative aliquot out of the whole before proceeding with an analysis) from the object of examination that contains the analyte and subjects it to a suitable *examination procedure* (process to obtain information about the analyte). The examination samples consist of the components of interest, the *analyte* (substance in a sample about which qualitative or quantitative information is needed) and all other components, known as the matrix. The process-relevant information is ascertained with the help of *literature and database searches*. Following the planning phase, the steps *analysis* (application and optimization of the suitable investigation procedure), *calibration* (analysis of concentration of the unknown samples), *assessment of the achieved results* (evaluation, interpretation and critical estimation of the reliability of the results), *validation* (checking the selected method for suitability) and the creation of *conclusive documentation* are conducted.

After that, a *judgment* of the various aspects is made together with the client regarding the overall requirements. Should the assessment prove unsatisfactory, corrective measures are taken or a new procedure is chosen.

4.2 Planning Phase

4.2.1 Analytical Problem

The exact determination of the problem to be analyzed (type of performance desired) is the prerequisite for the meaningful use of the analysis results at the end. Time and financial resources must be put to the best possible use. What triggers knowledge deficit on the part of a client are the specifications, requirements and basic conditions of internal or external involved affected parties, for example, specifications of the authorities, the requirements of specialist departments, quality assurance, corporate proposals and/or the implementation of rules and regulations (Figure 4.1).

The client requirements are formulated as an analytical problem together with the analyst. The client determines the expected use, specifies the solution specifications and names any restrictions. It is crucial for the analyst to have a bird's eye view of the social, economic, scientific and technical background of the problem. There should be a lively exchange between the parties at the beginning and the end of the process. The processed analytical problem now consists of the listing of the object to be examined, examination samples, the examination procedure to be used and the determination of the analyte and, very important, the specifications and limits (Table 4.1).

Table 4.1 Analytical problems (examples).

Analytical "Problem"	Object of Investigation	Sample	Examination Procedure	Analyte
Contamination of a lake	Lake of Zurich	Water sampling at different places at different times	Determination in compliance with EU norm	Pesticides
Hit and run driver	Car driver	Blood	Official procedure	Alcohol
Drug abuse	Confiscation of goods at the customs	Powder-like substance	Quick test	Opiates
Air pollution	Emissions from a factory	Air and suspended particulates	Instrumental trace analysis	Heavy metals

What is also important is to establish the boundary conditions of the system to be measured. Possibly they will have great influence on the analytical results and/or their interpretation. For example, if air pollution must be determined in a residential area, the results will be very different if not only the volatile components, but also the suspended particles are taken into account when expressing the pollution values.

4.2.2 Object of Investigation

Most analytical tasks are all about obtaining certain information about an object of investigation or an investigation system using a suitable analytical method. From these systems examination samples are extracted (aliquots in space and time), from which the desired analytes are qualitatively (presence of analytes = yes/no) or quantitatively (concentration of analytes) ascertained.

***system**, Definition according to IUPAC Compendium of Chemical Terminology 2nd Edition, 1997. Arbitrarily defined part of the universe, regardless of form or size, e.g., for clinical chemistry, patient, patient plasma, patient urine.*

4.2.3 Sample

The examination sample is a part of the object of investigation which must be representative and which must be holistically processed and analyzed in the laboratory. The results obtained are allotted to the system from which it was taken as a whole. The prerequisite for this allotment is to conduct correct sampling and suitable processing.

***sample**, Definition according to IUPAC Compendium of Chemical Terminology 2nd Edition, 1997. A portion of material selected from a larger quantity of material. The term needs to be qualified, e.g. bulk sample, representative sample, primary sample, bulked sample, test sample. The term "sample" implies the existence of a sampling error, i.e., the results obtained on the portions taken are only estimates of the concentration of a constituent or the quantity of a property present in the parent material. If there is no or negligible sampling error, the portion removed is a test portion, aliquot or specimen. The term "specimen" is used to denote a portion taken under conditions such that the sampling variability cannot be assessed (usually because the population is changing), and is assumed, for convenience, to be zero. The manner of selection of the sample should be prescribed in a sampling plan.*

How representative the sample is, is an essential prerequisite for the analytical process and must be clearly defined prior to analysis.

4.2.4 Sampling

The reliability and accuracy of analytical results is largely dependent on professional sampling. The final result of an analysis will never be any more reliable than the reliability of the sampling step.

A large number of standardized sample collection procedures exist, and are described in national and international norms.

Different technologies are used depending on the condition of each aggregate of the material to be sampled.

For example:

- Gases

  passive sampling sorption tube

  (numerous norms, for example, American Society for Testing and Materials (ASTM) E1370–96 (2008) Standard Guide for Air Sampling Strategies for Worker and Workplace Protection)

- Aerosols and dust particles

  sampling on filters, absorption in liquids, absorption in adsorption materials

  Most common collection media are filters.

  Filters are subsequently analyzed by gravimetric, chemical or biological methods.

Inertial separating devices, such as cyclones and impactors, classify particles based on their inertial properties (size). Classifying by size allows determination of the particle size distribution.

- Liquids

  parameter-specific sampling

  (numerous DIN-procedures, e. g. Water quality–Sampling–Part 3: Preservation and handling of water samples (ISO/DIS 5667-3); German version prEN ISO 5667-3:2010)

- Solids (floors, sludge, sediments, wastage, garbage)

  parameter-specific sampling

  (numerous DIN-procedures, e.g. DIN 38402-24 German standard methods for the examination of water, waste water and sludge–General information (group A)–Part 24: Guidance on sampling of suspended sediments (A 24))

4.2.4.1 **Types of Sampling**

Individual sample

An individual sample is a sample gained by a one-time extraction.

Independent sampling

Independent samples are those samples selected from the same population, or different populations, which have no effect on one another. That is, no correlation exists between the samples.

Random sampling

Random sampling is one (or several individual samples) taken to assess the momentary condition of the extract.

Or: a certain number of identical stacks of a charge, for example, 10 bottles of olive oil from one pallet.

Aggregate sampling (also known as composite sampling)

Aggregate sampling is a united sample from several individual samples

Average sampling

Average sampling is the name given to a special form of sampling individual samples by mixing these by hand, in compliance with given rules, or by mixing them automatically with sampling machines, continuously or intermittently, and whose composition is representative.

Sampling location

The sampling location and the concrete places from which samples are obtained are to be exactly specified. It must be possible to gain representative samples from these locations.

By means of corresponding pretreatment steps, the sample is made from the raw sample and subsequently undergoes measurement.

If necessary, sampling follows a preservation stage. Within the framework of quality assurance, as far as this is not specified in the analysis regulations, suitable possibilities of preservation must be sought and, in practice, checks must be made as to what period the samples to be analyzed can be stabilized for and which errors occur on a time basis. In so doing, attention must be paid to whether the sample is representative (sampling plan). It must be concentrated enough (content of analytes) and several samples must be extracted and analyzed separately.

sampling plan, *Definition according to IUPAC Compendium of Chemical Terminology 2nd Edition, 1997. A predetermined procedure for the selection, withdrawal, preservation, transportation and preparation of the portions to be removed from a population as samples. Summarizing the test values or observations from the selected portions yields an estimate for the concentration of an analyte or a value for a property determined with a calculable degree of uncertainty at a specified confidence level. A sampling plan includes the designation of the number, location and size of the portions, and instructions for the extent of compositing and for the reduction (in amount and fineness) of the portions to a laboratory sample and to test portions. It may also contain acceptance criteria. Some sampling plans do not include more than instructions for the statistical selection of portions to be removed. Such plans should properly be designated as "statistical sampling plans".*

A sampling protocol must ensure that the type of samples and their pretreatment can later be retraced. For this reason, all important activities, observations and external marginal conditions are to be documented.

4.2.4.2 **Sampling Errors**

Many sampling difficulties arise from inhomogeneities in the material under study; such inhomogeneities can occur in almost any mixture, including gases and liquids. As examples, the level of nitrogen oxides in the atmosphere is greatly affected by the distance from a highway, and concentrations of soluble substances in a lake or stream by location or depth. For mixtures of solids of moderate particle size, the problem of sampling can be particularly severe. One possibility for reduction of sampling error is to increase the sample size, because the greater the sample size, the closer your sample is to the actual population itself. If you take a sample that consists of the entire population you actually have no sampling error because you do not have a sample, you have the entire population.

sampling error, *Definition according to IUPAC Compendium of Chemical Terminology 2nd Edition, 1997. That part of the total error (the estimate from a sample minus the population value) associated with using only a fraction of the population and extrapolating to the whole, as distinct from analytical or test error. It arises from a lack of homogeneity in the parent population. In chemical analysis, the final test result reflects the value only as it exists in the test portion. It is usually assumed that no sampling error is introduced in preparing the test sample from the laboratory sample. Therefore, the sampling error is usually associated exclusively with the variability of the laboratory* *sample. Sampling error is determined by replication of the laboratory samples and their multiple analyses. Since sampling error is always associated with analytical error, it must be isolated by the statistical procedure of analysis of variance.*

4.2.4.3 **Sample Handling**

Goals of sample handling

- Homogenization of the sample (mixing, grinding, digesting, dissolving)

  For many measurement procedures, the sample must be liquid (exception: gas analysis).

  A lesser number of analytical measurement methods can detect analytes quantitatively in solids (solids NMR, IR from KBr pellets, surface analysis).

- Digestion (making the analytes available)

  Inorganic separation procedure, microwave digestion.

- Removal of interference (filtration, extraction, distillation, chromatography, etc.)

  Many modern physical measurement techniques, especially in the trace range, are sensitive to interfering substances that impair the sensor during measurement, or produce similar signals to the analyte (non-specific sensor). The separation of interferences is particularly important for trace analysis, where there is a probability that, in addition to the hundreds of small amounts of analyte in the matrix, hundreds of interfering substances are present. In proportion to the decreasing analyte content, the concentration of these interfering substances increases exponentially.

- Bringing the sample content into the working range of the method (concentration or dilution).

***sample handling**, Definition according to IUPAC Compendium of Chemical Terminology 2nd Edition, 1997. Any action applied to the sample before the analytical procedure. Such actions include the addition of preservatives, separation procedures, storage at low temperature, protection against light and irradiation, loading, etc.*

4.2.4.4 **Difficulties of Sample Processing**

Difficulties that may arise during sample processing are:

- Analyte losses (e.g., during separation of the matrix by thermal decomposition, etc.). Thus the sample contains less analyte than the sample under investigation. This difference should be either kept insignificantly small or quantifiable (and thus correctable).
- The addition of impure reagents, solvents, and so on and/or using contaminated equipment ("memory effect") leads to additional analyte in the sample. Both problems lead to systematic deviations which, once they have been identified, can be corrected accordingly in the measurement results.

Table 4.2 Core specifications for analytical results.

Application	Specification
Quality control of a product	Accuracy in a narrow concentration range
Screening analysis (yes/no)	Short analysis time
Limit compliance with federal regulations	Absolute accuracy
Forensic investigations	Documentation and full traceability

4.2.5 Examination Procedures

The examination process includes the stages of all the detailed work steps made to obtain analytical results for a defined analyte. Depending on the needs of the client (taking into account the requirements and limitations) an acceptable investigation procedure and an appropriate test method are now selected. As shown in Table 4.2, the analyst must choose a procedure that satisfies the requirements of the end user. It goes without saying that the analyst always wants to provide the most accurate analysis results which, however, also means time-consuming and costly services. However, this is not always desired by the client. That is why it is important to clearly define the specifications prior to each analytical process.

To achieve specifications in quality control for which a particular concentration range must be observed (min/max values), different investigation specifications are required than are required in, for example, a screening yes/no analysis. Screening analysis is used for high sample throughput or for rapid testing. Here, a method is preferred, which allows quick decisions on whether further investigations are required. If, for example, the limit of cadmium in water is set at $4\,mg\,l^{-1}$, the analysis must be quantitatively very accurate and provide reproducible results for the concentration value of $4\,mg\,l^{-1}$. The measurement procedure does not need to be accurate over a large linear operating range (e.g., 0.1–$40\,mg\,l^{-1}$) because if the limit is exceeded, it does not matter whether the measured values are 100% or 200% higher than the limits. In any case, the result is said to be positive. The same is true at a lower concentration (e.g., $1\,mg\,l^{-1}$). The fitness–for–purpose acceptance of the measurement uncertainty range is determined by the client.

In the case of forensic investigations it is mandatory to document all processes. Here, the traceability of all results is essential, since all the process steps could be viewed in court. To guarantee these specifications, the analytical laboratory is usually accredited and the achieved results are periodically substantiated in proficiency tests. Even the language and presentation of the documentation must suit the area of application. For example, what is clear to the chemist can be difficult for the lawyers to understand.

4.2.6
Analyte

The analyte or analytes are those substances in a sample on which information is to be gained, that is, which are to be analyzed. The analyte is a part of the analytical problem to be solved.

The information gained from the analysis permits the formulation of various degrees of quality of an analyte, for example, contamination of a body of water with cyanides:

- whether there is cyanide in the water
- total concentration of cyanide in the water
- various cyanides available (free, bound)
- concentration of the various cyanides
- exact nature of the cyanide, for example, the cyano group occurs in different types of gaseous, liquid or solid compounds.

analyte, *Definition according to IUPAC Compendium of Chemical Terminology 2nd Edition, 1997. The component of a system to be analyzed.*

4.2.7
Literature and Database Research

Once the framework has been established, the analyst requires relevant information to solve the problem. This is obtained primarily through searching the chemical literature.

What? You have to be clear about the subject under study and narrow it down. Under certain circumstances this is possible only after obtaining initial information. During the research and assessment of information it often becomes necessary to redefine the desired goal.

How much? It must be determined how much information you want to/should process and what depth, actuality and scope the final results (introductory information on a question, short presentation, term paper, thesis, etc.) should have.

When? It takes time to procure and evaluate documents. The completion date determines how much material can be processed, and obliges us to make a reasonable selection and do efficient research.

Where? For the most part, the evaluated chemical information is to be found only in the commercial products of journal publishers and database producers and is not accessible through search engines such as Google or free services such as Wikipedia. Successful research therefore requires knowledge of the sources, their function and structure, as well as of procedures that are conveyed in the following sections. Regarding the *where?* it must be clarified whether one's own institution (the library at a university or the documentation center in a company) has the appropriate source or whether a neighboring institution must be consulted (visit for research, interlibrary loan or copy job).

4.2.7.1 Types of Chemical Literature

Chemical literature can be divided into three types. Primary literature refers to documents in which the research results have been published for the first time. The most important types of documents in this category include journal articles, patents, theses and dissertations, conference and research reports. Secondary literature summarizes and makes primary literature accessible for research. Here the main categories of documents are papers that explore bibliographic information on the documents, abstracts, and terminology of the content, such as manuals that comprise cross references, reactions and properties. The most important papers and manuals in chemistry today are now available as databases. These database products make no distinction between papers and manuals. There are no absolute borders between them. Tertiary literature includes dictionaries, encyclopedias, and the individual monographs. What is typical of this type of literature is the high degree of abstraction that originates from the primary literature.

4.2.7.2 From the Question to the Document

Depending on the type of enquiry, there are three approaches to information retrieval:

1) If there are already quotations, for example, literature recommendations or references in books and magazine articles available, one then gets to the desired documents through research in library catalogues, magazines or patent directories (as described in Section 4.2.7.3).

2) Next you look for literature on a specific topic, such as an analytical method or a connection, its production and characteristics. Depending on the type of questions, databases are suitable for a detailed search for relevant primary literature, encyclopedias for training in a subject, library catalogues for tertiary literature, and search engines for pre-orientation of research on the Web. The result of the research is either the document itself or a list of quotations that can be then resolved. Appendix A lists the major databases and sources.

3) Information is constantly required, on a particular topic, for example, the latest magazine articles or patents on a particular specialist topic. This one uses information in databases or services from magazine publishers (see Section 4.2.7.8)

4.2.7.3 From the Quotation to the Document

To locate a document, the type of document must first be determined based on the nature of the citation. From this the suitable search engine and the manner of proceeding follow. The most important types of documents are:

- ***Books or Chapters in Books***

  Recognition characteristics: author(s) or publishers, title and edition, if necessary,

  Examples: (i) Book: Joseph Wang (1994) *Analytical Electrochemistry*, VCH Publishers, Inc. (ii) Chapter in a book: Price, G.D. ,Wood, I.G., and Akporiaye, D.E. (1992) in *The Prediction of Zeolite Structures in Modeling the Structure and Reactivity of Zeolites*, Ed. C. R. A. Catlow, Academic Press: London, p. 19.

  Procedure: Search in library catalog of the institution; author, title, and possibly year of publication.

- ***Journal Articles***

  Distinguishing features: Author(s), title of the article, journal title (often abbreviated), volume number, if applicable, issue number, page number of the first page or where on the page the information was found, year, if applicable, and possibly DOI (digital object identifier).

  Examples: (i) W.-C. Liao, J.A. Ho, *Analytical Chemistry*, **2009**, 81 (7), 2470, (ii) I. Sutherland, P. Hewitson, S. Ignatova, New 18-l process-scale counter-current chromatography centrifuge, *J. Chromatogr. A*, 1216(19), 4201, DOI:10.1016/j.chroma.2008.11.097.

  Procedure: As libraries collect journals only to the level of journal title and issue, the procedure takes place in two stages. In the first step, the journal title must be investigated (in the example: Anal. Chem = Analytical Chemistry, J. Chromatogr. A = Journal of Chromatography A) and looked for in the library catalog or journal directory. Many library catalogues cannot deal with journal abbreviations, but require the full title of the journal. Tools for translating from the short to the full title, for example, CASSI (Chemical Abstracts Service Source Index) or the CLICAPS catalog of the ETH Zurich, http://www.clicaps.ethz.ch/en/. Once the desired title and source has been found (print or online edition), the second step requires us to search directly in the journal for the year, volume, issue and page numbers, and, in the case of an available online journal, on the companion website of the publisher. When a DOI exists, you can jump directly to the online article by writing dx.doi.org/ in front of the browser's URL (for the example above: http://dx.doi.org/10.1016/j.chroma.2008.11.097).

  Example of a journal directory:

  Electronic Journals Library, http://rzblx1.uniregensburg.de/ezeit/index.phtml?bibid=ZHAW&colors=7&lang=en, with its convenient traffic light system indicating the availability of the electronic journal to the desired institution.

  Tip You can inform yourself in your own library or documentation center on the ordering and delivery of copies of journal articles not locally available.

- *Patents*

  Distinguishing features: Two digit country code, patent number, and, if applicable, kind code

  Examples: (i) WO2009045789 A2, (ii) DE2403138 (A1), (iii) US2781404

  Procedure: Patents can be searched for by using the patent server of the European Patent Office, ep.espacenet.com/, or the U.S. Patent Office, patft.uspto.gov/, including the country code and patent number.

Other types of documents and procedures are dealt with under http://www.infochembio.ethz.ch/en/search_strategies.html (accessed May 2012).

4.2.7.4 The Question of Topic

The choice of the search agent often depends on the depth of the topic: for a quick introduction to a specialist subject specialized dictionaries and encyclopedias like Römpp, Ullmann and Kirk-Othmer, universal encyclopedias like Wikipedia (Appendix A), or research using the "keyword" button in library catalogues for monographs and series can be recommended. However, if detailed and extensive research for journal articles (and sometimes even for patents) is required on chemicals topics, connections and reactions, there is no way to ignore the three following databases:

Chemical Abstracts

Chemical Abstracts is the largest and most important reference tool n the field of chemistry and comprises several databases (CAplus^{SM}, CAS Registry^{SM}, CASREACT®, CHEMCATS und CHEMLIST), for which there are several interfaces for making queries. Approximately 9000 journals and patents in the field of chemistry and related fields are constantly being assessed. Currently (April 2009) there is proof of the existence of around 30 million documents, 46 million organic and inorganic compounds, 61 million biosequences and 17 million reactions [10]. Companies often use the SciFinder software or the paid access via the online host STN. At universities [11] the easy to use software SciFinder Scholar and the browser-based SciFinder Web are widely used. Instructions for the use of SciFinder (Scholar or Web) can be found in the Chemical Abstracts Service [12] or in [13].

Beilstein and Gmelin

Beilstein and Gmelin are, after Chemical Abstracts, the most comprehensive reference for organic or inorganic and metal organic compounds and reactions. Both originated from the original manuals of the same name and are available as databases on software CrossFire Commander (Elsevier), or the web interface DiscoveryGate (Symyx) and Reaxys (Elsevier's successor product to CrossFire) [14]. As so-called factual databases, Beilstein and Gmelin contain not only compounds and reactions, but also several hundred million data values and information (literary information on spectra, physical and chemical properties, ecological, toxical and pharmaceutical data) organized in several hundred data fields which can also be researched like the structures and reactions.

Note: For licensing reasons, you should use the products SciFinder, Crossfire, DiscoveryGate and Reaxys only within your institution.

Tip

The online tutorials of the database manufacturers provide the service but not background information on the products, such as coverage of content, content gaps and limitations of the user interface. This useful information and several tips on how to research are available at courses in the libraries and at lectures of nearby universities, where you can sign up as an expert listener. Other sources are further education opportunities of professional organizations.

The following bibliographic databases enable research for authors and titles of journal articles as well as abstracts therefrom:

4.2.7.5 **Science Citation Index Expanded**

Interdisciplinary database of Thomson Reuters on natural sciences and biology, technology and medicine, which is also accessible through the licensed products ISI Web of Science (WoS) and ISI Web of Knowledge (WoK). They cover about 7100 journals published from 1900. As so-called citation databases WoS and WoK allow not only retrospective search for items, but also a prospective search for articles in which other articles are cited, and thus search for similar articles on a topic, do citation analysis and help to observe the competition. Instructions for use can be found in [12].

4.2.7.6 **Scopus**

Analogous to WoS, the interdisciplinary citation database of Elsevier with a wider coverage of journals (approx. 15 000 more European journals considered), but which covers a shorter period of time: Bibliographic data from 1960, citations 1996.

Tip

As a so-called Walk-in User (visitor) WoS, WoK and Scopus can be used in a library, even if you are not a member of the said library.

4.2.7.7 **Medline**

Medline from the National Library of Medicine contains some 16 million references in the biomedical field dating from 1949. Medline and other databases are freely accessible through PubMed and http://www.pubmed.org.

Tip

To learn whether an institution has licensed the above-mentioned products it is best to visit the respective database directory on the library's websites or the information center.

4.2.7.8 Current Information

The table of contents of desired journals can be obtained regularly by e-mail by registering with journal publishers such as ACS Publications, Wiley Interscience or Elsevier ScienceDirect using the button "Set E-Mail Alert", "e-Alert" or "alerts". Alternatively I recommend the RSS feeds of the publishers. Many database products also offer Alert services. The "My Alerts" feature of Scopus allows you to define searches as well as authors and documents that are cited. ISI Web of Knowledge contains the function "My Citation Alerts". In SciFinder and SciFinder Web searches can be filed as "keep-me-posted" profiles stored; the advantage here is that patents are also included. Information and in-company patent authorities often deposit research profiles (SDI, selected dissemination of information) with online hosts such as STN or offer their own patent watch services.

4.2.7.9 Specific Types of Documents Such As Norms and Patents

Norms Simply expressed, norms are rules of technology. They demand rationalization, facilitate quality assurance, serve workplace safety, and standardize test methods, (for example, in environmental protection), and facilitate in general communication between the economy, technology, science, administration and public life, to name just a few examples.

As norms are often required as an entire document for non-private use, they may not be copied in keeping with copyright law, but must be purchased at a norm publisher, for example, Beuth, http://www.beuth.de. Alternatively, it is possible to use them by purchasing a license to use the Perinorm database.

Patents For research requiring a brief overview Chemical Abstracts (via SciFinder Scholar or Web), Derwent World Patent Index (to be found in Web of Knowledge) and search aids on the websites of patent offices are ideal. However, if it is necessary to do comprehensive research into patents and their legal status for a planned patent application, this should definitely be entrusted to patent specialists or information brokers.

The rapid change of chemical information reflects the increase in the rate of change in the entire scientific community. Thus this script could be already partially outdated at the time of publication.

4.3 Analysis

4.3.1 Measurement

Measurement is defined as the signal response which results from the interrelation between the measured sample and the probe (sensor) in a detector.

Spectroscopic methods usually allow another formulation: in a spectrophotometer a reaction follows a provocation of the sample. This reaction is converted into a measurable signal in the detector. These signals are used to obtain a reliable (quantifiable) statement about an unknown measured quantity of an object of investigation.

The signals triggered by the measurement are also known as “raw data”. This raw data must in accordance with Good Laboratory Practice (GLP) guidelines always be available (at least 10 years) and form an integral part of the analysis documentation.

Measurements are conducted by means of analytical procedures determined in investigation procedures (see Section 4.2.4).

The defined requirements (limits) of the used analytical method are compared with the actual performance characteristics in order to determine the suitability of the method for its intended purpose. The requirements cannot be generalized and depend on the core requirements of the analysis results

4.3.2
Method Optimization

The laboratory’s quality assurance measures within the analytical work begin with the transfer of the raw sample to the test sample. In the simplest case, the measurement equipment is to be calibrated, standards with a known content and samples without the material to be tested, so-called blank values, are to be subjected to the normal measurement procedure and to statistical evaluation. In most cases, there are analysis specifications which determine the workflow of preliminary sampling, measurement, assessment, calculation and result expression. With the help of statistical interpretation of the compiled data it can be ensured that the measurement process is performed within the given framework. For example, blank values give information on contamination of reagents. In addition, standards show changes in the sensitivity of the measurement process. In spiked samples the recovery of a given amount of substance can be determined; by means of multiple determinations, measurement precision can be checked. The parameters exemplified here enable us to check the quality of the measurements, so that in the case of unsatisfactory results the measurement series can be discarded and the cause of the faulty quality can be investigated. Thus, within the analytical process, there is a feedback between the measurements and quality assurance measures which aims at “fulfilling” our “wishes” adequately. The selected analytical method is now used in a targeted manner. Even if, for example, the analysis by chromatography is a versatile method, in order to obtain meaningful results, a whole series of trial separation experiments must generally be conducted to obtain optimal method parameters. After optimizing the relevant (previously defined and desired) parameters the measurement campaign can begin.

The parameters to be determined can vary depending on corporate strategy and operational area. For example, a pharmaceutical laboratory investigates a drug for certain (known) impurities. If the synthesis is changed, new impurities can arise for which the existing method must be optimized. Environmental analytical laboratories are now confronted with trace analysis. Here achieving the lowest possible detection limit is certainly an important parameter. Also, not only analytical factors influence the selection of the target criteria but also, for example, economic aspects (high sample throughput) or ecological aspects (minimizing the solvent used).

> The specifications for the selected method and the optimization effort will depend on the framework conditions and the "wishes" of the client. It is the client who formulates the goal (acceptable measurement uncertainty of the results achieved, internal/external restrictions, dates, costs etc).

4.3.3
Calibration

The set of operations which establish, under specified conditions, the relationship between values indicated by the analytical instrument (e.g., absorbance, impulses per time, peak areas, currents or voltages) and the corresponding known values of an analyte is defined as calibration.

The construction of a calibration line should include seven or more calibration points. This time-consuming process can be shortened if applied after a "full" calibration procedure showing an acceptable calibration model over the range of interest.

Practical application:

- One-point calibration method (one standard)

  With this calibration it is not possible to indicate the blind signal and it should only be used when standard and test samples have comparable signals or when the measurement functions throughout the work area are definitely linear and the intercept (almost) zero.

- Two-point calibration method (two standards with different content)

  This type of calibration can be used under the following conditions:

  - The contents of the two standards do not differ greatly.
  - The sensitivity in the content range of the standards is continuous and relatively large.
  - The content of the samples lies between the content of the standards.

    It is possible to indicate the blind signal when the standard content is close to the detection limit.

- Multiple-point calibration method (More than two standards with different contents)

  Regarding this calibration, the contents of the standard solutions should be within the range of the samples to be measured. The determination of sensitivity and blank signal can be done in the form of a diagram or by linear regression (see Section 3.4.4).

4.3.3.1 **External Calibration**

In the case of external calibration, the relationship between two measurement sizes is determined, whereby one size is directly measurable. In general, the unknown factor is the concentration. Several samples with known concentrations are measured and a corresponding size, such as absorbance, is determined with the measuring process. From pairs of values a calibration curve is created with the help of which the unknown concentration can be determined.

- Prerequisites for an external calibration:
  - Test matrix and standard matrix must be comparable
  - Few systematic sources of deviation
  - High analysis reproducibility of the standard and the sample
- Advantages of an external calibration:
  - Many comparable samples can be analyzed without extra effort (routine operation)
- Disadvantages of an external calibration:
  - Systematic deviations are difficult to detect, system fluctuations falsify measurement values (for example, GC with an external standard)
- Lessening systematic deviations during calibration:
  - Matrix adjustment:

    The matrix of the standard solution and the sample is adjusted. The difficulty here is that the sample matrix is often not known.
  - Using the standard addition method:

    The matrix is modeled accurately, which can result in high effort.
  - Internal standard:

    A chemically related substance is added to the analyte in a known concentration. This may not be present in the sample and a relative response factor must be calculated. The internal standard and the analyte must be simultaneously determinable; the analysis must therefore be selective for both substances.

4.3.3.2 **Internal Standard (IStd)**

The IStd is used as a relative reference, and it is used in quantification. Because an internal standard method uses the ratio of responses for standard and analyte, quantitation does not depend on the amount of material injected. The ratio of the peak area of the target compound in the sample, or sample extract to the peak area of the internal standard in the sample or sample extract is compared to a similar ratio derived for each calibration standard. This ratio is termed the response factor (RF) or relative response factor (RRF).

$$\mathrm{RF} = \frac{A_{\mathrm{IStd}} \cdot C_{\mathrm{Sample}}}{C_{\mathrm{IStd}} \cdot A_{\mathrm{Sample}}} \tag{4.1}$$

where A_{Sample} is the area of the analyte, C_{Sample} the unknown concentration of the analyte in the sample, A_{IStd} the area of the internal standard, and C_{IStd} the concentration of the internal standard

$$C_{\mathrm{Sample}} = \mathrm{RF} \cdot C_{\mathrm{IStd}} \frac{A_{\mathrm{Sample}}}{A_{\mathrm{IStd}}} \tag{4.2}$$

internal standard (in chromatography)

A compound added to a sample in known concentration to facilitate the qualitative indentification and/or quantitative determination of the sample components. IUPAC Compendium of Chemical Terminology 2nd Edition, 1997.

4.3.3.3 **Standard Addition Method (Spiking Method)**

The sample matrix can lead to quantification interferences. The calibration must therefore be able to compensate such effects. In the method of standard addition, the calibration is not created separately with a man-made dilution series of standards (sample internal calibration). A defined quantity of the same substance as the analyte is added in separate and equally spaced amounts to the real investigation sample. Assuming that the analytical method is effectively linear over the required working range the standard addition method eliminates rotational effects with a negligible effect on precision.

4.3.3.4 **One-Time Addition**

The procedure corresponds to a calibration with only one standard, with the higher deviations as expected. After the addition, where appropriate, the dilution of the spiked sample is taken into account.

$$\frac{c_{\mathrm{A}}}{(c_{\mathrm{A}} + c_{\mathrm{S}})} = \frac{y_{\mathrm{A}}}{(y_{\mathrm{A}} + y_{\mathrm{S}})} \tag{4.3}$$

where c_{A} is the unknown concentration of the analyte in the sample, c_{S} the known concentration of the standard after adding to the sample volume, y_{A} the signal (absorbance) of the analyte in the sample, and $y_{\mathrm{A+S}}$ the signal from the sample with a standard addition

4.3.3.5 Multiple Additions

Here the real sample analysis is increased again and again. Usually the analyte concentration of the sample is estimated and increased in equal intervals up to twice the estimated concentration. From the individual measurements there now results a straight line which results in the content of the analyte sample being sought when extrapolating for $y = 0$.

The advantage of this method lies in the compensation (ideally elimination) of matrix effects, which can lead to proportionally systematic errors. Blank values (constant and systematic deviations) cannot be determined with the standard addition method

This method has proven itself during measurement in complex sample matrices.

In case of matrix interference, which is often dependent on concentration, the resulting calibration function is usually not linear.

Example:

Determining the Concentration of Phenol (Photometrically):

50 ml of a solution of unknown phenol concentration is spiked with a phenol standard solution and filled to 100 ml with water (Table 4.3, Figure 4.3).

Calculation:

The curve of the calibration line runs in accordance with the function.

Table 4.3 Absorbances of phenol solutions (measured at $\lambda = 510\,\text{nm}$).

Solution	Analysis sample	Spike 1	Spike 2	Spike 3	Spike 4
Spiking concentration in $\text{mg}_{\text{phenol}}/100\,\text{ml}$	0	1.0	1.5	2.0	2.5
Absorbance (Signal y)	0.114	0.221	0.271	0.314	0.363

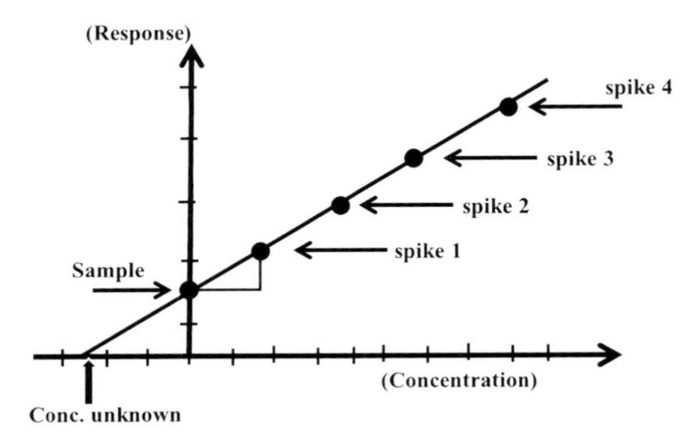

Figure 4.3 Standard addition procedure.

$$y = 0.099x + 0.1174$$

The concentration of the spiked sample (test sample) is obtained from the intersection of the calibration line with the abscissa:

$$y = 0 \quad 0.099 * \text{conc}\,(\text{mg}_{\text{phenol}}/100\ \text{ml}) + 0.1174$$

$$x = \frac{0.1174\ \text{mg}}{0.099 \cdot 100\ \text{ml}} \quad 1.18\ \text{mg phenol}/100\ \text{ml}$$

(i.e., 50 ml of the unknown sample contains 1.18 mg phenol).

The result:

Mass concentration of the unknown sample =

$$\frac{m_{\text{phenol}}}{V_{\text{sample}}} = \frac{1.18\ \text{mg}_{\text{phenol}}}{50\ \text{ml}} = \frac{23.6\ \text{mg}_{\text{phenol}}}{1}$$

4.3.3.6 **Recovery Standard (RStd)**

Depending on their intended purpose, recovery standards (RStd) allow the detection of sample losses (loss mechanisms) as well as compensation (computational compensation). Table 4.4 gives an overview of possible application areas of recovery standards and their corresponding name and function.

For the recovery reference standards to exercise these functions, the following requirements must be met (e.g., HPLC):

- If possible, similar chemical properties to those of the compounds to be quantified. Also suitable are, for example, compounds with the same basic structure but with different or additional functional groups which do not significantly affect the polarity.
- The RStd must neither react with the analyte nor with the matrix.
- A similar retention time to that of the sample compound(s) to be quantified.

Table 4.4 Use and designation of recovery reference standards.

Date of adding the standard	Description	Function
Prior to sample extraction	Extraction standard	Corrects for extraction and processing losses
Prior to extract processing	Processing standard	Corrects for processing losses
Prior to the quantification	Recovery standard	Allows the calculation of the losses that occurred in the previous steps

(Source: http://www.muttenz.ch/dl.php/de/46cc645b14b8f/QSKonzept-Oehme-deutsch-final.pdf, accessed May 2012).

4.4 Assessment

4.4.1 Quantification

The transition from the raw data to the actual analytical results is called assessment.

There is a functional relationship between the signal and the analysis result known as the analytical function. The analytical function can be a mathematical formula, a table of logical links or a diagram. Only in the case of a few analytical methods is the absolute size of all parameters of the analytical function known (absolute analysis method). For all other quantitative methods (relative procedure) these parameters and thus the course of the analysis function must be empirically ascertained by conducting sample measurements with known analyte content (standards). This process is called quantification.

4.4.1.1 Traceability

One of the greatest difficulties in analytical measurement technology is to determine the true content of standards (certification). It must be possible to base such content on a calibrated measurement method (for example, the weight of a sample of the analyte of high purity). It is, however, not always possible to trace back to weighable references of high purity. Many analytes (e.g., of a biological and biochemical nature) cannot be separated in their purest form. The sample matrix can affect the parameters of the measurement function (matrix interference). Should it not be possible to separate the matrix prior to measurement, this leads to a systematic deviation.

Remedy:

Interlaboratory testing (comparative laboratory studies) can be arranged in which several laboratories all analyze the same sample using the same method. The average of the results is then accepted as the best estimate for the right content (conventional true value CTV).

4.4.2 ("True") Content

It is the aim of every quantitative analysis method to determine the true content of the sample accurately, that is, to give an "exact measurement".

How much analyte is really in a sample, however, cannot be determined with certainty. If the content is measured with a relative method, one must first know the true content of the reference materials used for calibration.

However, an absolute method can also be affected by a systematic deviation, which is not apparent from the measurement results. By producing high-purity substances (primary standards) an attempt is made to counteract this problem. These primary standards are used to manufacture certified standard solutions, which are then analyzed with an absolute procedure. The “true” content thus obtained is used for validating other analytical methods.

Wherever there are no primary standards or certified standards available, the participation in interlaboratory comparisons is carried out if the effort is justified.

4.4.2.1 Terminology

Reference materials (RMs) are substances of sufficient homogeneity, of which one or more characteristic values is determined in such a manner that they can be used for the calibration of equipment, the assessment of measurement methods, or the allotment of material values. If reference materials have a certificate indicating the uncertainty and the associated confidence range, they are called certified reference materials (CRMs) [4].

reference material, *Definition according to IUPAC Compendium of Chemical Terminology 2nd Edition, 1997. A substance or mixture of substances, the composition of which is known within specified limits, and one or more of the properties of which is sufficiently well established to be used for the calibration of an apparatus, the assessment of a measuring method or for assigning values to materials. Reference materials are available from national laboratories in many countries (e.g., National Institute for Standards and Technology [NIST], USA; Community Bureau of Reference, UK).*

4.5 Validation

Definition (DIN EN ISO/IEC 17025:2000/1/)

Validation is the confirmation by examination and provision of objective evidence that the particular requirements are met for a specific intended use.

The analytical chemist is becoming increasingly governed by regulations controlling the validity and quality of the data being produced. To obtain maximum confidence in the results, attention should be paid to method development and validation procedures (Figure 4.4).

Validation is carried out after developing a new method, after changes to the application or working environment, and before using any published methods.

Especially in academic publications, the published methods are not always properly validated. Therefore a verification of performance should be carried out in order to meet the in-house required validation level.

As suitability verification and as an important part of quality control management, validation is subject to various factors that influence its quality.

The essential elements are:

- Employee qualification

  To achieve the goals of quality assurance, the manufacturers in every organization must have a sufficient number of competent and appropriately qualified staff.

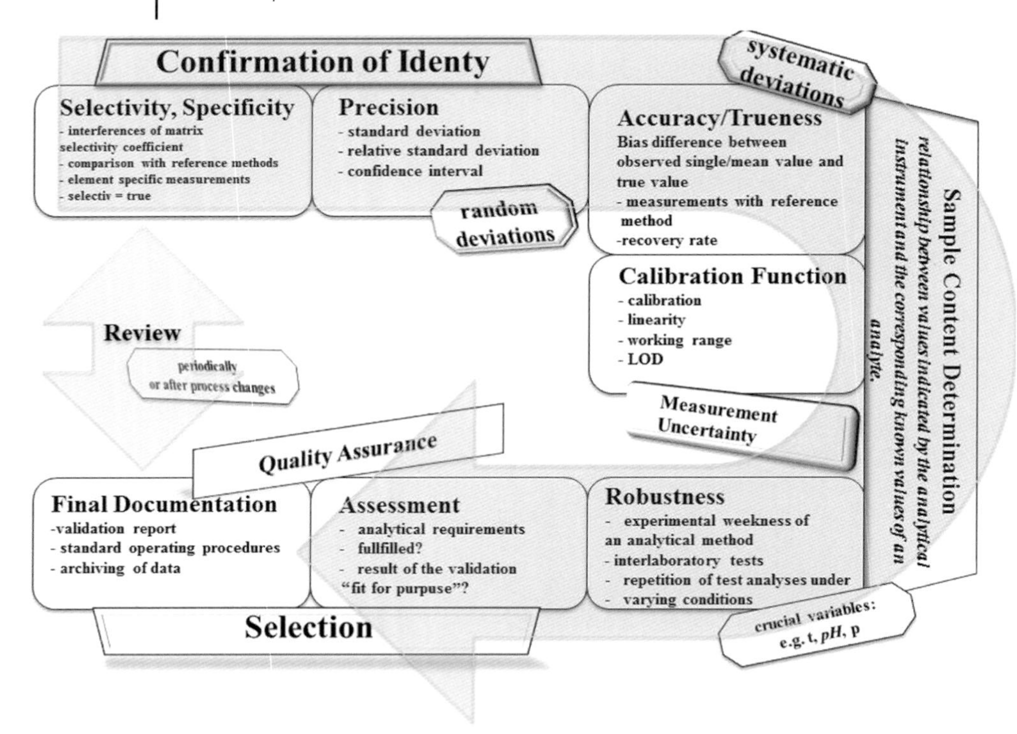

Figure 4.4 The phases of a validation.

- Qualified equipment

  The equipment must be checked periodically for proper functionality and adjusted as required.

- Suitable method

  In order to perform accurate measurements, the appropriate physical or chemical principle should be applied.

- Quality management

  The GLP is an officially recognized quality management system that places great value on documentation and archiving. Particular attention must be paid to the strict regulatory requirements, while taking care at the same time to effect implementation in a pragmatic and economic manner.

It is the goal of method validation to characterize the properties of a method, to determine its scope, area of application and limitations, and to identify any factors that can change the characteristics of the method (including the extent of the change).

Which analytes can determine a method, in which matrix, in the presence of which disorder and with which measurement uncertainty? The validation provides the answer.

A method validation includes the proof and documentation of the reliability and aptitude of a method for the appropriate solution of an analytical problem. It is conducted to ensure the quality of analytical results by way of a legal guarantee, or in response to market demand. A validation is appropriate following the development of a new method or when an existing method is modified.

The validation master plan (VMP) created prior to each validation describes the structure and organization of the workflow of a particular validation project, determines the individual projects and defines the responsibilities necessary for deviation and dispersion.

The validation plan is important for the analyst. It is a document describing the agreed measures and responsibilities, the prerequisites for validation and the limits of their scope as well as the reason for any exceptions.

The validation plan is a written proof that validation has been performed.

It should include the following:

- Purpose
- Scope
- Framework conditions
- Strategy
- Responsibilities
- Acceptance criteria
- Statistical evaluation, documentation
- Archiving of validation documents

4.5.1
Validation Elements

4.5.1.1 Selectivity/Specificity

Definition:

A method is said to be selective when different components are determined at the same time without mutual interference.

Selectivity coefficients are used to describe the selectivity of analytical methods. They show for example, the relative response of the method to analyte X as compared to analyte Y and can range from zero = no interference (not really applicable in practice because no analytical method is totally free of interferences) to higher units.

A method is said to be specific if it captures the component to be determined from other components present in the sample without falsification or distortion.

For example, the chloride content of a sample can be selectively proven in the presence of other halogenides or the sodium ion content can be selectively determined in the presence of other alkali and alkaline earth metal cations. The question is whether substances other than the analytes of interest also generate significant signals with the probe used.

Verification:

Measurement of standards of known concentration of analyte and interfering substances of interest, analysis of the same sample using other methods as far as possible.

selectivity, *Definition according to IUPAC Compendium of Chemical Terminology 2nd Edition, 1997.*

1) *(qualitative): The extent to which other substances interfere with the determination of a substance according to a given procedure.*

2) *(quantitative): A term used in conjunction with another substantive (e.g., constant, coefficient, index, factor, number) for the quantitative characterization of interferences.*

specific, *Definition according to IUPAC Compendium of Chemical Terminology 2nd Edition, 1997. A term which expresses qualitatively the extent to which other substances interfere with the determination of a substance according to a given procedure. Specific is considered to be the ultimate of selective, meaning that interferences are supposed to occur.*

- Precision (see Section 3.3.2)
- Accuracy/Trueness (see Section 3.3.3)
- Calibration Function (see Section 3.4.1)
- Linearity (see Section 3.4.3.5)

4.5.1.2 **Working Range**

The working range (scope) is the range of concentration of the analyte that can be adequately determined in the sample with an acceptable degree of precision, accuracy and linearity. For most analytical methods the working range is known from previous experience. When introducing a new method or measuring technique this range may have to be determined. It is determined by technical and practical considerations and confirmed by the validation.

4.5.1.3 **Detection, Determination and Quantitation Limit**

The concepts of the detection limit, determination limit and quantitation limit have been, and still are, most controversial in analytical chemistry. The multiple definitions and calculation methods proposed have contributed to this situation. Although in the last years, several international organizations, such as ISO, IUPAC, DIN and so on have tried to reach a consensus in their definitions and have issued guidelines for the estimation of this important parameter in chemical analysis, the subject is still a matter of scientific debate.

4.5.1.4 **German Standard DIN 32645 [15]**

At the lower end of the range of measurement three limits have been generally accepted, the names of which appear to be a semantic problem [16]. According

Figure 4.5 Measurement limits.
(Source of picture: Richard R. Rediske, Ph.D., Grand Valley State University).

to the German Standard DIN 32645, and in agreement with many authors, the lowest limit which deviates significantly from the signal obtained from blank measurements ("noise" signal) is defined as the detection limit. Direct and indirect methods have been proposed for the determination of this detection limit (Figure 4.5).

- *Limit of detection (LoD)*

  The limit of detection is equal to the smallest value significantly distinguishable from the blank sample that is received by insertion into the analysis function (critical value of the measured quantity). The limit of detection is used to evaluate a previously conducted qualitative analysis (*a posteriori* decision) and allows the inference of the presence or absence of a substance. In accordance with DIN 32645 [15] the limit of detection is defined as the lowest quantity which can be differentiated from a "zero" (blank) quantity with a stated probability. It is a decision limit for the presence of an analyte. Results below the detection limit do not indicate that the analyte is absent.

  - *International Conference for Harmonization* (ICH) defines the detection limit of an individual analytical procedure as the lowest amount of analyte in a sample which can be detected but not necessarily quantitated as an exact value.

– *The International Organization for Standardization* ISO [17] defines the LoD as the true net concentration (or quantity) of a component in the material subject to analysis that will lead, with a probability ($1 - \beta$), to the conclusion that the concentration (or quantity) of component in the material analyzed is greater than that of a blank sample. The International Union of Pure and Applied Chemistry (IUPAC) [18], in an earlier document, provided a similar definition and adopted the term "minimum detectable (true) value", as the equivalent to LoD.

limit of detection, *Definition according to IUPAC Compendium of Chemical Terminology 2nd Edition, 1997. The limit of detection, expressed as the concentration, cL, or the quantity, qL, is derived from the smallest measure, xL, that can be detected with reasonable certainty for a given analytical procedure. The value of xL is given by the equation:*

$$xL = xbi - + ksbi$$

where xbi is the mean of the blank measures, sbi is the standard deviation of the blank measures, and k is a numerical factor chosen according to the confidence level desired.

- ***Limit of determination or limit of identification***

  The limit of determination serves to assess a verification procedure (*a priori* decision) and specifies the minimum content that can be detected with a highly specified probability.

  The formulation that is frequently found in literature to the effect that "the limit of detection is the minimum content above which verification is possible with a specified probability of error" thus applies for an analytical procedure only to the determination limit in the meaning of this norm. Only when the limit of determination is less than that content, proof of which is required, is the analysis method suitable for this verification (Limit test).

- ***Limit of quantitation (LoQ)***

  To determine a substance is to determine its content, for example, as a mass fraction or mass concentration. Quantitation is therefore obligatory in all cases in connection with an associated figure and thus with a procedure with a quantitative result. Quantitation procedures thus presuppose a property of the component that can be quantified by measurement. A component is regarded as identifiable when its content can be determined with a relative uncertainty of outcome that meets the specified requirements. Thus, the limit of quantitation is also an *a priori* statement, that is, it provides information on the feasibility of future quantitative analysis.

- ***Blank sample***

  Sample that under ideal conditions does not contain the component to be verified or determined, but that otherwise agrees with the analysis sample.

- ***Blank value, limit of blank (LoB)***

  The arithmetic mean (average) of the measured values of the blank sample. The limit of blank is the highest *apparent* analyte concentration expected to be found when replicates of a blank sample containing no analyte are tested [19]. LoB is estimated by measuring replicates of a blank sample and calculating the mean result and the standard deviation.

- ***Critical value of the measurand***

  The critical value of the measurand (Kritischer Wert der Messgrösse) is that value which, at a defined probability, indicates that the signal measured is higher than the signal obtained from a blank sample.

  To determine the critical value of the measurand, the following requirements are necessary:

  – The analysis procedure

  – The matrix

  – The significance level P, the number n of the samples for determining the calibration function or the measurement values for ascertaining the blank value

  The matrices of the analysis sample and the blank should match as far as possible. In accordance with the consequences of a wrong decision, a significance level α of 0.05 or 0.01 (probability of error of 5% or 1%) is common.

4.5.1.5 **Calculations**

DIN 32645 distinguishes between direct methods for the limit of detection calculated from the uncertainty of measuring the blank (blank method, "Leerwertmethode") and indirect methods which derive the uncertainty of the blank from the uncertainty of low concentration samples by extrapolation (calibration line method, "Kalibriergeradenmethode").

Methods for Determining the Blank The results of content determination of a blank sample (blank) are fraught with uncertainty, which can be determined in two ways:

Direct Method (Blank Method) This approach is applicable whenever an appropriate blank sample is available. To calculate the detection and/or determination limit using the blank value determination, the slope of a calibration function is required. For this, calibration can be used across the entire work range. Theoretically, the blank value determination is to be given preference. However, as there are rarely ideal blank samples available in practice, recourse must be taken in such cases to the extrapolation method, which additionally takes into account the uncertainty of calibration.

Indirect Method (Extrapolation Method) For indirect methods blank solutions are stocked with calibration material in the lower range of the total "working range".

The function must be determined experimentally in a section of the analytical range which is close to the zero quantity if an indirect procedure is applied. As a rule of thumb a lower range from 0 to 10 times the detection limit is recommended.

In the case of the extrapolation method, the uncertainty of the blank value is only indirectly obtained from an extrapolation of regression data of a calibration (ordinate intercept). This is only conducted in the immediate vicinity of the limit of detection and is not to be confused with the "normal" calibration over the entire work area. The samples used for calibration can be determined by spiking the blank with the desired component.

As a rule, both methods provide not the same but equivalent results. In the case of significant differences, the result of the direct method is determining. What is determining for the detection and/or determination limit is not the value of the blank, but only its uncertainty. A constant blank value could always be subtracted from the sample signal. The limit of detection would then be equal to zero.

Strictly speaking, the dispersion of the blank and the analysis sample must be the same size for the calculation of the detection and determination limit from the dispersion range of several blank sample measurements to apply. This is not generally realized, since the dispersion of the measured values is influenced not only by real existing analyte content, but also by other factors.

Definitions

x_{NG} limit of detection, for flash estimate $x(_{NG})$

x_{EG} limit of determination, for flash estimate of $x(_{EG})$

x_{BG} quantification limit, for flash estimate $x(_{BG})$

k (e.g., $k = 5$, corresponding to a relative uncertainty of results ranging from 20% to the specified level of significance)

$\phi_{n,\alpha}$ factor Φ in flash estimate of the limit of detection (Appendix B, B.3)

$\phi_{n,\frac{\alpha}{2}}$ factor Φ in flash estimate of the limit of quantitation (Appendix B, B.3)

n number of measurements

α level of significance

S_L blank signal dispersion/standard deviation of the measured values of the blank

a intercept of the calibration lines

b slope of calibration line

$\hat{y}_i$ estimated functional value of the calibration

y_i measured value of the calibration sample

y_L measurement value of the blank

$s_{y,x}$ residual standard deviation of calibration data

$$s_{y,x} = \sqrt{\frac{\sum_{i=1}^{n} \hat{y}_i - y_i^2}{n-2}} \tag{4.4}$$

s_{x0} process standard deviation (in the case of variance homogeneity)

$$s_{x0} = \frac{s_{y,x}}{b} \tag{4.5}$$

Flash Estimates The flash estimates of the detection-, determination- and quantitation limits can be interpreted as a multiple of the process standard deviation $s_{x0} \approx s_L/b$.

- *Flash estimate limit of detection*

$$x_{(NG)} = \phi_{n,\alpha} \cdot \frac{s_L}{b} \quad \text{direct method (blank method)} \tag{4.6}$$

$$x_{(NG)} = 1.2 \cdot \phi_{n,\alpha} \cdot s_{x0} \quad \text{indirect method (extrapolation method)} \tag{4.7a}$$

$$x_{(NG)} = 4 \cdot s_{x0} \quad \text{indirect method (extrapolation method)} \tag{4.7b}$$

$$x_{(NG)} = 1.2 \times 3 \times s_{xo} \approx 4 \cdot s_{xo} \tag{4.4a}$$

Factor $\Phi\, n, \alpha = 3$ for 10 measurements ($n = 10$) and an accuracy of 99% or an error probability of 1% ($\alpha = 0.01$) (See Appendix B, B.3)

- *Flash estimate limit of determination*

$$x_{(EG)} = 2 \cdot x_{(NG)} \quad \text{direct method (blank method)} \tag{4.8}$$

$$x_{(EG)} = 2 \cdot x_{(NG)} \quad \text{indirect method (extrapolation method)} \tag{4.9}$$

- *Flash estimate limit of quantitation*

$$x_{(BG)} = k \cdot \phi_{n,\alpha} \cdot \frac{s_L}{b} \quad \text{direct method (blank value determination)} \tag{4.10}$$

$$x_{(BG)} = 1.2 \cdot k \cdot \phi_{n,\alpha/2} \cdot s_{xo} \quad \text{indirect method (extrapolation method)} \tag{4.11a}$$

$$x_{(BG)} = 12 \cdot s_{x0} \quad \text{indirect method (extrapolation method)} \tag{4.11b}$$

$$x_{(NG)} = 1.2 \times 3 \times 3.4 s_{xo} \approx 12 \cdot s_{xo} \tag{4.8a}$$

Factor $\Phi n,\alpha = 3.4$ with 10 measurements ($n = 10$) and an accuracy of 99.5% or an error probability of 0.5% ($\alpha = 0.005$). (See Appendix B, B.3)

The indirect method (calibration line method) and the direct method (blank method) do not provide the same results as a rule, but equivalent results. In the case of significant differences, the result of the direct method is determining [15].

Confidence Ranges The verification and detection limits can be determined from the multiple of a standard deviation. Since a confidence range can be calculated for a standard deviation (see DIN 53804-1), according to the law of error propagation confidence ranges can also be ascertained for verification and detection limits. For this, the values are multiplied by the factors κ_u and κ_o (see Appendix B, B.4).

- *Confidence range of the limit of detection*

$$K_U \cdot x_{(NG)} \leq x_{(NG)} \leq K_O \cdot x_{(NG)} \tag{4.12}$$

- *Confidence range of the limit of quantitation*

$$K_U \cdot x_{(BG)} \leq x_{(BG)} \leq K_O \cdot x_{(BG)} \tag{4.13}$$

K_U Factor for the lower limit
K_O Factor for the upper limit

For the limit of detection ($V = n - 1$ degree of freedom) applies for determination, and for quantitation limits ($V = n - 2$ degree of freedom) applies.

4.5.1.6 Limit of Detection according to Kaiser

Especially in chromatography, one of the most widely used methods to determine the limit of detection is known as 3σ (according to Kaiser and Specker [20]). The basic approach is as follows. Seven or eight replicates of a blank are analyzed by the analytical method, the responses are converted into concentration units, and the standard deviation is calculated (noise).

noise, *Definition according to IUPAC Compendium of Chemical Terminology 2nd Edition, 1997. The random fluctuations occurring in a signal that are inherent in the combination of instrument and method.*

This statistic is multiplied by 3, and the result is the detection limit (Figure 4.6). The advantages of this procedure are that it is easy to collect the data and it is easy to calculate. This classic definition has been described as follows:

$$x_S = \bar{x}_{Bl} + 3\sigma_{Bl} \tag{4.14}$$

where x_S is the measurement signal at the limit of detection, $\bar{x}_{Bl}$ the mean value of the blind signal, and σ_{Bl} the standard deviation of the blind values.

- **Limit of Determination**

The practice-oriented limit of determination, also known as the *Kaiser*'s guarantee limit, is defined as the concentration at which the received signal is x_C:

$$x_C = \bar{x}_{Bl} + 6\sigma_{Bl} \tag{4.15}$$

- **Limit of Quantitation**

The limit of quantitation is defined roughly in a simplified manner:

$$x_C = \bar{x}_{Bl} + 10\sigma_{Bl} \tag{4.16}$$

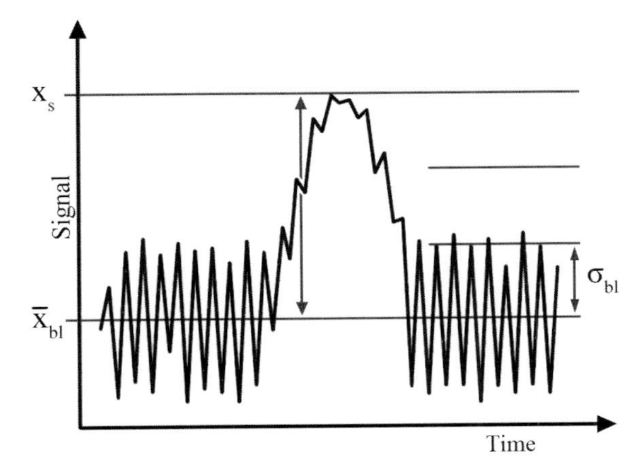

Figure 4.6 Signal/noise ratio (*S/N* ratio).

The meaning of "*detection limits*" is perhaps clear to all, in a qualitative sense. That is, the detection limit is commonly accepted as the smallest amount or concentration of a particular substance that can be reliably detected in a given type of sample or medium by a specific measurement process. Within such a general definition, however, lurk many pitfalls in terminology, understanding, and formulation [21].

However, what remains different are the recommendations for estimating the currently valid parameters each time, especially the "true" blank value standard deviation σ_{BL}, whereby these differences are often task-and process-related.

Which procedure should be preferred?

Signal/Noise ratio

Determination through the signal/noise ratio is very common in chromatography and it is (still) accepted in official literature.

Calibration-Based LoD (DIN 32645)

The blank value determination and the extrapolation method are supported by many software programs and, as an "official" method, they enjoy wide acceptance.

Due to the multiple definitions and calculation methods proposed in the literature it is crucial that the analytical chemist strictly defines the limit values investigated, how they were calculated and which measuring procedure was used.

4.5.1.7 **Robustness**

Robustness is the ability of a system to maintain its function even in the face of fluctuating environmental conditions (e.g., fluctuations in temperature, change of personnel, biodegradability of matrix and analyte). It is often useful to specify what the system is supposed to withstand (for example, to withstand changes in the environmental temperature or faulty operation). The robustness of the method with respect to a boundary condition can be checked by (periodic) repetition of test analyses under varying conditions. The robustness influences the uncertainty of a result, the influence on susceptibility to interference by changed conditions (analysis parameters, equipment, laboratories, etc.). The robustness is to ensure that the quality of the data of smaller fluctuations is independent during process execution. The robustness can sometimes be determined with the help of proficiency testing, the implementation of which is under discussion in external quality control. In the in-house laboratory, the robustness of a procedure can be tested within permissible limits by varying the experimental parameters.

4.5.2
Using the Computer

A spreadsheet program helps to automate the various steps of calculation. Microsoft Office Excel provides useful tools for visualization and modeling of data. The Excel software can be complemented with additional macros (EasyStat).

However, several programs offer far more. These are independent software programs that were developed specifically for validation, quality assurance and statistical process control. They are quite suitable for practical application in the analytical laboratory when statistical data security is required. They are also partly network compatible applications that can be used under the current Windows operating systems. The following are some examples:

XLSTAT http://www.xlstat.com/en/products/ (accessed May 2012)

XLSTAT is a modular statistical software. XLSTAT relies on Excel for the input of data and the display of results, but the computations are done using autonomous software components. The use of Excel as an interface makes XLSTAT a user-friendly and highly efficient statistical and multivariate data analysis package.

Winstat WinSTAT is a statistics Add-In for ExcelMicrosoft® (http://www.winstat.com/)

Analyse-it http://www.analyse-it.com/products/standard/

Transforms Microsoft Excel into a comprehensive, easy-to-use statistical analysis package, with powerful statistics like ANOVA, linear and polynominal regression.

PAST http://folk.uio.no/ohammer/past/

PAST is a free, easy-to-use data analysis package originally aimed at paleontology but now also popular in many other fields. It includes common statistical, plotting and modeling functions.

ProLab http://quodata.de/index.php?id=18&L=1 (accessed May 2012)

ProLab is a software program for planning, organizing, realizing and analyzing interlaboratory tests.

Typical features of a "simple" statistical software (cost approx. €/$ 100–200)

- Test for homogeneity of variances
  - F (*Fisher*)-test for homogeneity of variances according to *Bartlett* (up to 10 groups)
  - F-test according to *Mandel*
  - F-test according to *Funk*
- R/s test for normal distribution
- *Grubbs* test for outliers (see Section 3.3.7.2)
- Calculation of the precision in and between the series based on ISO 5725 Ascertaining the precision within and between the series
- The *t*-tests determine whether a value lies within the confidence interval of the target value
  - *t*-Test for comparing two mean values
  - *t*-Test for comparison of measured values with a target value
- Calculation of analytical characteristics according to DIN 32645
- Calculation of the recovery rate
  - Linearity test according to *Mandel*

4.6 Final Documentation

At the end of the validation process, the analysis results, including their assessment, are recorded in a report or method validation protocol and discussed in connection with the client. The traceability of measurement results must be ensured and it must be ensured that the measurements can be repeated at a later date. Critical aspects of the method are presented in detail and recommendations for routine users are developed. The completely validated method is recorded in a Validation Report and in separate Standard Operating Procedures (SOP).

Care should also be taken to ensure that non-chemists (lawyers, doctors, entrepreneurs) can also comprehend the language and presentation of the final documentation. On demand translated copies have to be made to order.

If the result of the validation is considered satisfactory by the client meeting the required target values, the validation is completed and can be declared as fit for purpose. The validation can now be released by Quality Assurance and the validation documents can be archived.

In case the validation is not satisfactory, corrective measures are decided on by the analyst and client followed by reassessment of the target aims, and the process starts again.

4.6.1 Review

Conducting periodic reviews on processes that have already been validated is a critical aspect in any validated environment. It is important to review a system, or process, for changes, both physical and procedural, in order to verify that a system, or process, is operating as expected. For the most part, management of periodic reviews is achieved using spreadsheets, paper-driven tools or software that automatically alert user groups of any upcoming, or delayed periodic review tasks. This in turn equates to reliance upon a single user, or group of users, to ensure that periodic reviews are completed on time and documented accordingly.

Failing to plan is *planning* to fail.

5
Example of a Validation Strategy

In this chapter we will use a concrete example to demonstrate how a validation strategy is developed. However, first let me make some comments.

The maximum scope of a validation includes accuracy, precision, linearity, recovery, selectivity, robustness, copping-out tests, limit of detection and determination limits, methodology, and process stability. The actual effort required depends on the nature and purpose of the analysis. Terms such as detection, determination and quantification limits are often defined differently in analytical/chemical literature and in standards (see Section 4.5.1.3) and, to some extent, various mathematical and statistical aspects are applied for their calculation. There are also a number of guidelines and recommendations for the interpretation and definition of measurement uncertainty (see Section 3.3.7). In validation it is, therefore, imperative to define the terms used.

This chapter aims to help to convey a basic knowledge in the area of validation in analytical chemistry, while outlining different approaches. More than the mathematical details, it is the conceptual interpretation of the terms that should be in the foreground. I have, therefore, consciously restricted myself here to presenting the simplest approach of linear regression with constant variance. The evaluation procedures in compliance with DIN 32645 [15] are used to calculate the limits of detection, determination and quantitation. An extension to more complex model approaches is then merely a question of applying a more comprehensive array of mathematical instruments. This book deliberately places value on the application of simple estimation techniques that can be quickly applied and implemented in the course of laboratory classes. The workflow leading to a solution remains visible to students and a later application of automated calculation methods using statistical software is facilitated.

The students should not only be introduced to the different instrumental chemical analytical methods and their applications, but also they should realize that analytical results are of little value without the necessary statistical evaluation.

Essentially what should be conveyed is the following:

- A heightened degree of personal quality consciousness
- The safe handling of analytical and statistical measurement data and their correct evaluation and interpretation
- A deepened understanding of quality in the development of analytical methods

Practical Instrumental Analysis: Methods, Quality Assurance and Laboratory Management, First Edition.
Sergio Petrozzi.

For detailed explanations I recommend the works of Wenclawiak and De Bievre [22, 23], in which there is a very good and detailed didactic description of the topic of method validation.

In the following example and in the projects dealt with in Chapter 8, there are three types of notes, which are each characterized by a distinctive icon:

= **Customary in Practice**

The accepted deviations and measurement ranges specified here are guidelines, which are usually aspired to in practice, many of them being guidelines of the Food and Drug Administration, Surveillance and Evaluating Authorities for Food and Drugs in the U.S. (FDA) [24] and the International Conference on Harmonization (ICH) that is promoting international harmonization by bringing together representatives from the three ICH regions (EU, Japan and USA) to discuss and establish common guidelines [25]. The acceptance of these deviations can vary from company to company, industry to industry and according to needs, and must be established prior to conducting validation. Most industries have their own specific guidelines which must be regarded as mandatory and binding.

Generally, these guidelines and acceptance criteria are treated confidentially.

= **Practical Implementation**

= **Result**

5.1 Determination of Phenol in Industrial Waste Water

Requirements and conditions of the method to be validated:

Analytical Problem

The phenol concentration from the waste water plant of a synthetic resin plant should be determined routinely and an appropriate examination procedure should be validated.

Examination Procedure

In principle, several methods are available. According to the literature, gas chromatography allows the achievement of low detection limits (<0.01 mg phenol/l). The chromatographic conditions and parameters are known from literature search and will be implemented.

Content of the Validation Plan

The validation plan includes acceptance criteria, all tests and measurements to be conducted, validation parameters and rules governing the evaluation and guidelines for their documentation.

Table 5.1 shows the client specifications for the analytical method.

Table 5.1 Client specifications for the analytical method.

Validation parameters	Acceptance criteria
A. Confirmation of Identity	
1) Selectivity	If the accuracy/trueness of the method has been proven by use of standard reference material, processing of the selectivity can be waived.
	Additional identification of phenol (relative retention time of comparison with a test chromatogram = 1.00). Visual comparison of peak shape, peak width, baseline resolution.
2) Precision (repeatability, within-run)	The coefficient of variation (CV) calculated from single runs should be <10%.
B. Sample Content Determination	
1) Accuracy	
Verification	
1.1) Recovery Rate (RR)	The recovery rate should be < ±8%.
1.2) Verification using a second, independent method	An Environmental Protection Agency (EPA) method is to be used.
2) Calibration Function — Calibration	The most frequently expected values are about 4.0 mg $_{phenol}$ l^{-1}. Therefore, linearity should be investigated in a working range of 0.5–10 mg $_{phenol}$ l^{-1}.
— Linearity	Linearity test is made on a visual basis.
— Working range	In addition, the coefficient of determination (R^2) should not be less than 0.95.
3) Detection, Determination and Quantitation Limits	The limits are calculated in accordance with the DIN 32645 norm (flash estimates). A limit of detection of 0.10 mg $_{phenol}$ l^{-1} is aimed at.
4) Measurement Uncertainty	– Sources of uncertainty are to be identified.
	– The uncertainty of the two employed independent methods is to be determined.
5) Robustness	
Verification	
Timely Process Stability	Measurement results of control samples within 3 days measured morning and afternoon are to be compared.
	Discrepancies in content should be less than CV 5%.
Change of method parameters	System suitability (SST) requirements are to be checked for variations of the carrier gas flow, the initial oven temperature and the split ratio.
C. Selection	
1) Final Documentation	The documentation should also include a German summary.

5.1.1 Confirmation of Identity

5.1.1.1 Selectivity

The statement about selectivity is gathered from a plausibility evaluation:

Since selectivity is one of the requirements for trueness, an accurate reference method is also selective.

The analytical method chosen can be considered as specific for the determination of the phenol content in the selected matrix (waste water produced from a specified industrial process).

In addition, the chromatographic retention time is checked with a certified phenol-standard.

5.1.1.2 Precision (see Section 3.3.2)

Determination of The Repeatability (Within-Run Precision)

Ten measurements ($n = 10$) of the same control sample (e.g., concentration $2.32 \pm 0\ 21\ \mathrm{mg}_{\mathrm{phenol}}\ \mathrm{l}^{-1}$) are performed. Statistical parameters, such as mean ($\bar{x}$), standard deviation (s), relative standard deviation and coefficient of variation (CV), are calculated.

The following mass concentrations ($\mathrm{mg}_{\mathrm{phenol}}\ \mathrm{l}^{-1}$) were measured:
2.25, 2.20, 2.28, 2.28, 2.40, 2.21, 2.25, 2.27, 2.68, 2.35.

The following values result:

$$\text{Mean} \quad \bar{x} = 2.317\ \mathrm{mg}_{\mathrm{phenol}}\ \mathrm{l}^{-1} \tag{3.1}$$

$$\text{Standard deviation} \quad s = 0.141\ \mathrm{mg}_{\mathrm{phenol}}\ \mathrm{l}^{-1} \tag{3.3}$$

$$\text{Relative standard deviation} \quad \mathrm{RSD} = \frac{0.141\ \mathrm{mg}_{\mathrm{phenol}}\ \mathrm{l}^{-1}}{2.317\ \mathrm{mg}_{\mathrm{phenol}}\ \mathrm{l}^{-1}} \tag{3.6}$$

$$\text{Coefficient of variation} \quad \mathrm{CV} = \frac{0.141}{2.317} \cdot 100\% = 6.1\% \tag{3.7}$$

Concentration control sample (new measurements): $2.32 \pm 0.14\ \mathrm{mg}_{\mathrm{phenol}}\ \mathrm{l}^{-1}$ = CV 6.1%

Concentration control sample (previously measured): $2.32 \pm 0.21\ \mathrm{mg}_{\mathrm{phenol}}\ \mathrm{l}^{-1}$ = CV 9.1%.

Although the mean is the same, the dispersion of values is different. In this case, the precision has improved.

> The coefficient of variation is suitable for comparing the precision of a series of measured values. This coefficient allows comparisons between values differing in order of magnitude [26]. The acceptance criteria of the deviations depend largelyon the specifications at the time of formulating the problem.
>
> Thus, in quality control conducted in the pharmaceutical industry, CV-values of 1–2% are customary. In environmental analysis values of 10% are quite acceptable.

Outlier Test–the 4-s Barrier (see Section 3.3.7.1) The simple test often used in practice is conducted here on 10 weighings (repeat measurements on one sample).

Ten weighings were conducted and the following results obtained (x_i mg_{phenol}):

30, 28, **43**, 31, 31, 32, 31, 30, 32, 31.

Among the given weighings, the value $x_i = 43$ mg is suspected to be an outlier. After this value is eliminated, the result is:

$$\bar{x} = 31.9\,\text{mg} \quad (3.1)$$

$$s = 4.07\,\text{mg}\ (n = 10) \quad (3.3)$$

$$\text{check: } 31.9 +/- (2.5 * 4.07)\,\text{mg} \quad (3.20)$$

smallest accepted value: 21.72 mg

largest accepted value: 42.08 mg

$x_i = 43$ mg can therefore be regarded as an outlier

5.1.2 Sample Content Determination

5.1.2.1 Accuracy (see Section 3.3.3)

If the dispersion of the method lies within the stated specifications, the method is considered to be suitable for the defined concentration range.

Recovery Rate (RR) (see Section 3.3.3.6)

Solution 1	100 ml aqueous phenol solution	($m_{\text{phenol}} = 3$ mg)
	1 ml standard solution	($m_{\text{phenol}} = 1$ mg)
	101 ml solution $_{\text{total}}$	($m_{\text{phenol}} = 4$ mg)

$\text{Signal}_1 = 410\,000$ counts

Solution 2	100 ml aqueous phenol solution	($m_{\text{phenol}} = 3$ mg)
	1 ml water	($m_{\text{phenol}} = 0$ mg)
	101 ml solution $_{\text{total}}$	($m_{\text{phenol}} = 3$ mg)

$\text{Signal}_2 = 305\,000$ counts

Solution 3	100 ml water	($m_{\text{phenol}} = 0\,\text{mg}$)
	1 ml standard solution	($m_{\text{phenol}} = 1\,\text{mg}$)
	101 ml solution $_{\text{total}}$	($m_{\text{phenol}} = 1\,\text{mg}$)

$\text{Signal}_3 = 100\,000$ counts

$$\text{RR}(\%) = \frac{(410\,000 - 305\,000)}{100000} 100\% \tag{3.10}$$

RR = 105%

What are the internal requirements for the RR?

Ideally, the recovery rate is 100%, that is, the total amount of analyte spiked can be recovered.

Experience shows that in practice deviations of the recovery rate of ± 8% are accepted.

If matrix effects are observed, a correction factor is introduced.

Method of Comparison If no certified reference material nor sufficiently well characterized inter-laboratory testing material is available, the accuracy can be determined with a second validated method (reference method). Thereby, the analysis procedure of the reference method must differ from this method to be validated.

As a method of comparison the photometric method for phenol quantification according to the Environmental Protection Agency (*EPA*) Method 9065 is used (color reaction with 4-aminoantipyrine) [27].

5.1.2.2 Calibration Function – Calibration, Linearity, Working Range (see Section 3.4)

The linear working range is determined by means of a calibration line. Standards with equidistant concentrations are measured under repeated conditions. Every calibration begins with the selection of a preliminary working range. This depends on:

- The practice-related application target

  Experience shows that the expected concentrations of phenol are about 2.5–4.0 mg $_{\text{phenol}}\,\text{l}^{-1}$ (measured factory effluent values over a long period of time).

  Since the mid-range should be approximately equal to the most frequently expected sample concentration, the preliminary measurement range between 1.0 and 10.0 mg l^{-1} is selected.

- Requirements for technically feasible options
  - The laboratory staff are familiar with the method.
  - The analytical chromatography column recommended in the literature is purchased.
  - The chemicals used (reference compounds) are of known quality.
- Linearity of the selected method

  The linear relationship is known from the literature.

"Normal" Calibration After the *preliminary* working range has been set, 10 standard solutions ($n = 10$) with a concentration of 1.0 to 10 mg $_{\text{phenol}}$ l^{-1} are determined. Then the calibration function s is calculated from the 10 calibration data pairs, in compliance with the rules of the simple linear regression.

Calibration function

$$y = 4470.8x + 2028.8 \quad R^2 = 0.99 \tag{3.21}$$

Without forced zero value passage

- **Linearity**

From the calibration line (pair values, Table 5.2) it is visually apparent that the linear regression analysis can be performed quite reasonably, since the points can be described quite well by a straight line.

The correlation coefficient of linear regression here is taken as a measure of linearity and in this series is 0.99. By common criteria the correlation coefficient obtained here is considered to be sufficient, so that a linear relationship can be accepted.

Table 5.2 Measured values "normal" calibration (preliminary measurement range).

Mass concentration (mg $_{\text{phenol}}$ l^{-1})	Area *F*
1.0	7160
2.0	12 340
3.0	12 000
4.0	19 580
5.0	25 300
6.0	29 000
7.0	35 233
8.0	36 568
9.0	43 000
10.0	46 000

- **Working range**

The working range has been set definitively between 1 and 10 mg $_{phenol}$ l^{-1}, since the linearity in this range was considered to be satisfactory.

- In technical applications a correlation coefficient of at least 0.90 is frequently requested in order to be able to accept a linear relationship. In biological applications one generally works with essentially smaller correlation coefficients due to the system-inherent variations in the measured variables.
- Should it be visually evident that the signal/concentration relationship is clearly nonlinear, a suitable function must be selected or another area of concentration must be tested (the system may be saturated or overloaded).
- If the dispersion of the method is within the requirements set out in the specifications, the method is considered to be appropriate for the defined concentration range.

5.1.2.3 Determination of the Detection, Determination and Quantitation Limits according to DIN 32645 – Direct and Indirect Method [15] (see Section 4.5.1.4)

After linearity has been determined in the desired working range, the detection, determination and quantitation limits are ascertained.

The experimental determination of the detection, determination and quantitation limits takes place by the blank value method or the calibration line method (DIN 32645).

- **Determination of the blank value**

Individual measurements of a sample assumed to be phenol-free are taken to determine the blank value.

It specifies:

$n = 10$

$k = 3$ (corresponding to a relative uncertainty of results of 33%, $\alpha = 0.01$)

The following measured values (area F) were determined:

2003, 1901, 2212, 1976, 2279, 1853, 2165, 2108, 2368, 1943.

From these data calculations were made using Equations (3.1) and (3.3):

s_L standard deviation of the measured values of the blank sample 172 F
$\bar{y}_L$ blank value = arithmetic mean of the measured values of the blanks 2081 F

Flash Estimates Based on Blank Method – Direct Method

- **Limit of detection**

$$x_{(NG)} = 3 \cdot \frac{172F}{9662 \cdot F\,l\,mg^{-1}} \tag{4.6}$$

b (9662 $F\,l\,mg^{-1}$) known from calibration data

$$\text{LoD} = 0.053\ \text{mg}_{\text{phenol}}\ \text{l}^{-1}$$

The confidence range of the limit of detection is determined by the factors from Appendix B, Table B.4:

Degrees of freedom (n-2)	9
Factor for K_u	0.69
Factor for K_o	1.83

$$\begin{aligned} \text{Confidence range} \quad & 0.69 \times 0.053\ \text{mg l}^{-1} \leq x_{(\text{NG})} \leq 1.83 \times 0.053\ \text{mg l}^{-1} \\ & 0.037\ \text{mg l}^{-1} \leq x_{(\text{NG})} \leq 0.097\ \text{mg l}^{-1} \end{aligned} \tag{4.12}$$

$$0.037\ \text{mg l}^{-1} \leq 0.053\ \text{mg}_{\text{phenol}}\ \text{l}^{-1} \leq 0.097\ \text{mg l}^{-1}$$

- **Limit of determination (LoD)**

$$\text{LoD} = 2 * 0.053\ \text{mg l}^{-1} = 0.11\ \text{mg l}^{-1} \tag{4.8}$$

- **Limit of quantitation (LoQ)**

$$\text{LoQ} = 3 * 0.053\ \text{mg l}^{-1} = 0.16\ \text{mg l}^{-1} \tag{4.10}$$

The confidence range of the quantitation limit is determined by the factors from Appendix B, Table B.4:

Degrees of freedom (n-2)	8
Factor for K_u	0.68
Factor for K_o	1.92

$$\begin{aligned} \text{Confidence range} \quad & 0.68 \times 0.16\ \text{mg l}^{-1} \leq x_{(\text{EG})} \leq 1.92 \times 0.16\ \text{mg l}^{-1} \\ & 0.11\ \text{mg l}^{-1} \leq x_{(\text{EG})} \leq 0.31\ \text{mg l}^{-1} \end{aligned} \tag{4.13}$$

$$0.11\ \text{mg l}^{-1} \leq 0.16\ \text{mg}_{\text{phenol}}\ \text{l}^{-1} \leq 0.31\ \text{mg l}^{-1}$$

Flash Estimates according to Calibration Line Method–Indirect Method (DIN 32645) This regression line is not to be confused with the "normal" calibration over the entire working range. The samples used should have a very low concentration.

Ten standard solutions are measured for calculating the calibration lines (indirect method; Table 5.3).

It specifies:

$$n = 10$$

$$k = 3\ (\text{corresponding to a relative result uncertainty of } 33\%, \alpha = 0.01)$$

The residual standard deviation ($s_{y,x}$) of the measured values of the calibration was calculated according to Equation (4.4):

$$s_{y,x} = \sqrt{36977} = 192\ F$$

Table 5.3 Measured values using calibration line method.

Concentration β (mg l^{-1}) (x)	F (counts) (y)	F (counts) calculate	$\bar{y}_1 - y_i$	$(\bar{y}_1 - y_i)^2$	$\frac{(\bar{y}_1 - y_i)^2}{(n-2)}$
0.05	3060	2964.0	−96.0	9223.0	1152.9
0.10	3522	3447.1	−74.9	5615.9	702.0
0.15	3707	3930.2	223.2	49799.3	6224.9
0.20	4280	4413.3	133.3	17756.8	2219.6
0.25	5058	4896.4	−161.6	26130.2	3266.3
0.30	5510	5379.4	−130.6	17043.7	2130.5
0.35	5703	5862.5	159.5	25454.8	3181.8
0.40	6205	6345.6	140.6	19780.3	2472.5
0.45	7156	6828.7	−327.3	107099.5	13387.4
0.50	7178	7311.8	133.8	17912.2	2239.0
					Σ = 36977

The uncertainty of the blank is now obtained by extrapolating the regression data (*x*, *y*) (22).

Slope (sensitivity of the analytical method) 9662 $F^* \text{mg}_{\text{phenol}} \text{l}^{-1}$

Intercept (=blank) 2481 *F*

The process standard deviation (s_{x0}) is determined by regression:

$$s_{x0} = \frac{192}{9662} = 0.0199 \text{ mg l}^{-1} \quad (4.5)$$

- **Limit of detection**

$$\text{LoD} = 4 * 0.0199 \text{ mg l}^{-1} = 0.08 \text{ mg}_{\text{phenol}} \text{ l}^{-1} \quad (4.7\text{b})$$

- **Limit of determination**

$$\text{LoDetermination} = 2 * 0.08 = 0.16 \text{ mg}_{\text{phenol}} \text{ l}^{-1} \quad (4.9)$$

- **Limit of quantitation**

$$\text{LoQ} = 12 * 0.0199 \text{ mg l}^{-1} = 0.23 \text{ mg}_{\text{phenol}} \text{ l}^{-1} \quad (4.11\text{b})$$

The confidence level is calculated analogously to the direct method.

If blank values are unusually high, every single step of the process must be examined to find the causes. For this purpose, the individual components are examined: the devices, the test bench, the solvents and the chemicals.

In routine operations, the calibration is checked prior to each analytical series run and after a maximum of 10 injections. For this purpose, two calibration solutions are successively injected, the analyte concentrations of which are 20% or 80% of the upper limit of the linear working range selected for each analyte. (Compared with internal control limits or bracketing = 1 blank, 1 standard, 10 samples, 1 blank, 1 standard, 10 samples, etc.)

A wording like ". . . the detection limit for lead is $0.1\,ng\,ml^{-1}$" is therefore to be avoided. On the other hand, it would be correct to say, for example, ".. the limit of detection for the determination of lead in drinking water according to DIN 32645 (flash estimate, calibration line method – indirect method) is $0.1\,ng\,ml^{-1}$".

5.1.2.4 Measurement Uncertainty (see Section 3.3.5)

The measurement uncertainty results from experimentally determined uncertainties and/or estimated uncertainties. It affects the complete measurement process. If the result refers to a homogenized sample, the measurement uncertainty only comprises the analytical part. Otherwise the pre-analytical part must also be taken into consideration.

Standard Measurement Uncertainty (Gas Chromatographic Method)

Uncertainty of the result from the precision test expressed as an empirical standard deviation (easiest way).

The series of measurements for precision control results in:

$$\pm u = 0.141\,mg_{phenol}\,l^{-1} \qquad (3.11)$$

Combined Measurement Uncertainty

The uncertainties of three partial steps (weighing, dilution to the mark and measurement process) of the phenol determination can be ascertained:

1) Preparation of the standard solutions
   Weighing:

Tara:	31.0234 g	weighed
Gross:	36.1284 g	weighed
Analyte (phenol):	5.1050 g	calculated

Specifications of the balance:
   (i) QA-certificate (test weight up to 50 g) $s = 0.07\,mg$
   (ii) Calibration certificate when ascertaining a weight by subtraction ± 0.1 mg (with P 95%) $s = 0.051\,mg$

$$\pm u_{weighing} = \sqrt{0.07^2 + 0.051^2} = 0.087\,mg = 0.02\% \qquad (3.14b)$$

2) Filling up to the final volume
   (i) Uncertainty of the volume given
   Manufacturer's note on (tolerance): 250 ± 0.15 ml

$$s = \frac{0.15\ \text{ml}}{\sqrt{3}} = 0.087\ \text{ml} \tag{3.18}$$

(ii) Deviation when filling to the mark (10× filled)

$s = 0.012$ ml

(iii) Temperature difference between the volumetric flask and the solution during calibration

Temperature fluctuation ± 3 K (with 95% confidence level)

Volume expansion coefficient $2.1 \cdot 10^{-4}\ \text{K}^{-1}$

$s = 0.080$ ml

$$\pm u_{\text{filling up}} = \sqrt{0.087^2 + 0.012^2 + 0.080^2} = 0.12\ \text{ml} = 0.05\% \tag{3.14b}$$

As the measurement uncertainties of the weighing and filling up to the mark are far below the 10% of the measurement uncertainty of the quantitation (CV = 6.1%), they are not taken into consideration.

Expanded Measurement Uncertainty The measurement uncertainty calculated here is an expanded standard measurement uncertainty, using a coverage factor 2, which results in a confidence level of approximately 95%.

$$\pm U = 2 * 0.141 = 0.282\ \text{mg}_{\text{phenol}}\ \text{l}^{-1} \tag{3.16}$$

Standard Measurement Uncertainty of the Alternative Method As an alternative analytical method, phenol is quantified photometrically with a one-point calibration:

First the sensitivity is determined:

$$\text{sensitivity} = 0.416\ \text{abs.} * (\text{mg}_{\text{phenol}}\ \text{l}^{-1})^{-1} \quad \text{(standard deviation} = 0.032\ \text{abs.)}$$

The signal response of the sample is 0.203 abs. The standard deviation of this signal lies at 0.013 abs. (known from experience):

Calculation of the mass concentration:

$$\beta(\text{phenol}) = \frac{\text{abs}_{\text{sample}}}{\text{sensitivity}} = \frac{0.203}{0.416} = 0.49\ \text{mg}_{\text{phenol}}\ \text{l}^{-1}$$

$$\pm u_{\text{photometrically}} = 0.49\sqrt{\left(\frac{0.032}{0.416}\right)^2 + \left(\frac{0.013}{0.203}\right)^2} = 0.049\ \text{mg}\ \text{l}^{-1} \tag{3.15b}$$

The effort expended on ascertaining the measurement uncertainty depends on the analytical problem and is defined by the client. The quantitation of the measurement uncertainty must only be as accurate as necessary, that is, adapted to the measurement problem and the client requirements.

5.1.2.5 Robustness (see Section 4.5.1.7)

- Check of the timely stability of the analytical method measuring standard solutions and control samples: Measurements within 3 days measured morning and afternoon were compared.
- Method parameter variations of +/−10 kPa in the carrier gas flow, +/−10 °C in the initial oven temperature and +/−50% in the split ratio were investigated and compared.

The obtained results indicated that the studied variations do not cause any significant changes in system suitability and the method is regarded as robust.

If one factor significantly affects the results, the acceptability limits can be determined by further tests and the limits should be newly defined in the test instructions.

Standard Form for Reporting the Analytical Results Reporting an analysis result without specifying its range of uncertainty (confidence level, measurement uncertainty) is incomplete, since the quality of the analytical result is not obvious. Therefore, at least n and P (95% = 2 × standard deviation) must be indicated.

$$\text{Phenol content of the sample: } (426 \pm 42)\ \text{mg l}^{-1}\ (n = 10, P\ 95\%) \qquad (3.19)$$

Additionally, the following should be specified:

- Complete identity of the sample
- Date of the sample entry in the laboratory
- Analyst in charge
- Method used
- All deviations from the named procedure, which possibly could have influenced the analytical procedure
- All peculiarities arising during measurement
- Working range

Significant Figures

As a rule of thumb, only one significant figure is sufficient to report uncertainty in laboratory classes as deviations are seldom exactly known and can, therefore, only be estimated (see Section 3.3.8).

5.1.3 Selection

5.1.3.1 Final Documentation (see Section 4.6)

At the end of the validation process, the analysis results, including their assessment, are recorded in a report and discussed with the client. The traceability of measurement results must be ensured and it must be ensured that the measurements can be repeated at a later date. Critical aspects of the method are presented in detail and recommendations for routine users are developed. Attention must also be paid to ensure that the language and layout of the final documentation can also be comprehended by non-chemists.

Due to the fulfillment of the established specification limits, the quality assurance laboratory can recommend the implementation of the tested procedure for determining phenol in industrial waste water. This recommendation provides the basis for decision-making regarding the procurement of five new gas chromatographs.

A one-page summary in German will be added to the final documentation.

Validation is not a destination, it is a never ending journey.

6 Organizational and Practical Procedures in the Teaching Laboratory Program

6.1 Goals

The students should be able to optimize and validate instrumental analytical measurement methods for a specific analytical problem based on theoretical knowledge and a study of the literature. They should have knowledge of the physical and chemical basics of modern measurement technology and the statistical aids. The theoretical principles worked out in class should be implemented accordingly in practice. Furthermore, the students are to learn and practice the basics of quality assurance, statistical data evaluation and project management.

It is not about becoming an expert on a measurement instrument, but rather it is about receiving an overview of important instrumental analytical methods and practicing with practical examples the (principally technology independent) work technique of the analytical chemist when developing and validating methods.

Hereby the following aspects are in focus:

- Problem analysis and work planning
- Sample processing
- Creating of measurement plans for validating a method
- The use of suitable working techniques
- The independent processing of approaches to solutions, in order to solve (almost) any problems that may arise
- Writing laboratory reports, which comply with the guidelines of Good Laboratory Practice

Practical Instrumental Analysis: Methods, Quality Assurance and Laboratory Management, First Edition.
Sergio Petrozzi.

6.2
Safety in the Laboratory Class

Security instructions prior to the start of the laboratory class are an integral part of the training. Students are obliged to attend. Whoever misses the sessions cannot participate in the internship.

- Immediately before the start of the practical laboratory work, a commitment (Appendix C) is completed and submitted to the assistant.
- The laboratory rules must be observed. In all laboratories, the protective goggles and protective clothing (lab coat) must be worn.
- Experimental work outside the official laboratory hours is only allowed with written approval of the laboratory responsible.
- Experimental work over lunchtime is allowed only if at least two students are present.
- Experiments that run overnight (exceptional cases) are to be marked by placing a plaque on the device and a completed form signed by the laboratory responsible is to be hung on the laboratory door.
- The refrigerator should only store chemicals/samples, which bear the date, description of the content and the name of the owner.
- Particular caution should be exercised in the proper handling of waste (heavy metals, solvents, toxic substances) (Appendix F).

6.3
Experimental Project Workflow

6.3.1
Preparation

For the very short laboratory time available for the processing of the experimental tasks, it is important for the students to familiarize themselves from the very beginning of the project with the technologies used.

- Prior to starting the practical work, the glassware, chemicals and device accessories should be checked for completeness. Depending on the analytical task, safety data sheets of the chemicals used are also to be completed.
- Prior to commencement of the experiment, each student should inform themselves about the physical and chemical properties as well as the potential hazards of the chemicals and substances in use for potential analysis (*R*/*S*-phrases, see Appendix F). It is also necessary to be thoroughly informed about the experimental set-up of the experiment to be carried out. In case of doubt, the supervising assistant or lecturer should be consulted before the start of the experiment.

- Only after submitting the safety data sheets to the respective laboratory assistant in charge of the experiment, will the detailed experimental task be issued and discussed.
- After that, a rough work plan is created, which is constantly updated during execution. Time should be planned for unexpected problems. The plan is to be attached to the report.
- During the practical work placement a bound laboratory logbook (A4 or A5 booklet, not a collection of loose leaves) is to be kept, in order to be able to retrace the steps executed.
- At the end of each day of internship the workplace is to be tidied so that neither chemicals nor glassware are around the devices, in the weighing chambers or on the working space.
- At the end of an experiment the equipment used must be cleaned and any defects in the equipment or accessories are to be reported immediately to the assistant in charge, and missing components must be replaced immediately.

6.3.2 Laboratory Notebook

The laboratory notebook is still the most significant evidence of practical laboratory work. Each group keeps a lab logbook. It includes all activities in chronological order and without gaps (including weighing reports, etc.). The journal is the basis for evaluation and reporting.

Key Points of a Laboratory Notebook

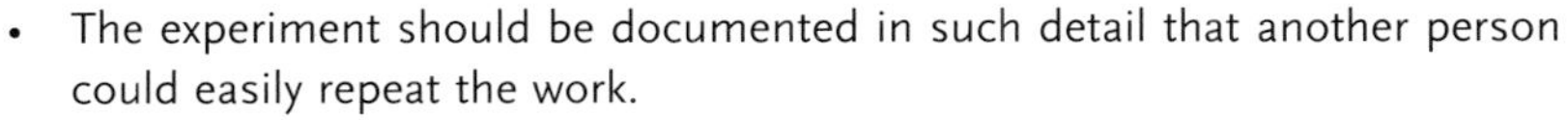

- The experiment should be documented in such detail that another person could easily repeat the work.
- The Laboratory Notebook says exactly what was done, how and when.
- The full documentation of *all* experimental observations (descriptive comments) should not be missed from any laboratory evaluation.
- The traceability of raw data must be ensured by providing accurate documentation.
- All numerical data are to be assessed using statistical methods and how they have been calculated must be evident (formulary).

Writing a Laboratory Notebook The following information is essential (Figures 6.1 and 6.2):

- Date
- Team

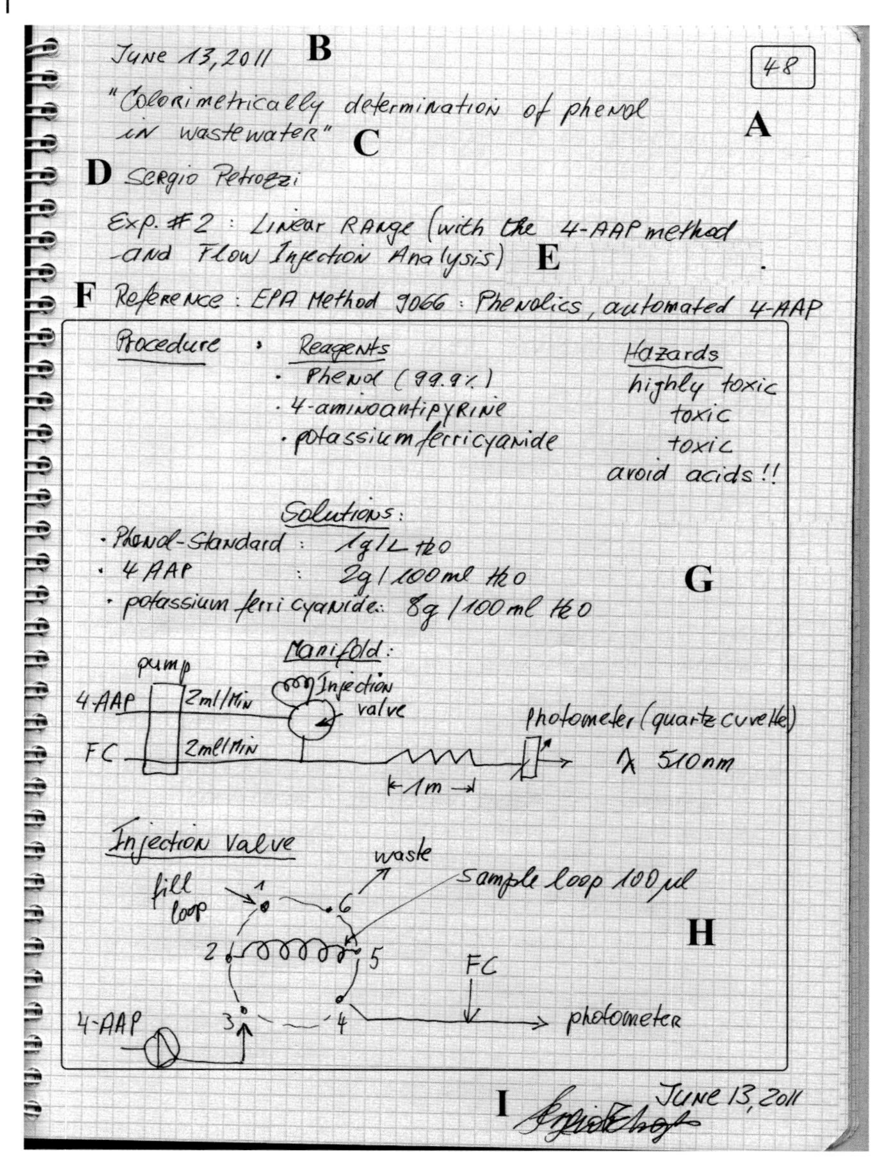

Figure 6.1 Laboratory logbook. **A** page number, **B** date, **C** title/statement of purpose, **D** team member, **E** objective of experiment, **F** background, **G** description of procedure, **H** important details, **I** signature.

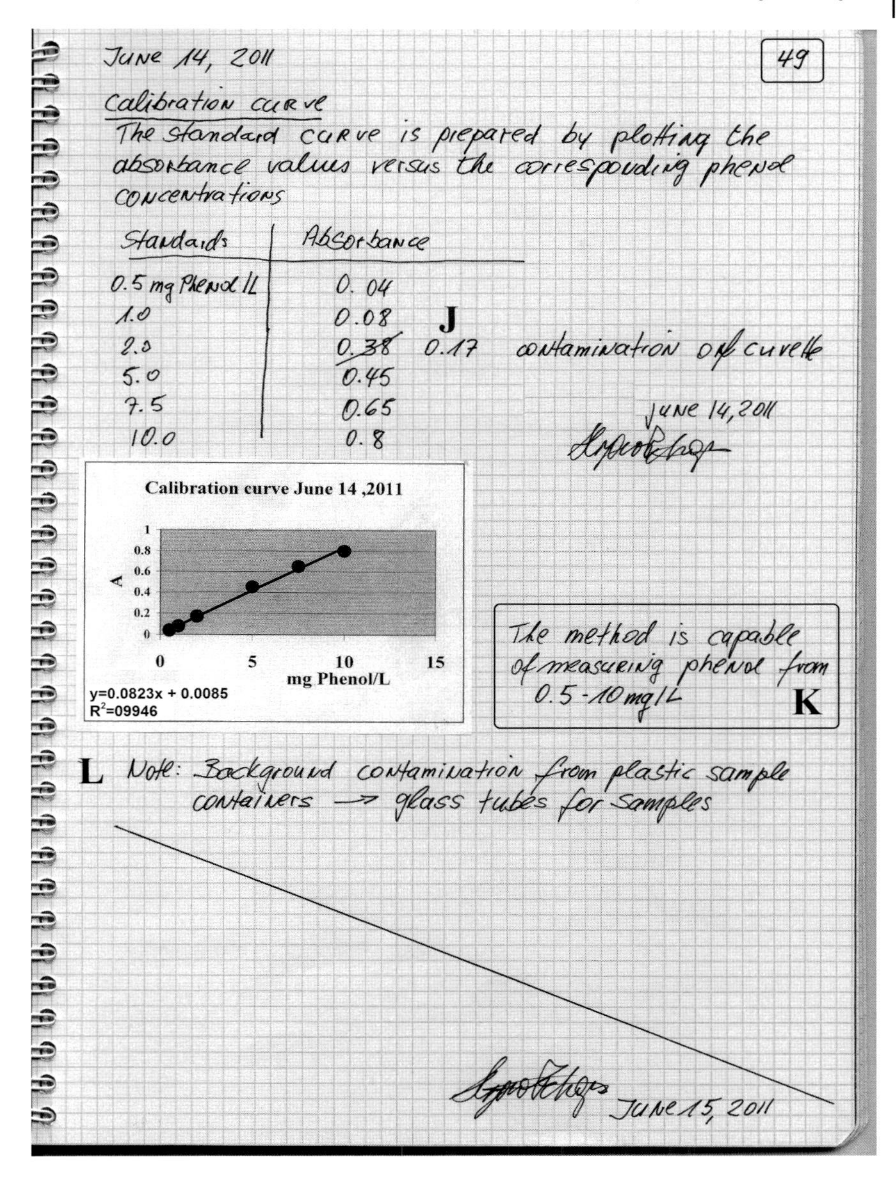

Figure 6.2 Laboratory logbook. **J** error crossed off with explanation, **K** computation of data and answer to the question posed by the experiment, **L** important information.

- Objective(s)

  Statement of the problem or task

- Bibliographical references
- Experimental plan
- Safety comments (special hazards)

  Safety! Part of chemical education is the instruction in handling potentially hazardous materials safely. There are still plenty of hazards around, and you should take these into account when planning the experiment. It may affect the quantities you use, or whether a fume cupboard is needed or not, and many other things. One needs to make a *risk assessment*. Standard practical exercises will have been assessed by teachers, but this does not remove the need to consider safety for undertaken experiments.

- Environmental protection, waste disposal
- Chemicals used, approach (g, ml, moles), purity
- Very detailed information on the apparatus used

  For example: The static mixer of the FIA unit consisted of a low pressure liquid chromatography column (length 50 mm, bed volume 0.35 ml) filled with inert material. The peristaltic pump was a 4-channel Ismatec MS-4 Reglo with a low pulsation pump-head fitted with 8 rollers operated with color-coded Ismatec® pump tubes. Tubes i.d. 2.06 mm (color code purple/purple) were used to propel the solutions. A 6-port Rheodyne Teflon rotary valve Type 50 fitted with a PEEK sample loop was used as the injector. The spectrophotometric readings were made by means of a single-beam UV/VIS Perkin Elmer lambda 1 spectrophotometer. A Hellma® flow-through quartz cuvette was used as the detection system. The optical path-length of each cuvette was 10 mm and the chamber volume was 80 μl. The photometer was interfaced with a PC through an IOtech (Personal Daq/3000™) Multifunction Data Acquisition Module. The software was DASYLab® (Data Acquisition System Laboratory) running under Windows XP.

- Procedures
- All manipulations and observations

The observations made and the data recorded will lead to the acceptance or rejection of the stated hypothesis, and will decide what future experiments may be done. The observations and data are therefore central to the whole exercise [1].

They need to be:

- recorded honestly.
- recorded as you go along, in the notebook, in ink, immediately.

- do not trust to memory, even for a minute or so – someone talks to you, and that data are forgotten.
- do not trust to memory; you do not want your mind occupied with trivial things and small details. You need to keep the overall experimental plan in mind.
- do not use odd scraps of paper or the edge of your lab coat to record data.
- the raw data are precious. Information on where raw data are stored is essential. Kanare [2] suggests that the data are treated with the care you would bestow on a family heirloom.
- the data must be recorded as completely as is possible. Do not worry too much about interpreting the data as you go along, and do not worry if some of the observations appear banal. For an example of how omission of even the simplest things can dramatically affect the outcome of an experiment, go to http://www.rod.beavon.clara.net/detail.htm.
- use good penmanship. Take care with numbers – never over-write, always cross out erroneous material with a single line and re-write the correct data.
- Interpretation of experimental data
- Discussion and conclusion

  write any calculations out clearly, showing all the steps and using units throughout
- relate your results to your hypothesis – do they support or refute it? Comparisons must be as quantitative as possible. Of course a simple analysis practical will only produce a result.
- record any ideas you have, however brief – if you don't write them down, you will forget them
- your conclusions should state
  - what you found out;
  - whether the hypothesis was supported or not, if appropriate;
  - the error limits on your answer(s); a quantitative assessment of error should be made if possible, so that you can decide whether the use of a measuring cylinder rather than a pipette, say, really did make any meaningful difference to the result;
  - suggestions for improvement in experimental design, if appropriate; the error analysis will be useful here.
  - what to do next, if appropriate.

6.4
Reports

The structure and style of writing should be based on publications, Bachelor's theses and so on, and should be written as follows:

- **Cover Sheet**

  The cover sheet should include the following information:

  - Title of the experiment
  - Summary
  - Date of the experiment
  - Name of the group members/group numbers

- **Title**

  The title should be as concise as possible. The purpose of the investigation, the object of investigation and investigation parameters should be apparent.

- **Summary/Abstract**

  The summary is a short version of the report and allows assessment of the relevance of the report without someone having to read the entire report. Analytical/scientific problem studied, procedures, results and conclusions are included in short form. Thereby, the summary must be independently readable, that is, there should be no references to formulae, diagrams and tables in the lab report.

Example:

Analytical problem:
Development of an analytical method for the quantification of total polyphenols in coffee using a flow injection analysis (FIA) apparatus with colorimetric detection.

Goals of the investigation:

1) Selection of a suitable method for the colorimetric determination of total polyphenols in coffee.
2) Optimization of the FIA method to determine the concentration of total polyphenols in coffee.
3) Validation of the method.
4) Suitability of the method with real samples examined.

The analytical problem and the related goal(s) of the investigation defined in the task are mentioned in the summary/abstract as well as the major results and conclusions.

Summary/Abstract
Analytical problem. A FIA procedure with colorimetric detection was developed for the determination of the total polyphenol content (TPP) directly from coffee brews. Goal 1. A comparison of three official accepted colorimetric methods based on different reactions was performed to select the most suitable colorimetric assay. Goal 2. The Folin–Ciocalteu reaction was found to be the superior one and therefore selected for the investigations of real samples. Goal 3. The FIA design adopted was characterized by its simple base, relatively inexpensive equipment and great capacity for achieving results that are excellent in view of rapidity, accuracy, reproducibility and precision. Key parameters were optimized for sensitivity, linearity, working range and matrix effects, and the method was successfully tested against false-positive adsorption from non-polyphenol components. Goal 4. Finally, the assay was applied to brews from two different coffees (light roast). Results. First, *Robusta* coffee showed 30% higher content of TPP compared to *Arabica* coffee. Secondly, coffees roasted to three different time–temperature roasting profiles were compared. High-temperature roasting and a short time showed higher TPP content in the brew than low-temperature roasting. Conclusion. An observation that indicates the potential for optimizing the roasting process to increase polyphenol content.

- **Motivation**

In the motivation, the problem is described and the basics are illustrated. For specific tasks, important points must be mentioned. Attention should be drawn to important publications in the field and it should contain enough background information. A brief but clear outline of the measuring principle should be included.

- **Description of practical work**

This section should include all information necessary to replicate the described work (equipment/key chemicals/methods/instruments). Also, it should be apparent that the methods used are appropriate to enable correct inferences.

The following questions must be answered:

- In what sequence were the experiments conducted?
- Which parameters were varied (table settings)?
- Boundary conditions (pressure, temperature, humidity, etc.)
- Reasons for procedure (why not otherwise?)
- Procedure for experimental evaluation (how to evaluate?)
- Possible error potentials (what "incidental issues" should one pay attention to?)

Since there is generally a laboratory manual for laboratory experiments, the representation of the experimental set-up and testing methodology should not exceed a maximum of two pages and it should not be a too detailed account of the activities carried out during the laboratory class.

- **List of Chemicals and Device Parameters**

List all chemicals used with quality data (puriss, p.a, etc.) and source (company). List device parameters so completely that the measurement could be repeated based on this information (see Laboratory Notebook)

- **Wet Chemistry**

Whole regulations should not be copied. Reference should be made to the source of such regulations and the workflow should be described in as detailed a manner as possible. In case of deviations from the rule with regard to *what, why* something was done differently, this must be explained exactly and justified.

- **Evaluations/Results**

The results are described and trends are recorded. Tables and illustrations are numbered and captioned and referenced in the text. All final results must be specified with dispersion. Raw data with a clear reference to measured curves from which they were extrapolated, are attached in the appendix. In the measured curves mark the spot where something has been extrapolated. Represent data and evaluations in tabular form as far as possible. Always specify measurement units. The discussion of test results should include a critical examination of the experimental results. These may include:

- A comparison of measurement series
- A comparison between theory and practice
- Meaningful improvement measures (laboratory apparatus, methodology, etc.)

- **Writing Unit Symbols and Names, and Expressing the Values of Quantities**

Bureau International des Poids et Mesures, The International System of Units (SI), 8th Edition, 2006

Organisation Intergouvernementale de la Convention du Mètre

General principles for the writing of unit symbols and numbers were first given by the 9th CGPM (1948, Resolution 7). These were subsequently elaborated by ISO, IEC (International Electrotechnical Commission), and other international bodies. As a consequence, there now exists a general consensus on how unit symbols and names, including prefix symbols and names, as well as quantity symbols should be written and used, and how the values of quantities should be expressed. Compliance with these rules and style conventions, the most important of which are presented in this chapter, supports the readability of scientific and technical papers (See Appendix D).

- **Validation and Statistical Evaluation of Results (see Section 6.4)**

A quantitative statistical evaluation must always be conducted in order to assess the test results.

- **Discussion and Outlook**

The discussion includes the interpretation of the results. Conclusions derived from the results are substantiated with models and hypotheses. Any interpretation of the results should be explained. The analyst's own observations and experiences (both positive and negative) should be described. The methods should be assessed (including suitable or what, where not suitable).

It should be stated which targets were not achieved and why. Possible follow-up work and suggestions for improvement should be listed.

- **Reference List**

Citations should be numbered consecutively and should provide a unique text reference.

The cited references are to be given in standard format.

- **Appendix**

The appendix should include everything that disturbs the reading flow in the main part, but which, however, is necessary for the documentation and traceability of the test results. Typical attachments include program listings, extensive data tables and diagrams, derivations from formulae. Material from the appendix is not necessary for a general understanding of the work

- **General Conditions of Business**

- The work should be understandable on the basis of the report.
- Describe your own work in past tense, results are always reported in present.
- Do not start a sentence with a numeral.
- Spell check the document thoroughly.

What should be avoided?

- The "I/we" and "you" form.
- An excessively large number of sentences with that: "It is known that" becomes "as we all know".
- Unnecessary fillers such as: yes, exactly, after all, but, certainly, in this regard, possibly, to a certain extent.
- Excessive cumulatives with: very, especially, absolutely, extremely.
- Adjectives like big, small, much, few (relative statements).

- **Source References**

It is a basic principle of scientific documentation, to clearly distinguish between services provided by the author himself and those that represent the intellectual property of a third party. This distinction is mainly made by specifying the sources and references used. The sources are to be cited in such a way as to enable the reader to obtain each source.

- **Tables**

Structured information is presented in tables. In addition to data, qualitative characteristics can be compared. Very often tables can be used to summarize a matter more clearly than in text. It is always necessary to explain or interpret, respectively.

- **Figures and Diagrams**

The following information must be considered in the design of diagrams and illustrations:

- Graph axes are to be labeled with numbers and units.
- A meaningful subtitle should be selected for each chart.
- Illustrations and diagrams must be numbered.
- Each diagram must be referenced in the text at least once.
- Illustrations and diagrams are often easier to understand than long text passages.
- Use 3D charts only when absolutely necessary.
- If you use images from outside sources, the sources must be specified and referenced.

- **External Form**

The external form often has a decisive influence on the willingness to read and the verdict of the reader. It makes a good impression when it gives a clear overview of the information.

Grading Grading using a form (suggestion: Evaluation Guideline–Appendix D) has proven its practicality, as it ensures a certain objectivity. The grade is calculated from the assessment of theoretical understanding, practical skills and reporting.

References

1 http://www.rod.beavon.clara.net/lab_book.htm (accessed July 2012).
2 Kanare, H.M. (1985) *Writing the Laboratory Notebook*, American Chemical Society, Washington, DC, ISBN 0 8412 0933 2.

> Those who write clearly have readers, those who write obscurely have commentators.
>
> *Abraham Lincoln*

7
Literature

7.1
Cited Literature

1 Shewhart, W. (1986) *Statistical Method from the Viewpoint of Quality Control*, Dover Publications Inc.

2 Mikosch, T. (1998) *Elementary Stochastic Calculus with Fin*, World Scientific Pub, ISBN-10: 981-02-3543-7.

3 (2006) DIN-ISO: 3534-1:2006: *Statistics – Vocabulary and Symbols – Part 1*: General statistical terms and terms used in probability.

4 Barwick, V., Burke, S., Lawn, R., Roper, P., and Walker, R. (2001) *Applications of Reference Materials in Analytical Chemistry*, Laboratory of the Gouvernment Chemist, Teddington, UK, ISBN-0-85404-448-5.

5 Czichos, H., Saito, T., and Smith, L. (2006) *Springer Handbook of Materials Measurement Methods*, Springer Science + Business Media Inc., ISBN-13: 978-3-540-20785-6.

6 Stephenson, W.K. (2009) *Journal of Chemical Education*, **86** (8), 933–935.

7 Grubbs, F.E., and Beck, G. (1972) *Technometrics*, **14**, 847.

8 Anscombe, F.J. (1973) Graphs in statistical analysis. *The American Statistician*, **27** (1), 17–21.

9 Kellner, R., Mermet, J.M., Otto, M., Valcárcel, M., and Widmer, H.M. (2004) *Analytical Chemistry – A Modern Approach to Analytical Science*, Wiley-VCH Verlag GmbH, Weinheim, ISBN-3-527-30590-4.

10 Current figures, http://www.cas.org/index.html (accessed May 2012).

11 Consortium licences: e.g., the Consortium of the Swiss Academic Libraries, http://lib.consortium.ch/products_categories_lizenzen.php?lang=2 (accessed May 2012).

12 http://www.cas.org/support/academic/index.html (accessed May 2012).

13 Keese, R., Brändle, M.P., and Toube, T.P. (2006) *Practical Organic Synthesis: A Student's Guide*, Wiley, Chichester.

14 https://www.reaxys.com/info/ (accessed May 2012).

15 (November 2008). *DIN 32645* Chemical Analysis – decision limit, detection limit and determination limit under repeatability conditions – terms, methods, evaluation.

16 Haeckel, R., Hänecke, P., Hänecke, M., and Haeckel, H. (1998) *Journal of Laboratory Medicine*, **22** (5), 273–280.

17 ISO 11843-1 (1997) *Capability of Detection. Part 1: Terms and Definitions*, ISO, Geneva.

18 IUPAC (1995) Nomenclature in evaluation of analytical methods including detection and quantification capabilities. *Pure and Applied Chemistry*, **67**, 1699–1723.

19 Clinical and Laboratory Standards Institute (2004). Protocols for determination of limits of detection and limits of quantitation approved guideline. Wayne, PA: CLSI; CLSI document EP17.

20 Kaiser, H., and Specker, H. (1956) *Fresenius' Journal of Analytical Chemistry*, **149**, 46–66.

Practical Instrumental Analysis: Methods, Quality Assurance and Laboratory Management, First Edition.
Sergio Petrozzi.

21 Currie, L.A. (1997) Detection: International update, and some emerging dilemmas involving calibration, the blank, and multiple detection decisions. *Chemometrics and Intelligent Laboratory Systems*, **37**, 151–181.

22 Wenclawiak, B.W. (2010) *Quality Assurance in Analytical Chemistry: Training and Teaching*, Springer-Verlag, ISBN-13: 978-3-642-13608-5.

23 De Bievre, P., and Guenzler, H. (2005) *Validation in Chemical Measurement*, Springer Science+Business Media Inc., ISBN-13: 978-3-540-35988-3.

24 U.S. Department of Health and Human Services, Food and Drug Administration Guidance for Industry, Q2B Validation of Analytical Procedures: Methodology. (November 1996).

25 (October 1994) International Conference on Harmonisation of Technical requirements for Registration of Pharmaceuticals for Human Use, ICH Harmonised Tripartite Guideline, *Validation of Analytical Procedures: Text and Methodology Q2 (R1)*.

26 Danzer, K. (2007) *Analytical Chemistry: Theoretical and Metrological Fundamentals*, Springer Science+Business Media Inc, ISBN 3-540-20788-0.

27 (September 1986) U.S. Environmental Protection Agency (EPA) Method 9065 Phenolics (Spectrophotometric, manual 4-AAP with distillation).

7.2 Recommended Norms (Selection)

7.2.1 Calibration

DIN 38402 (Part 51), German standard methods for the examination of water, waste water and sludge; general information (group A); calibration of analytical methods, evaluation of analytical results and linear calibration functions used to determine the performance characteristics of analytical methods (A 51).

DIN 53804 (Part 1), Statistical evaluation – Part 1: Continuous characteristics.

DIN ISO 11843-1, Capability of detection – Part 1: Terms and definitions (ISO 11843-1:**1997**, including Technical Corrigendum 1:**2003**).

DIN ISO 11843-2, Capability of detection – Part 2: Methodology in the linear calibration case (ISO 11843-2:**2000**.

DIN V ENV 13005, Guide to the expression of uncertainty in measurement.

EA 4/16 "EA guidelines on the expression of uncertainty in quantitative testing", Dez. **2003**.

EUROLAB Technical Report 1/**2002** "*Measurement Uncertainty in Testing*".

NORDTEST Report TR 537 (2003) Handbook for calculation of measurement uncertainty in environmental laboratories; Nordtest, Finland.

7.2.2 Inter-laboratory Tests

DIN 38402 A41, German standard methods for the examination of water, waste water and sludge; general information (group A); inter-laboratory tests; planning and organization (A 41).

7.2.3 Quality Management

DIN 55350, Concepts for quality management – Part 11.

ISO 8402, Application of statistics. Guide for the selection of an acceptance sampling system, scheme or plan for inspection of discrete items in lots.

ISO 9000, Quality management systems – Fundamentals and vocabulary (ISO 9000:2000).

ISO 9004, Managing for the sustained success of an organization – A quality management approach (ISO 9004:**2009**).

7.3 Suggested Books (Selection)

Baiulescu, G.E., Stefan, R.-I., and Aboul-Enein, H.Y. (2000). *Quality and Reliability in Analytical Chemistry*, CRC Press, ISBN-10: 0849323762.

Konieczka, P., and Namiesnik, J. (2009). Quality Assurance and Quality Control in the Analytical Chemical Laboratory: A Practical Approach (Repost), ISBN: 1420082701.

Krull, I.S., and Swartz, M.E. (2011). *Handbook of Analytical Validation*, CRC Press, ISBN: 9780824706890.

Lira, I. (2002). *Evaluating the Measurement Uncertainty: Fundamentals and Practical Guidance*, CRC Press, ISBN-10: 0750308400.

Mullins, E. (2003) *Statistics for the Quality Control Chemistry Laboratory Book*, Royal Society of Chemistry, ISBN-10: 0854046712.

Swartz, M.E., and Krull, I.S. (1997). *Analytical Method Development and Validation*, CRC Press, ISBN: 9780824701154.

Vetter, T.W., and Guthrie, W. F. (2011) *Uncertainty in Analytical Chemistry: Practical Tools and Applications*, CRC Press.

Books are like friends, they should be few and well-chosen.

8 Projects

Comment I

Old hands in everyday life and qualified analytical chemists would rightly argue that the chapters here described cover only the most essential principles of the analysis process and the application of statistical evaluation methods. However, as a vocational teacher working with newcomers to analysis, one finds that in addition to the understanding of basic theoretical principles, the ability to consider the project as a whole and interpret it in practice is a major goal of the training program.

Practical relevance means that the reproducibility of the analyses must be ensured, the calculation of key figures defined, the correct application of statistical weighting to results applied, and details of acceptance criteria observed. Last but not least, craftsmanship in relation to analytical work needs practice: this includes the writing of laboratory notebooks, and writing reports and presentations.

Comment II

The apparatus, devices and software applications described here are representative of similar instruments, accessories or computer programs from other sources (*make, manufacturer, version, type*). So instead of a peristaltic pump, a piston pump can be used, or for chromatographic analysis several programs are equally suitable.

Comment III

The globally harmonized system (GHS) partially provides for *completely new* pictograms (see Appendix F.12). New hazard symbols or hazard pictograms replace the current symbols to date. There are also partly new classification criteria and limits.

The new icons are in the form of a red-bordered diamond with a black pictogram on a white background and pictorial warnings of the dangers. The presently valid symbols (and those symbols used in this book on an orange-green background)

Practical Instrumental Analysis: Methods, Quality Assurance and Laboratory Management, First Edition.
Sergio Petrozzi.

will be replaced by the new symbols as soon as they are introduced. The new GHS-designation will be binding for materials with effect from 1.12.2010 and for mixtures with effect from 1.6.2015. Until then, the classification under the old and new system must be carried out using safety data sheets. In addition to the pictograms the level of danger is described by using one of two signal words: "Danger" or "Warning".

Notes on the Project Phases

- **Literature Research**

  The works of the authors D.A. Skoog and J.J. Leary (*Principles of Instrumental Analysis*), D.A. Skoog, F.J. Holler, S.R. Crouch (*Fundamentals of Analytical Chemistry*), R. Kellner, J.M. Mermet, M. Otto, M. Valcárcel, H.M. Widmer (*Analytical Chemistry*) and D. Harris (*Quantitative Chemical Analysis*) have become indispensable reference works for students of chemistry, pharmaceuticals and biotechnology, and provide the analytical knowledge in an easy-to-understand form with an eye for the essentials. Their books are well explained and deal comprehensively with (almost) all subject areas of analytical chemistry, including practical examples from the practice of modern analysis.

  These books complement each other and belong in their own specialist book collection that I would recommend to every chemist and student of chemistry.

- **Bibliographical Research (General Topics)**

Skoog, D.A., and Leary, J.J. (2006) *Principles of Instrumental Analysis*, 6th edn, Brooks Cole, ISBN-10: 0495012017, ISBN-13: 978-0495012016.

Skoog, D.A., Holler, F.J., and Crouch, S.R. (2003) *Fundamentals of Analytical Chemistry*, 8th edn, Brooks Cole, ISBN-13: 978-0-030-35523-3.

Keller, R., *et al.* (2004) *Analytical Chemistry: A Modern Approach to Analytical Science*, 2nd edn, John Wiley & Sons, ISBN-13: 978-3-52730590-2.

Harris, D. (2010) *Quantitative Chemical Analysis*, 8th edn, Palgrave Macmillan, ISBN-13: 978-1-429-23989-9.

Ahuja, S., Ahuja, C. (eds) (2006) *Modern Instrumental Analysis*, Elsevier, Calabash, NC, USA, Neil Jespersen, Chemistry Department, St. John's University, Jamaica, NY, USA, ISBN-13: 978-0-444-52259-7.

- **Bibliographical Research (Specific Topics)**

  This literature deals in detail with a specific analytical method. The proposed works are not limited to the function, but also critically examine the performance of each method and name typical sources of error.

- **Keywords**

  The recommended keywords should be sought in the specialist literature and understood prior to beginning the projects. They point to important connections between the methods and are essential for understanding a measurement principle and optimizing an analytical method.

- **Project Start**

  Before beginning the actual project, it is important to conduct a brief meeting with the students. Essentially, this meeting is to discuss the project process and solve any possibly existing discrepancies (tasks) and/or problems (*fit for purpose*).

- **Introduction**

  In the introductory part of each project's essential theoretical foundations and instrumental details are presented in short form for quick reference.

- **Questions**

  The attaching of solutions to the problems posed was deliberately omitted as the answers will also be graded. Since the same projects are assigned several times per semester, this forces students to produce an independent solution.

  At the report meeting the answers should be dealt with individually.

8.1 Chromatography

8.1.1 Gas Chromatography–GC, Project: Tanker Accident

8.1.1.1 Analytical Problem

After a collision at night near the town of Bratislava between two waterway craft, around 120 tonnes of diesel and light gas oil spilled into the river Danube. During the collision between the inland tanker "Caramel" and the bulk carrier "Valentine", four of the Caramel's eight tanks were seriously damaged. After the collision the 82 meter long ship, which had been carrying 900 tonnes of diesel, was tied up at the sluice in Bratislava, where fire and water police sealed the leaks and pumped out the remaining fuel. The 95 meter long "Valentine" later docked at the Port of Vienna. Why the accident occurred is still unclear. One day later an investigator re-examined the inland tanker, which had meanwhile been moved to a shipyard. The Environment Agency emphasized that oil barriers would remain in place to protect two nature reserves. They also stated that polluted shorelines would again be checked and cleaned if necessary. On behalf of the Ministry for the Environment, the State Institute for Water Pollution Control took hundreds of representative samples of water and had them tested for their gas oil content. The results will show whether further adverse effects can be expected for flora and fauna and how much contamination is present. A task force has been appointed to optimize, validate and implement a suitable method for the rapid detection and measurement of gas oil.

Object of Investigation/Investigation System

The river Danube

Samples

Water samples (taken at various locations)

Examination Procedure

Gas chromatography (GC) is one of the most widely used techniques in modern analytical chemistry. In its basic form GC is a technique capable of separating individual components contained in a mixture. The separation takes place in the vapor phase. The detector measures the quantity of the components that exit the column. The breadth of application of gas chromatographic analysis ranges from gaseous compound analysis (e.g., also with portable gas chromatographs) to the analysis of non-volatile substances after derivatization and after defined pyrolysis. With capillary GC, the highest separation efficiency in all chromatography can be achieved. The importance of gas chromatography in quality control and process control in the chemical and drug industry, in environmental pollution investigations, and in clinical analysis is critical.

Analytes

n-Alkanes, light gas oil/diesel oil

Keywords

Sample preparation by solid–liquid extraction, absolute and relative analysis methods, calibration with external standard, capillary gas chromatography, composition of petroleum fractions, naturally occurring hydrocarbons, flame ionization detector (FID), temperature programming in GC, characteristic values in GC.

Bibliographical Research (Specific Topics)

McNair, H.M., and Miller, J.M. (2009) *Basic Gas Chromatography*, 2nd edn, Wiley-Interscience, ISBN-13: 978-0470439548.

Rood, D. (2007) *The Troubleshooting and Maintenance Guide for Gas Chromatographers*, 4th Revised and Updated Edition, Wiley-VCH, ISBN-13: 978-3527313730.

Grob, R.L. (2004) *Modern Practice of Gas Chromatography*, 4th edn, Wiley-Interscience, ISBN-13: 978-0471229834.

Project Process/Task

(A) Assessing the separation efficiency of a GC column

Using the GC test procedure for capillary columns [1, 2] ascertain the number of theoretical plates N and the separation efficiency of the columns used.

(B) Optimization of the temperature program

Conduct optimization experiments in order to achieve the best possible separation of an alkane mixture. The temperature program serves as an optimization parameter. Analyze a test mixture at first isothermally, then by varying the heating rate.

(C) Identification of the major components of gas oil (qualitative analysis)

Using the previously optimized temperature program (Task B), analyze a gas oil standard, and then an alkane standard (n-decane, n-tetradecane, n-hexadecane and n-octadecane). Thereby identify the dominant peaks of the n-alkanes based on their retention time.

(D) Determination of the gas oil content in different "contaminated" water samples (quantitative analysis)

D.1 Calculate the extraction efficiency of the method.

D.2 Using the percentage area method calculate the relative proportions of the major n-alkanes after qualitative identification from task B. Calculate the sum of "non-n-alkanes".

D.3 Determine the sensitivity of the method.

(E) Determine the amount of gas oil in different water samples

(F) Project reporting

The results are to be submitted in the form of a report (or presentation). Comparisons with bibliographical references are to be made.

8.1.1.2
Introduction

Gas chromatography is a powerful method for separating mixtures of volatile components. The required sample sizes are relatively small (<1 ng/components). The high separation efficiency attainable allows the separation of very complex mixtures. The components of a gas chromatograph are illustrated in Figure 8.1. The operation of the GC is simple. A gas, a vaporizable liquid or a solid without decomposition is introduced into a sample injection port via a septum and volatilized. There the substances are transported by a carrier gas through a thermostated column, where the chromatographic process takes place at a fixed temperature. After leaving the column, the in-time separated substances then pass through a detector that displays each single element on a chart recorder, integrator or personal computer.

During the whole procedure the system is kept at a temperature sufficient to maintain distribution equilibrium for each analyte between the vapor phase and the condensed phase, but at the same time the samples must be sufficiently stable at the given temperature in order to avoid decomposition during the chromatographic process. In some instances, the volatility and the thermal

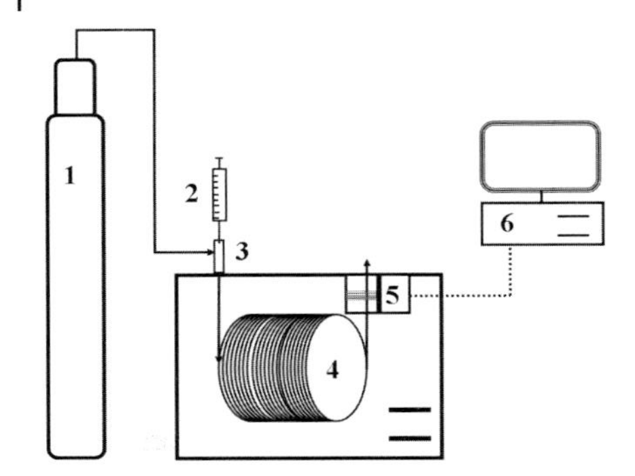

Figure 8.1 Schematic diagram of a gas chromatograph. **1** carrier gas supply, **2** syringe, **3** injection port with septum, **4** column, **5** detector, **6** data acquisition and analysis system.

stability of the sample components are enhanced by making chemical derivatives of them and analyzing them in the derivative form. The GC is always the method of choice when the substances are to be analyzed in samples that can easily evaporate. If time-consuming sample preparation and preparations for the separation of matrix components or derivatization are required, frequently liquid chromatography can be used advantageously.

In gas chromatography adsorbing liquids (partition columns) or adsorbing solids are used as stationary phases. A basic distinction is made between the two types of columns:

Packed columns

The packed column contains an inert carrier material on which the stationary phase is coated (e.g., silica particles). Most packed columns used in practice are made of metal. However, in certain applications, such as analysis of unstable biological substances and pesticides, columns made of glass tubing may be preferable.

Capillary columns

The capillary (open tubular columns, OTC) consist of long thin tubes (glass, quartz). A distinction is again made between two types of capillary columns: the wall-coated open tubular (WCOT) columns and the porous layer open tubular (PLOT) columns. The stationary phase in WCOT columns is a liquid film coated to the deactivated wall of the column. These are the most widely used columns in gas chromatography. In the PLOT columns the stationary phase is a solid substance that is coated to the column wall.

Instrumentation and Techniques

Injector

The injector is used to inject the sample into the separation system. The sample mixture is usually dissolved in a volatile solvent and inserted into the injector using a syringe. Basically, two types of injectors can be distinguished. In the case of the on-column injector, the syringe needle is inserted into the separation column and the sample solution is thus delivered directly into the capillary. In the case of the split injector, however, the sample solution is injected into the evaporation chamber through a septum (a membrane made of soft plastic or rubber). The sample vaporizes due to the prevailing temperature there (usually 200–300 °C) and a vapor cloud arises, which is transported into the column by the carrier gas. The sample quantity that reaches the column can be regulated with the help of a split.

Carrier Gas

The carrier gas (mobile phase, carrier) is led to the injector from gas cylinders by means of a pressure reducing valve. The most commonly used carrier gases are hydrogen, nitrogen and helium. In gas chromatography an inert mobile phase is used, in contrast to liquid chromatography. The separation is only due to the nature of the stationary phase.

The practical gas chromatographer is mainly interested in three problems:

- Which carrier gas gives the best performance?
- Which carrier gas permits fast analysis?
- How to achieve the fastest analysis in a given system?

Since the principles of measurement of the individual detector are different, the carrier gas used has to be selected accordingly.

With few exceptions, hydrogen is the most suitable carrier gas for capillary chromatography. Because the optimum velocity for hydrogen is higher than that for helium or nitrogen, hydrogen can provide better separation in less time than either of the other gases.

The great disadvantage using hydrogen is that it forms explosive mixtures with air at concentrations 4–10%.

Helium is the second best carrier, and is popular because no safety precautions are needed.

When using packed columns nitrogen could be the carrier gas of choice.

Carrier gases for chromatographs must be of high purity.

Injection Techniques

Introducing and evaporating the sample into the gas chromatographic system is one of the most important steps in analysis. Gaseous samples can be directly entered into the system by injection. Liquid and solid samples are injected after dilution with a gas chromatography non-interfering solvent. The dissolved sample and the solvent have to be evaporated instantaneously after injection and carried to the column with a minimum of band spreading.

There are mainly two different types of sample injectors: the on-column-injector, where the sample is directly injected onto the column, and the split injector. With this latter technique the sample is first injected in an evaporation area, before the sample is pushed onto the column with the carrier gas. A splitter divides the injected and vaporized sample into two parts. One part is directed into the column and the other (usually the larger) is vented. Split injection is preferred when working with samples with high analyte concentrations in order to prevent column and detector overloading.

For the on-column-injector, there is only one method to bring the sample onto the column, the direct one, where a certain volume is injected onto the column. With a split injector, there are mainly four different ways to inject the sample onto the column.

The first method is the *filled-needle* technique. The sample is held in both the needle and the syringe barrel, and is injected rapidly or slowly, depending on the injection mode.

The second method is the *cold-needle* technique. The sample is drawn completely into the syringe barrel and the needle filled with air. The sample is injected as soon as the needle is fully inserted into the injection port.

The third method is the *hot-needle* technique. After the sample, some air is taken up until the air can be seen. Then, the needle is put through the septum. After about five seconds, the sample in the barrel is injected.

The fourth technique is the *solvent flush/sandwich* technique, where the syringe is first filled with pure solvent, then air is taken in until the needle is full. Then the needle is filled with the sample. The last layer is air again. The sample is then injected as in the hot-needle technique.

Stationary Phases/Columns

By virtue of its chemical properties, the stationary phase influences the retention behavior of chemical compounds. An almost infinite number of stationary phases are commercial available. A detailed listing of these phases is beyond the scope of this text book. Here we give just a few examples.

Important criteria for selecting columns are:

- Application (e.g., trace analysis)
- Column material (type of sample)
- Film thickness (resolution)
- Column diameter (limit of detection)
- Column length (performance)
- Carrier gas flow rate
- Size of sample (e.g., when amount of sample is limited)

In GC the separation of two analytes occurs due to differences in their interaction with the stationary phase, therefore a phase must be chosen that matches the properties of the sample, for example, separation on a nonpolar column is based predominately on different boiling points.

Another important criterion for classifying the selectivity of the stationary phase is its polarity. Nonpolar compounds interact selectively with nonpolar stationary phases and are retained according to their polarity, that is, the polarity affects the distribution coefficient K.

An example of a nonpolar stationary phase is Apiezon L (a mixture of saturated hydrocarbons with relative molecular weights of up to 15 000). This phase is suitable for the separation of nonpolar mixtures, such as saturated and unsaturated hydrocarbons and halogenated compounds. The classic nonpolar phases include polysiloxane-phases. The phase, which has long been on the market, is called a classical phase. Classical phases are characterized by good reproducibility and a long lifecycle.

A stationary phase made of polyethylene glycol (e.g., Carbowax) is a very polar stationary phase which is used in the separation of alcohols, esters, ketones, aldehydes, and so on.

Oven

Apart from the film thickness and chemical composition of the stationary phase, the influence of temperature on the retention behavior is of great importance. The vapor pressure of a compound and thus its distribution coefficient K is temperature dependent. Separations of mixtures with a large volatility range are, therefore, best when conducted with a temperature gradient (temperature program). The temperature is continuously increased during the separation, and thus the distribution coefficients of the components still in the separation system are decreased. The result is a rapid elution of higher boiling components, better resolution and more superior quantitative evaluation.

Detector

The most widely used detector in gas chromatography is the flame ionization detector (FID) (Figure 8.2). Compared to other detectors, it has the advantage of greater sensitivity when used in a very wide range of applications. The FID is sensitive to carbon-containing compounds and hardly limited to certain classes of substances. The eluents are burned in a diffusion flame, whereby ions are formed by subsequent oxidation with oxygen, which is fed into the flame. The ions are accelerated in an electric field (approx. 300 V) and the resulting current is measured. The FID responds very badly to small heteroatom-containing organic molecules such as CO, CS_2, COS, CH_2O, HCOOH, HCN or $HCONH_2$ and fails to register almost all the inorganic components.

The thermal conductivity detector (TCD) is also very versatile, but it is less sensitive than the FID. It is mainly used for the detection in separations of inorganic gases in packed columns. The electron capture detector (ECD) is a selective detector, which responds to components with electron capture properties (certain molecules containing heteroatoms such as Cl, Br, O or N). The scope of application is therefore restricted. However, the sensitivity is very high. This detector is used for the selective detection of certain components (e.g., some pesticides).

Temperature as a Gas Chromatographic Parameter

The capillary is surrounded by an oven, which can either hold a constant temperature or heat up at a certain rate (Figure 8.3). The oven is needed because the retention time of a certain sample is dependent on the temperature, the used column and the pressure. If the temperature is too high, the substances elute too fast and the resulting peaks are not separated, if the temperature is too low, the

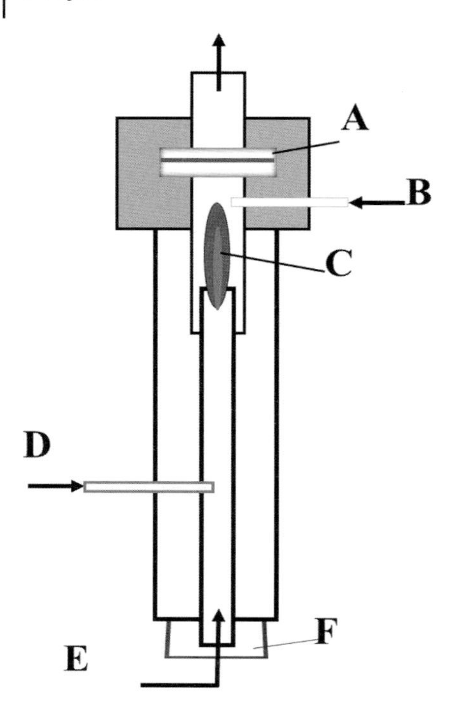

Figure 8.2 Typical flame ionization detector (FID). **A** collector electrode, **B** hydrogen, **C** burner jet/gas flame, **D** air, **E** carrier gas with sample, **F** column connection.

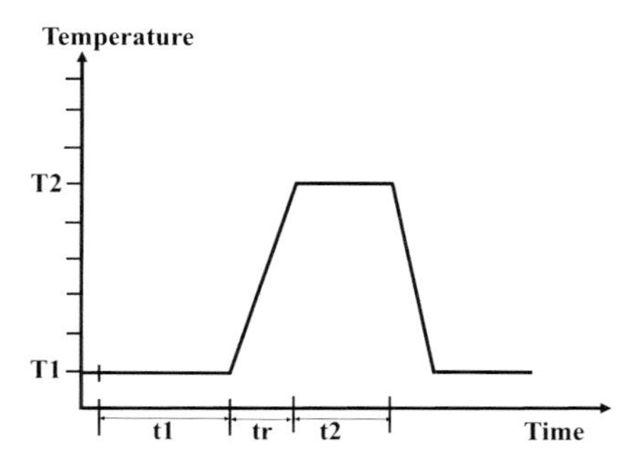

Figure 8.3 Temperature programmed GC-analysis. **tr** temperature ramp from T1 to T2 (°C min^{-1}).

analysis takes too much long. The temperature, which the column assumes in the oven, is in the thermodynamic sense a gas chromatographic parameter.

With a given stationary phase, it influences the distribution coefficient of the components to be separated. The basic rule is:Retention volumes (V_R) decrease exponentially with increasing temperature (T).

$$\log V_R \sim 1/T \tag{8.1}$$

The distribution coefficients of various classes of substances show characteristic temperature dependences in a given stationary phase, depending on their structure. For the separation of certain material pairs there are, therefore, optimal column temperatures that need to be adjusted in the analysis. At these temperatures the differences between two distribution coefficients are greatest.

It may well be the case that a better separation is achieved with increasing temperature. Normally, however, greater differences in the distribution coefficients result by lowering the column temperature.

Optimization of the separation system (see also Section 8.1.3)

Separation Efficiency of a Column

The separation efficiency of a column can, among other things, be measured by how broad a substance peak becomes when passing through the column and is expressed as the number of theoretical plates (N) or the height equivalent to one theoretical plate (H or HETP). The larger the number of theoretical plates (the smaller the value of HETP), the better the separation efficiency of a particular column.

Theoretical Plate Number (N)

The most common measure of efficiency is the theoretical plate number (N). This is defined by Equation (8.2) and is measured from a test chromatogram (Figure 8.4).

The concept of the theoretical plate (derived from the distillation process) is an important model for describing chromatographic separation. The classical theory of chromatography by Martin and Synge (Nobel Prize 1952) assumes that the chromatographic column is divided into a number of discs. These discs are known as separation stages or theoretical plates.

Separation stage or plate height is the term given to the length of a chromatographic system, within which the distribution equilibrium is adjusted.

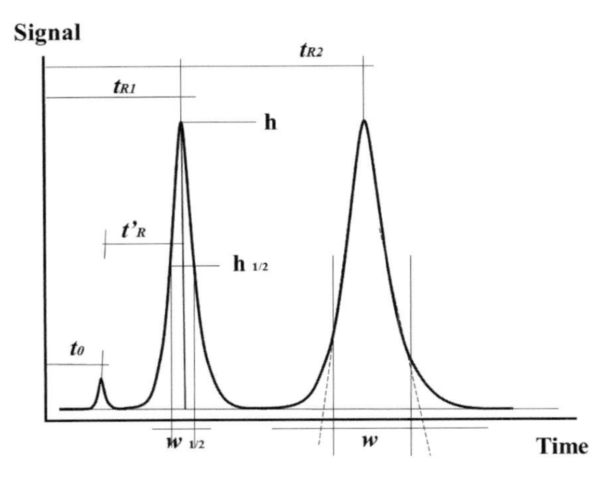

Figure 8.4 Typical GC chromatogram.

The number of theoretical plates N is a further characteristic of the separation system. The higher the number of theoretical plates, the more powerful (efficient) the chromatographic system is (Figure 8.5).

The number of theoretical plates increases with increasing column length. In order to compare columns of different lengths, however, yet another parameter has been defined, namely the plate height (H).

$$N = 16\left(\frac{t_R}{W}\right)^2 = 5.54\left(\frac{t_R}{W_{1/2}}\right)^2 \tag{8.2}$$

$$H = \frac{L}{N} \tag{8.3}$$

N Number of theoretical plates
L Column length (m)
W Peak width (s)

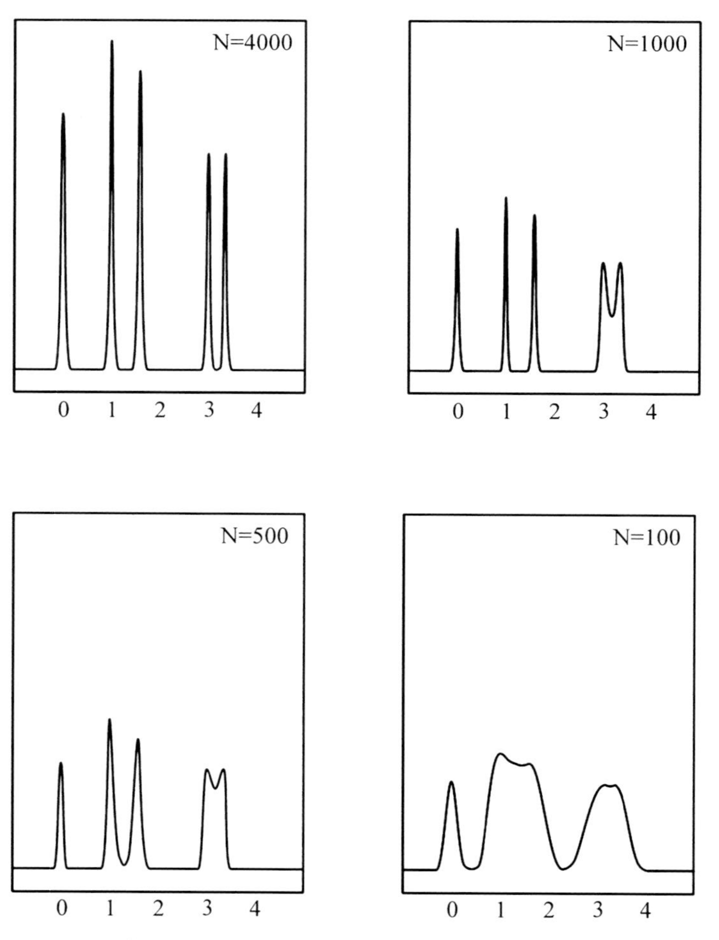

Figure 8.5 Influence of the number of theoretical plates (N) on the resolution.

$W_{½}$ Peak width at half height (s)
H Plate height (m)
t_R Retention time (s)

The number of theoretical plates of a column is calculated from the retention time and peak width. The smaller the plate height, the greater the efficiency of the column is.

> ***Plate number N (in chromatography)***
>
> *The value of 5.545 stands for 8 ln2. These expressions assume a Gaussian (symmetrical) peak. In these expressions the units for the quantities inside the brackets must be consistent so that their ratio is dimensionless: that is, if the numerator is a volume, then peak width must also be expressed in terms of volume. In former nomenclature the expression "number of theoretical plates" or "theoretical plate number" was used for the same term. For simplification, the present name is suggested. PAC, 1993, 65, 819 (Nomenclature for chromatography (IUPAC Recommendations 1993)), p. 847.*

Influence of Flow Velocity: "The *van Deemter* Equation" [3]

While flowing through a chromatography system (at the start still "infinitely" small) the substance peak is increasingly dispersed. This happens primarily due to three mechanisms:

(A) In a column packing different analyte molecules go varying long distances between the stationary phase and therefore progress at differing speeds ("Eddy-diffusivity"). This influence is independent of the flow velocity and the mobile phase. While the van Deemter equation captures those parameters that influence the theoretical plate degree H in the case of packed columns, the *Golay* equation describes the theoretically achievable degree of separation for capillaries.

(B) In the "normal" diffusion in the mobile phase there are always analyte molecules swimming against the stream. A statistical distribution of molecules around their "average" arises. This effect increases with decreasing flow velocity.

(C) A chromatographic system is in dynamic equilibrium. As the mobile phase is moving continuously, the system has to restore this equilibrium continuously resulting in slightly shifted concentration profiles of sample components between mobile and stationary phase.

The van Deemter equation takes all three factors into account. It describes the dependence of theoretical plate height (H) on the flow velocity (u) ($\mathrm{m\,s^{-1}}$):

$$H = A + \frac{B}{u} + C \cdot u \tag{8.4}$$

A Eddy diffusion
B Longitudinal diffusion coefficient ($\mathrm{m^2\,s^{-1}}$)
C Mass transfer term.

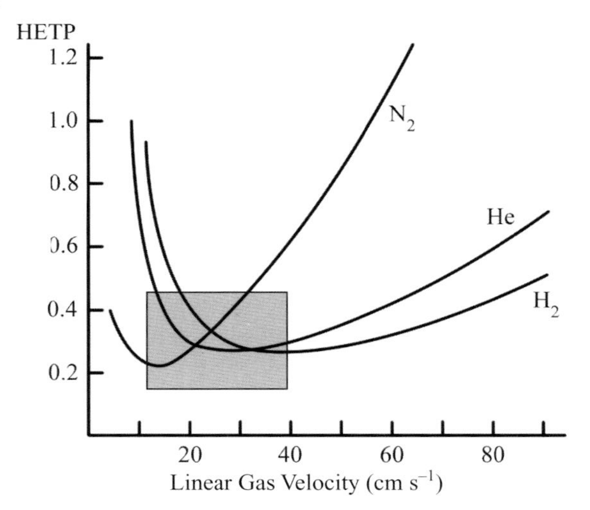

Figure 8.6 van Deemter diagram (correct Golay plot) for three carrier gases (conditions: 30 m fused silica capillary column, 0.32 mm internal diameter, covered with 0.25 μm methyl silicone at 120 °C) (Gray area = optimal range).

If this function is plotted versus u, the diagram shown in Figure 8.6 results.

Obviously H is minimal for a certain flow velocity, that is, separation efficiency is at its best. This optimum flow velocity must be determined experimentally for a certain column. It also depends on the type of carrier gas.

Flow rates are usually recommended, which are 1.25 to 1.5 higher than the minima determined. This is particularly important in analyses with a temperature program. Since the viscosity of flowing gases increases with increasing temperature, the flow velocity decreases at a constant column pressure.

Since, in the time measured, the unretained solute (i.e., methane) passes the length of the column, the average linear gas velocity ($\bar{u}$) can be calculated by dividing the column length (L) (cm) by the retention time of the unretained solute (t_{methane}) (s):

$$\bar{u} = \frac{L}{t_{\text{methane}}} \tag{8.5}$$

In addition to providing information needed to calculate the linear gas velocity, injections of methane show the quality of the column installation. If the system is properly connected, the peak will be symmetrical. Dead volumes, leaks or obstructions will produce tailing peaks.

Separation Number (SN)

The separation number is probably the most realistic way to express the separation efficiency of a column [4]. The SN is the number of completely separated compounds that can be eluted between two sequential peaks of a homologous series, usually n-paraffins differing by one CH_2 unit. The SN can be used with temperature programmed analyses, for which N is meaningless.

$$\mathrm{SN} = \frac{(t_{R2} - t_{R1})}{(W_{1/2_1} + W_{1/2_2})} - 1 \tag{8.6}$$

t_R Retention time (s)
$W_{1/2}$ Peak width at half height (s)
1 and 2 Sequential homologs.

Determination of Extraction Efficiency (Sample Processing Prior to Analysis)

All chromatographic separation methods are based on a heterogeneous distribution of equilibrium: the sample components are distributed between two phases in which the analytes can be adsorbed or dissolved, respectively.

A distinction is made between differing types of distribution: solid–liquid, liquid–liquid, solid–gas, liquid and gas.

In order to change from one medium to another, the molecules must go through a phase limit. For this reason, here one also speaks of a multiple phase or heterogeneous balance. The thermodynamic description of heterogeneous equilibria is by far more complicated than when only one phase is involved. In spite of this, most phenomena can be explained with a very simple model.

The equilibrium position is described by the Nernst distribution law, which is analogous to the Law of Mass Action (LMA) defined in homogeneous chemical equilibrium reactions:

$$K = \frac{c_{II}(x)}{c_I(x)} \tag{8.7}$$

K Nernst distribution coefficient
$c_{I,II}(x)$ concentration of x in medium I, or II, respectively.

If one medium is a solid, then x will only be adsorbed on the surface, and the equilibrium position is, logically, not dependent on the concentration of x in this phase, but on the loading density, and hence the surface of the solid (which can, e.g., be controlled by the particle size). In this case, the Nernst theorem is formulated differently:

$$k' = \frac{n_{\text{solid}}(x)}{n_{\text{liquid/gas}}(x)} = \frac{m_{\text{solid}}(x)}{m_{\text{liquid/gas}}(x)} \tag{8.8}$$

k' Retention (capacity) factor
$n_{\text{solid}}(x)$ Mass of analyte x in the stationary phase
$n_{\text{liquid/gas}}(x)$ Mass of analyte x in the liquid/gaseous phase
$m\ (x)$ Mass of analyte x

As a substance is distributed between the two phases, that is, the size of K depends primarily on the following factors (see also further below "Influence of the equilibrium constants"):

- Solubility of x in I and II
- Boiling point of x, if a phase is gaseous
- Interactions between x and the surface of a solid medium
- The capacity factor k also depends on the volume of the liquid phase, or, in chromatography, on the flow rate.

While this residual, and thus dwindling relative share α keeps decreasing with a growing number of extraction cycles (n) it never reduces to zero.

$$\alpha_n = \frac{m_{I,n}}{m_{\text{tot}}} \tag{8.9}$$

n Number of extraction cycles
$m_{\text{I},n}$ Mass of analyte in phase I
m_{tot} Total quantity of analyte in the sample

It is, therefore, important to be able to estimate how many cycles it takes until the quality requirements of the current analysis problem have been sufficiently met. If there is a sample preparation from only one extraction, the extraction efficiency corresponds precisely to the recovery rate.

Calculation of the extraction efficiency E_n

$$E_n(\%) = 100 * (1 - \alpha_n) \tag{8.10}$$

Liquid–liquid extractions

$$\alpha_n = \left(1 + K.\frac{V_{\text{II}}}{V_{\text{I}}}\right)^{-n} \tag{8.11}$$

Solid–liquid extractions

$$\alpha_n = \left(1 + \frac{1}{k'}\right)^{-n} \tag{8.12}$$

n Number of extraction cycles
k' Capacity factor
V_{I}, V_{II} Volume of phase I or II, respectively

Quantification Methods for Complex Mixtures

Area Percent Method

If all peaks in a chromatogram are integrated (with diluted solutions without the solvent peak) and the sum of all surfaces is set to 100%, then the percentage share of a single component of interest in the mixture can be calculated by calculating the percentage share of its surface in relation to the entire surface. This method does not require a calibration, and only the retention time of the peaks of interest must be known. This simple method is also used in the quality control of solvents, (so-called "GC-purity"): purity of the solvent = 100% minus the sum of all relative peak areas of the impurities visible in the chromatogram.

External Standard

This method first determines the sensitivity for selected components by analyzing standard solutions. With the analysis function thus found sample injections can be evaluated, and in this way the absolute content of the same components (and only these) can be determined in the samples.

Reference to Representative Individual Component

The FID is characterized by the fact that there is relatively little variation in sensitivity for organic compounds within a homologous series, if calculated in surface counts/mass. This condition is exploited to simplify quantitative analysis of complex mixtures by using only one (representative for the group of the desired analyte) external standard to determine sensitivity and by then transferring this sensitivity to all components of the desired group. In order to indicate this assessment method, the result should be formulated as follows: "mass fraction of n-alkane (measured as n-tetradecane): (15 ± 2) g/100 g".

8.1.1.3
Material and Methods

Equipment

GC Auto System (Perkin Elmer), two-channel gas chromatograph with temperature program and split injector. (An automatic sample injector can be connected.)

Stationary phase:	**polydimethylsiloxane with 5% phenyl instead of methyl groups**
Column length:	30 m
Inner diameter:	0.25 mm
Film thickness:	0.2 µm
Centrifuge	
Gas cylinders	H_2, He

Chemicals

n-hexane	CAS- number 110-54-3	F, Xn, N
n-decane	CAS- number 124-18-5	Xn
n-undecane	CAS- number 1120-21-4	Xn
n-nodecane	CAS- number 112-40-3	Xn
n-tetradecane	CAS- number 211-096-0	Xn

n-hexadecane	CAS- number 208-878-9	Xi	
n-octadecane	CAS- number 593-45-3	Xi	
isooctane (puriss)	CAS- number 540-84-1	F, Xn, N	
nonanal	CAS- number 124-19-6	Xi	
1-octanol	CAS- number 111-87-5	Xi	
2, 3-butanediol	CAS- number 513-85-9		–
2, 6-dimethylphenol	CAS- number 576-26-1	T, N	
2-ethylhexanoic acid	CAS- number 149-57-5	Xn	
dicyclohexylamine	CAS- number 101-83-7	C	
2, 6-dimethylaniline	CAS- number 121-69-7	T, N	
methyl esters of fatty acids – C10-C12			–
volatile gas oil / diesel oil		Xn, N	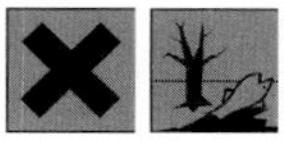

Reagents

Gas Oil Standard Solution

$\beta = 4\,\text{mg}$ diesel/mL isooctan puriss p.a.

Alkane Standard Solution

Each $\beta = 5\,\text{g/l}$ n-decane, n-tetradecane, n-hexadecane and n-octadecane

Test Mixtures I and II (Modified) According to Grob [1]

Test Mixture I contains the following substances without dodecane.

Test Mixture II contains the following substances without undecane.

Decane, 2,6-dimethylaniline, undecane, 2,6-dimethylphenol, dodecane, 2-ethylhexanoic acid, nonanal, dicyclohexylamine, 1-octanol, methyl ester of the fatty acids C_{10}–C_{12}, 2,3-butanediol.

For each $\beta = 500\,\text{mg}\,\text{l}^{-1}$ in n-hexane (or methylene chloride).

Experiment

(A) Assessing the separation efficiency of a GC column

The test mixtures I and II by Grob allow simple quality re-inspection of polar and nonpolar capillary columns, respectively.

Since the duration of the original method is too long, the method should be meaningfully shortened.

Determine the separation efficiency of the column and the reproducibility of the analytical method.

Proposal:

Inject 1 μL of the test solution with the following temperature program:

Start at 50 °C, wait 2 min, then heat to 220 °C at a rate of 10 °C min^{-1}, finally heat for 5 min at 250 °C.

(B) Optimization of the temperature program

Analyze the alkane standard solution at various different temperatures. The following oven temperatures should be selected:

- **Isothermal**

  100 °C, 120 °C, 130 °C, 140 °C, 150 °C (reference values)

- **Varying the heating rate**

  Inject 1 μL of the test solution with the following temperature program:

  Start at 150 °C, heat at 5 °C min^{-1} to 200 °C, then heat at 20 °C min^{-1} to 260 °C, hold for 10 min at 260 °C until no more peaks come out of the column.

What still does not suit you in the appearance of the chromatogram? Consider how to solve the following problems by skillfully changing the temperature program and adjust the program accordingly.

Problem:	low-boiling analytes disappear in the solvent peak
Problem:	poor resolution of the first peaks of the chromatogram
Problem:	poor resolution of the peaks in the main part of the chromatogram (during the temperature ramp)

(C) Identification of the major components of gas oil (qualitative analysis)

By using the previously optimized method (B), analyze 1 μL of gas oil standard and identify the dominant peaks of n-alkanes in the gas oil on the basis of the retention times.

(D) Determination of the gas oil content in different "contaminated" water samples. (Quantitative Analysis)

Sample work-up

D.1 Determination of extraction efficiency

Take four clean centrifuge tubes that can be sealed, weigh approximately 0.5 g of water sample accurately into three tubes and add 10 ml of isooctane to each of the samples (A and B) as an extraction solvent. The third sample (sample C) is spiked with 20 μl of the undiluted gas oil standard and then made up to 10 ml with solvent. 10 mL of pure extraction solvent is added to the fourth centrifuge tube (blank D).

- All four samples are manually shaken for approximately 2 min and then centrifuged.
- The supernatant clear liquid from each of the four samples is taken, filtered, and then 20 μm is injected into the GC with a syringe.

Calculate the extraction efficiencies.

Quantification

D.2 Percentage area method

Calculate the relative proportions of the major n-alkanes from the resultant chromatograms (C) of the gas oil standard.

Use the area percentage method for this.

How great is the sum of all "non-n-alkanes"?

D.3 Sensitivity

Calculate the mass ratio of n-alkanes in the gas oil in relation to tetradecane from the chromatograms. By what percentage does this value change if one of the other n-alkane standards is used as a reference?

Calculate the sensitivity E from the chromatogram of the standards according to the following four methods:

- Signal = total area of all gas oil peaks, $E = S/$mass of gas oil per injection
- Signal = area octadecane peak, E is defined as above
- Signal = area tetradecane peak, E is defined as above
- Signal = height of tetradecane peak, E is defined as above

(E) Determination of contamination in various water samples

After extraction D.1 determine the content of the various n-alkanes in unknown water samples.

8.1.1.4
Questions

(1) Explain: retention time, partition ratio, number of theoretical plates.
(2) What mechanism leads to peak broadening at a flow rate that is too low and which leads to peak broadening at a u that is too high?
(3) Why is it that Eddy diffusion does not trigger any peak broadening in capillary gas chromatography (parameter A in the van Deemter equation)?
(4) What are the possible advantages and disadvantages of the different carrier gases?
(5) What are the advantages and disadvantages of the various injection techniques?
(6) Explain when to run an isothermal analysis and when to run a temperature programmed analysis.
(7) List the major instrumental components of a gas chromatographic system.

References

1 Grob, K. Jr., Grob, G., and Grob, K. (1978) *Journal of Chromatography*, **156**, 517.
2 Grob, K. Jr., Grob, G., and Grob, K. *Journal of Chromatography*, **219** (1981) 13.
3 van Deemter, J.J., Zuiderweg, F.J., and Klinkenberg, A. (1956) Longitudinal diffusion and resistance to mass transfer as causes of nonideality in chromatography. *Chemical and Engineering Science*, **5** (6), 3869–3882.
4 Guide 853 (1988) *Troubleshooting Guide for Capillary Gas Chromatography*, Supelco Inc.

8.1.2
Gas Chromatography Coupled with Mass Selective Detection – GC/MS, Project: "Circumvention of the Formerly Mandatory Declaration of Fragrances in Perfumes?"

8.1.2.1
Analytical Problem

The history of perfume began in the ancient civilizations of Egypt and India. The various applications of aromatic substances were always considered a great source of inspiration – inspiration by means of inhaling them. Western culture became familiar with the different fragrant raw materials and mixtures of the Orient only through the crusades. The basic components of a perfume are primarily alcohol (80%), distilled water and dissolved natural essences (essential oils of vegetable or animal origin), and, increasingly, synthetic fragrances. The technical term for natural oils and synthetic perfumes (chemicals) is "fragrance" and the most common fragrances are now synthesized in large quantities. Fragrances are the major cause of contact allergy, after nickel. The scientific advisory committee of the EU (Scientific Committee on Cosmetic Products and Non-Food Products – SCCNFP) has declared 26 fragrances to be particularly strongly allergenic. These substances must, according to the revised EU Cosmetics Directive 76/768/EEC, be declared if they exceed a certain concentration in cosmetic products. Although this provision for consumer protection is positive, perfume manufacturers tend to still use allergenic substances, but at concentrations below the maximum allowed, or replace them with substances that do not have to be declared. Currently, the EC is negotiating a reform of EU chemical legislation (Registration, Evaluation and Authorisation of Chemicals – REACH). The aim is to provide appropriate analytical procedures that enable fragrances to be identified and quantified by independent agencies.

Object of Investigation/Investigation System

Fragrances

Samples

Perfumes

Examination Procedure

Mass spectrometry is a technique to detect and quantify selectively an amount of an unknown analyte. It is also used to determine the elemental composition of a given compound.

After a chromatographic separation technique, such as gas chromatography, the mass selective detector (MSD) enables the specific identification of substances by electrically charging the sample molecules, accelerating them through a magnetic field, breaking the molecules into charged fragments and detecting the different mass/charge ratio (m/z). Due to its high sensitivity and selectivity the MS has become indispensable in trace analysis. The combination of gas chromatography/mass spectrometry (GC-MS) into a single analytical instrumentation is widely used in aroma and environmental analytics and in toxicology, where the characterization and quantification of unknown compounds has to be achieved. Further applications in the area of chemistry are: molar mass, constitution analysis, isotope and elementary analysis.

Analyte

Limonene

Keywords

Ionization, fragmentation, fragment ion, total ion chromatogram (TIC), single ion monitoring (SIM), mass analyzer, headspace analysis, solid phase micro extraction (SPME)

Bibliographical Research (Specific Topics)

Hübschmann, H.J. (2009) *Handbook of GC/MS: Fundamentals and Applications*, Wiley-VCH, ISBN-13: 978-3-52731427-0.

Ekman, R., Silberring, J., Westman-Brinkmalm, A., and Kraj, A. (2008) *Mass Spectrometry (Instrumentation,*

Interpretation and Applications), Wiley & Sons, Hoboken NJ, online library doi: 10.1002/9780470395813.ch4.

Sparkman, O.D. (2010) *Gas Chromatography and Mass Spectrometry: A Practical Guide*, Academic Press, ISBN-13: 9780123736284.

Project Process/Task

(A) Optimization of the chromatographic separation of a mixed standard

A.1 Find out through bibliographical study the molecular structure of the analytes in the mixed standard (toluene, benzyl chloride, dodecane, methyl benzoate).Consider in what order and under which conditions they can be separated.

A.2 By testing different temperature programs, a suitable method for the separation of the substances has to be optimized.

(B) Identification of the components in the mixture by standard spectral library

Identify the separated components by a database search.

(C) Interpretation of mass spectra

C.1 Interpret the mass spectra of the results with a database search.

C.2 Look for suitable mass fragments for a SIM analysis

(D) Determination of the composition of a perfume

D.1 By testing different temperature programs, a suitable method for the separation of the perfume components is to be developed.

D.2 Identification of the components in the perfume with the spectral library.

D.3 Quantitation of limonene using single ion monitoring (SIM). The results are subsequently to be compared with the literature.

(E) Project reporting

The results are to be submitted in the form of a report (or presentation). Comparisons with bibliographical references are to be made.

8.1.2.2
Introduction

Principle

In gas chromatography (GC) the compounds to be analyzed are vaporized and transported through the column with the mobile phase and separated. The separation methods have already been dealt with in Section 8.1.1, here the focus will be on sample introduction methods and mass spectrometry coupled to gas chromatography. For separation of volatile components from the non-volatile or semi-volatile matrix, the headspace technique (HS) can be used in GC analysis. After

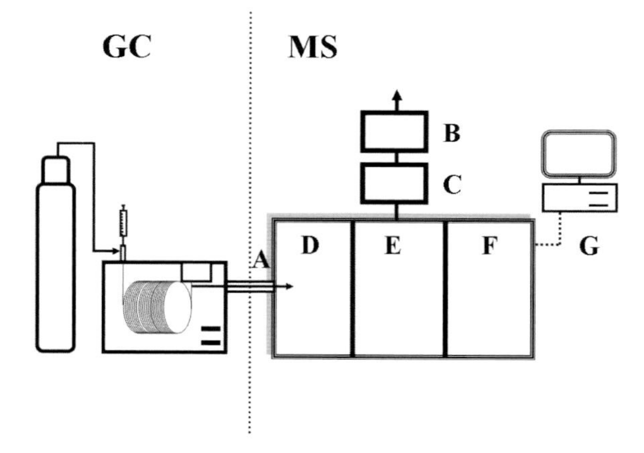

Figure 8.7 Schematic illustration of a GC/MS-system. **GC** (separation according to polarity and boiling point), **A** interface GC/MS, **MS** identification and quantification, **B** rough/**C** high vacuum pump, **D** ion source (ionization/(fragment)ion generation and focusing, acceleration and bundling of the ions. **E** mass filter (sorting the ions according to mass-to-charge relationship) and detection ("counting" the abundances and m/z values of ions arising from a certain mass), **F** data recording (acquisition of chromatograms and spectra) and data processing (interpretation of spectra).

separation by gas chromatographic techniques the use of coupling techniques, such as mass spectrometry, can help to identify and quantify the separated substances. The combination of a GC with a mass spectrometer (MS) as a detector is the most common method with which to identify fragrances. After the separation, the gaseous compounds are led into an ionization cell in which they collide with electrons (Figure 8.7). As a rule, the electrons have about 70 eV of kinetic energy. The remaining energy is stored in the molecule and leads to fragmentation. The charged molecular ion and the resulting fragment molecules are accelerated and focused, and enter into a mass filter. Only ions of a particular mass-to-charge ratio (m/z), that is, the ions are sorted according to mass, reach the detector through the mass filter and at synchronised intervals. The detector records the frequency (isotope abundance) of a particular m/z-ratio, converts this to an electrical signal and displays it as a spectrum in the function of the m/z-ratio. The analytes are identified through comparison with spectra from a database or by the retention time and spectra of a standard. The quantification is done by reconstructing the chromatogram and its area integration.

Headspace Technique

The headspace technique exploits the fact that in a closed, static system, an equilibrium between the liquid phase (sample matrix) and the overlying gas phase containing the volatile substance is reached after a determined time. This process can be facilitated through heating and agitation. After equilibration, a gas aliquot is withdrawn and injected directly into the GC for separation of all the volatile components. Most consumer products and biological samples are composed of a wide variety of compounds that differ in molecular weight, polarity, and volatility, therefore headspace sampling is the fastest and cleanest method for analyzing volatile organic compounds for complex samples.

Solid Phase Micro Extraction (SPME)

In SPME the fiber coating removes the compounds from the sample by absorption in the case of liquid coatings or adsorption in the case of solid coatings. Traditional sample preparation methods try to completely remove the analytes of interest from the sample. SPME does not work this way. With SPME, the amount of analyte removed by the fiber is proportional to the concentration of the compound in the sample. This is true when the fiber and the sample reach equilibrium or as long as the sampling parameters are carefully controlled even before equilibrium is reached. The ability to use SPME quantitatively before equilibrium is reached allows much shorter sampling times with this fast, economical, and versatile technique.

The efficiency of the extraction depends on the type of coating. There are various coatings used for different applications. For general applications, polymer coatings of polydimethylsiloxane or polyacrylates are used. The volume of the coating is also important. The larger the volume, the more sensitive the method. However, the fiber becomes more susceptible to damage, the greater the quantity of coating applied.

After extraction, the fiber is inserted into the injector of the equipment, in which the desorption process takes place, and the volatile compounds are delivered to the GC column (Figure 8.8).

Interface for GC/MS-Coupling

The interface between GC and MS plays an important role in the overall efficiency of the instrument. It must be capable of providing an inert pathway from the column to the ion source and serves as a molecular separator, so that the MS vacuum pump is not overloaded with carrier gas. The separators substantially reduce the pressure of the eluent stream from the GC that enters the mass spectrometer.

One divides mainly between the jet separator and the effusion separator. In the case of the jet separator the carrier gas is pressed through a nozzle and shoots toward an orifice in the wall of an adjoining chamber on its way to the ion source. The low mass carrier gas molecules diffuse away from the jet axis, leaving the sample stream enriched with the heavier analyte molecules.

The effusion separator consists of a sintered glass tube through which the gas passes at high speed. The heavier sample analyte molecules are enriched in the center of the gas jet while the carrier gas is pumped away. Usually a Ryhage separator is two stages, although all-glass single stage units are used, particularly in combination with a quadruple mass spectrometer.

Vacuum System

The vacuum system of an MSD consists of two devices. The first is a mechanical pump, which reduces the vacuum in the system to values around 10^{-1} to 10^{-2} Torr.

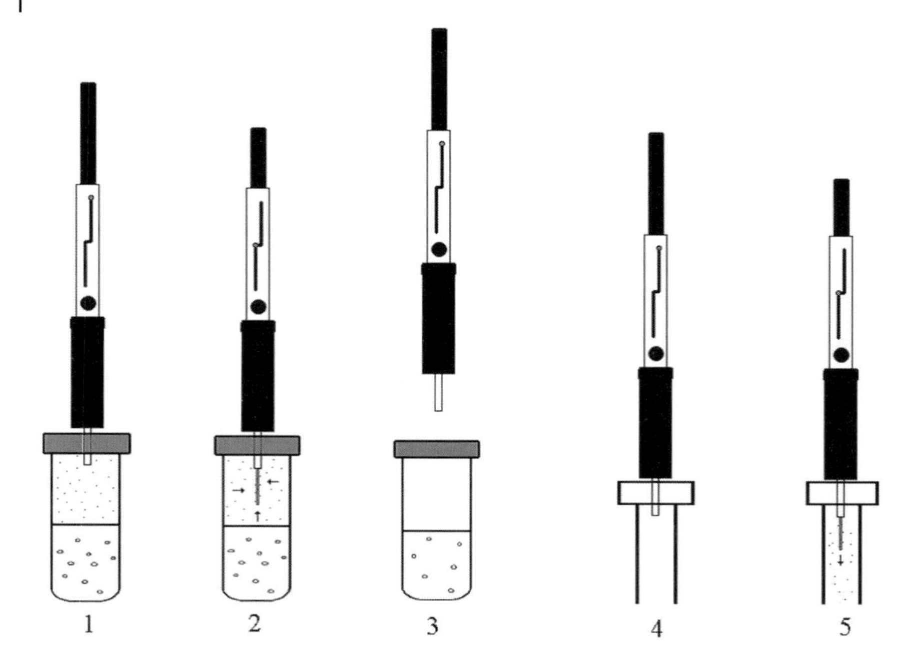

Figure 8.8 Solid phase micro extraction. **1** piercing sample container, **2** exposing SPME fiber/adsorbing analytes, **3** retracting fiber/withdraw needle, **4** piercing septum in GC inlet (or introducing needle into SPME/HPLC interface), **5** exposing SPME fiber/desorbing analytes (Supelco, Bulletin 923 A).

It also serves as a forepump for high vacuum. The high vacuum pump – an oil diffusion pump or turbomolecular pump – lowers the vacuum system to values around 10^{-5} to 10^{-6} Torr.

The generated high vacuum facilitates the production of free ions and electrons in the gas phase, it prevents collisions between ions and of ions with other particles (these reactions would lead to a complicated spectrum) and increases sensitivity (nitrogen and oxygen ions from air would oversaturate the detector).

Ion Source

The classical method to generate ions is the bombardement of the (previously evaporated) sample with electrons in the gas phase (electron impact ionization). However, this technique is limited by two marginal conditions, since (i) the substance must be vaporizable without decomposing, and (ii) the ions formed must be stable enough to reach the detector. Not until the late 1980s was it possible to establish so-called "soft ionization" techniques, which could compensate for these disadvantages. The most essential of these techniques is the matrix-assisted laser desorption and ionization (MALDI) process.

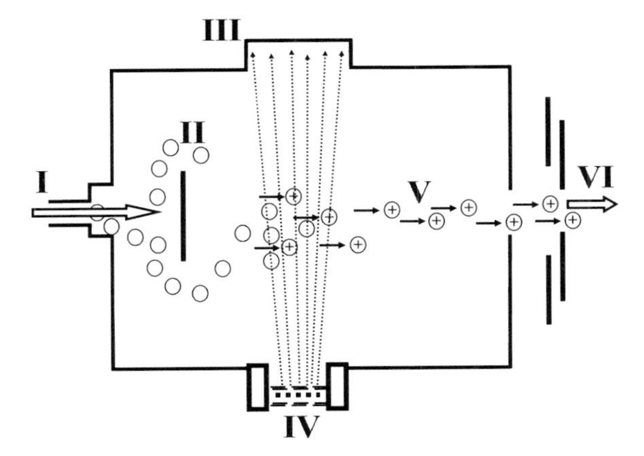

Figure 8.9 Schematic representation of an EI source. **I** sample from GC (transfer line), **II** molecules in gaseous state, **III** trap electrode, **IV** heated filament, **V** electron beam, **VI** ions.

Electron Impact Ionization (EI) [1]

In order to achieve fragmentation of a molecule, there must be oxidation or reduction processes involved. Therefore the molecules that enter the ionization chamber in the gaseous state through the sample inlet are bombarded with electrons. A heated filament in the ionization chamber, which is under high vacuum pressure ($p < 10^{-5}$ Pa), is producing electrons (Figure 8.9). With the help of magnets, the electrons are accelerated, resulting in an electron beam that radiates across the ionization chamber. Since the electrons have a very small mass, the molecular ions are not impaired by the required magnetic field strength. The electrons are then collected in a trap. Through acceleration the electrons can reach energies of 50–100 eV. If an electron now collides with a molecule, the energy is absorbed and the molecule is excited. If the stored energy (internal energy) is now greater than the activation energy for a decomposition reaction, decomposition reactions take place, leading to the formation of daughter ions and neutral particles. If the daughter ions still possess enough excess energy, they can decompose to other fragment ions. The most frequently occurring decomposition reaction is the formation of simply positively charged molecular ions. It occurs 1000 times more frequently than the negatively charged molecular ion. Double and triple charged ions are observed only very rarely, which is why neither multi-charged nor negative molecule ions are dealt with in detail here. Due to the high internal energy of the molecular ion, which is achieved in the EI mode, it already decomposes in the ionization chamber to a great extent, so that mainly fragment ions reach the detector.

Chemical Ionization (CI) [2]

Chemical ionization is based on the same principle as EI. An advantage of CI is the increase in selectivity, suppression of interferences from the sample matrix, identification of position isomers in ring systems, and determination of substituents on double bonds. In CI the analyte reacts with the ions of a reagent gas, for example, with methane due to a large excess of CH_4 molecules in the chamber, mainly the CH_4 is ionized when reaction gas and analyte are bombarded with electrons In a second step, a charge transfer to the analyte molecule occurs. The chosen reaction gas must not react with itself. The most common reaction gases for chemical ionization are methane, isobutene, and ammonia. In contrast to EI, CI is a more gentle ionization method, whereby the analyte molecular ions last longer and remain detectable.

Matrix Assisted Laser Desorption and Ionization (MALDI) [3]

The MALDI technique uses a laser to vaporize the sample. Thereby, especially nitrogen lasers (emitting $\lambda = 337$ nm) are used, as they are relatively inexpensive and work more or less without problems on a daily basis. By far, not all samples absorb at the emitting wavelength of 337 nm. For this reason, the following "trick" is used: the addition of a suitable "matrix" to the sample being tested. This is a (usually small, organic) molecule that absorbs the energy of the laser at the wavelength indicated. Thus the sample gets "hot" and evaporates and the analyte molecules are "dragged along". Bombarding the substance to be tested and the matrix, or an added auxiliary medium (e.g., an organic acid), or ions already present from the beginning, ions are formed between the substance to be examined and the matrix leads to the formation of ions that are first accelerated in an electric field and subsequently in a field-free space left on its own. This triggers separation in accordance with the *m*/*z*-ratio: light ions reach the detector faster than heavier ones and can, therefore, be separated (Figure 8.10). Since the mass separation is very much dependent on the traveled space, a higher mass resolution is reached by extending the flight route. For the manufacture of reasonable small equipment, a special "passive reflector" (reflector-detector) is used, which prolongs the flight route and the analyte ions travel the same distance twice.

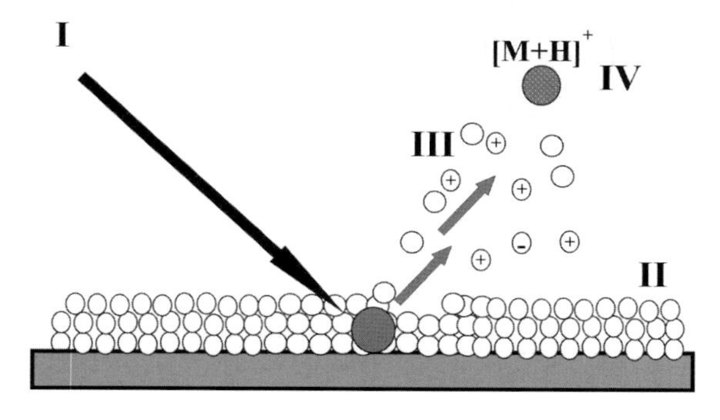

Figure 8.10 Diagram of MALDI-procedures. **I** Laser λ 337 nm (high-energy primary beam) **II** matrix molecules, **III** desorbed particles (ions, neutrons, electrons), **IV** intact biomolecule.

In contrast to EI and CI, MALDI is a pulsated ionization technique. As a soft ionization method, MALDI provides information about the molecular weight, results in little or no fragmentation, and is suitable for particularly large biomolecules and polymers. The sensitivity is strictly dependent on the type of sample substances. Suitable polar and ionic connections (e.g., metal complexes) can be detected by means of MALDI-TOF.

Focus

The molecular and fragment ions, which are formed in the ionization chamber are accelerated in the ion source in the direction of an analyzer. They are then influenced by a magnetic field so that the ions seem to be concentrated in a converging manner when leaving the magnetic field. Bundling and focusing the ion beam is further achieved by charged apertures (lens). The final velocity is reached at the exit slit, which leads directly to the analyzer. The non-ionized molecules are removed from the system with the vacuum pump.

Mass Filter (see also LC-MS)

Quadrupoles

After focusing, only selected ions reach the detector. That means only certain mass ratios m/z pass through the analyzer. A mass filter is used for this mass separation. For example, in gas chromatography quadrupole mass filters (Figure 8.11) are primarily used. The separation is achieved by deflecting the masses through electrical fields. These are four hyperbolic (parallel arranged) metal rods to each of which a DC voltage and an adjustable AC voltage are applied at two opposite ends. Depending on its mass, each ion behaves differently in the generated magnetic field. This property is utilized in the quadrupole mass filter. By changing the frequency of the AC voltage, the ions of particular masses start to oscillate and

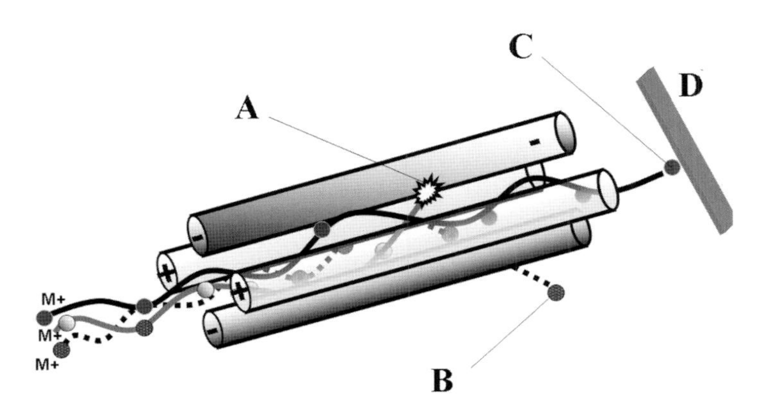

Figure 8.11 Work principle of a quadrupole. **A** discharge of a high mass, **B** discharge of a low mass, **C** filter a specific (desired) mass, **D** detector.

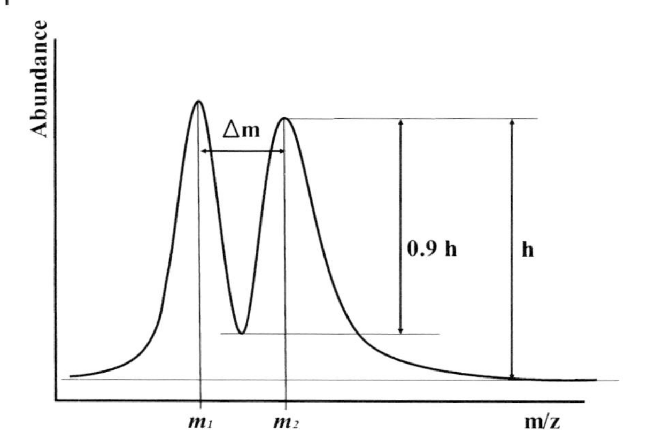

Figure 8.12 Resolution in mass spectrometry.

cross the mass filter on stable paths and reach the detector. All other masses are more strongly deflected and are, therefore, removed from the system prior to entering the detector. The registered mass is proportional to the frequency. Thus, the whole mass spectrum can be recorded.

The quadrupole mass filter is inexpensive and is very prone to disturbances due to its simple structure. The rapid acquisition of mass spectra and good reproducibility lead to frequent use in GC-MS coupling with both ionization modes, EI and CI.

Other major mass analyzers are dealt with in Section 8.1.4.

Resolution between Masses

Only when a sufficiently high resolution of adjacent signals is achieved, can isotopic analysis be performed. The larger the masses, the greater can be the interference. In general, two mass signals are considered to be resolved if the distance from the minimum between two adjacent masses to the baseline is equal to or less than 10% of the peak height (Figure 8.12).

In mass spectrometry, the resolution between two signals is defined as follows:

$$A = \frac{m}{\Delta m} \tag{8.13}$$

A: Resolution
m: Mass (m/z)
Δm: Distance from the adjacent mass (m/z)

The usual mass analyzers in GC-mass spectrometry have a resolution of 1000, that is, they can distinguish between ions of mass m/z 1000 and m/z 1001.

High-resolution mass analyzers can achieve a resolution of 150 000. In that case, the masses m/z 1000.000 and m/z 1000.007 can be distinguished.

Detection

High Energy Dynode (HED)

The detector is expected to measure the intensity of the ions and translate this into an electronic signal. The most widely used GC-MS detector is the high-energy dynode detector that elaborates electron multipliers (*EM*). Electron multipliers are dynodes that are arranged behind each other. The dynodes are essentially alloys of alkali or alkaline earth metals in combination with more precious metals. An ion coming from the mass filter hits a dynode, which in turn emits a multiple of electrons (secondary ions) from its surface. The secondary ions hit the opposite positively charged dynode causing other electrons to be emitted. Thus, a cascade of electrons is triggered. At the end of the detector there is an anode, where the electrons are discharged. The resulting current is recorded as a signal, which is proportional to the amount of molecular or fragment ions.

Electron Multiplier (EM)

An electron multiplier is designed to focus the ions from the quadrupole on the HED, which operates at −10 kV. When the ion beam hits the HED, the generated electrons are separated and are attracted by the more positive EM (in relation to HED). At the opening of the horn-shaped EM, a voltage of up to −3 kV is applied versus 0 V at its other end. The electrons emitted from the HED hit the EM and cascade through the horn, thereby releasing more electrons. At the end of the horn, the current produced by the electrons is transmitted to the signal amplifier card (see Figure 8.13). This system increases sensitivity by generating more and more electrons (i.e., a stronger current) and by reducing contamination by charged ions.

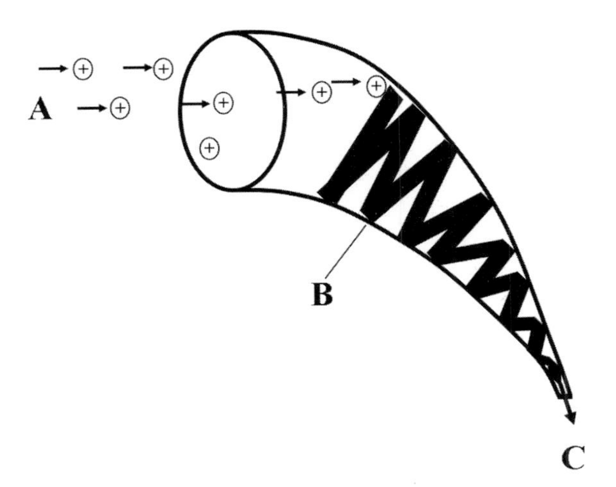

Figure 8.13 Functionality of the electron multiplier. **A** incoming ions, **B** cascading ions, **C** signal output.

Data Acquisition and Processing

Full Scan Mode (Total Ion Chromatogram, TIC)

In chromatography coupled with mass spectrometry data from measurements are presented in three dimensions. Data are recorded in dependence on time, intensity and mass number (x = time, y = isotope abundance, $z = m/z$).

For each retention time a corresponding mass spectrum is recorded. In addition, the intensity of all ions that occur in a mass spectrum at a given retention time, is integrated. This results in the intensity of the chromatogram, the so-called total ion chromatogram. The intensity and sensitivity of the signals are thus dependent on the processing time of the ions, the scan rate and the mass range. The larger the mass range, the smaller the sensitivity, because it takes longer to scan all the masses. For this reason, it is advantageous to adapt the mass range of the scan to the wanted masses (if these are known). The scan mode is used mainly for qualitative measurements.

Single Ion Monitoring (SIM)

If only a few, specific masses are being sought, sensitivity can be equally increased by SIM (single ion monitoring). In this process only specified masses (molecular ion or a fragment) are recorded with the mass spectrometer and it is not necessary to record the entire spectrum. This approach is particularly advantageous for quantification purposes. As with other detectors (e.g., FID), quantifications can be conducted with single ion monitoring. The mass spectra at the corresponding retention time can then be consulted to identify the substances. The quantification is done by integrating all peaks that result from individual, pre-determined mass fragments. A big advantage of this data collection is that the SIM mode can compensate for insufficient chromatographic resolution of the peaks in question. In this way quantification is possible despite overlapping peaks.

The most important criterion for a sensitive quantitative determination using SIM is the selection of one (or more) suitable fragment ion(s), which is/are specific for a substance and thus display high intensities in the mass spectrum.

Representation of the Spectra

The mass spectrum reflects the frequency quantity of the ions as a function of their mass/charge ratio m/z. The most intense signal is called the base peak and is assigned an intensity of 100% (random assignment). The data of ions are displayed as a line spectrum.

Analytical Information in the Mass Spectrum:

Molecular ion:	information on molar mass
Isotope peak:	information on elemental composition
Fragment ion:	typical for the molecular structure
Difference (mass defect):	typical for the molecular structure
Isotope abundances:	frequency of occurring isotopes

A mass spectrum of a substance contains various analytical information on the composition and structure of the substance analyzed.

The Molecular Ion [M^+]

The molecular ion M^+ results from the removal of an electron from a molecule.

The molecular ion peak is of importance for the following applications:

Molar mass determination
Determination of elemental composition
Isotope analysis
Determination of the ionization potential

However, whether a particular peak in the mass spectrum is really the molecular ion peak, must be evaluated first by critical consideration of the following points.

No thermal or catalytic decomposition of the molecule prior to ionization has occurred.
Molecular ion peak must contain all the elements found in the fragments → elemental composition of all ions in the spectrum.
Fragments must be in relation to the molecular ion peak through a suitable mass difference.
Molecular mass of derivatives and the starting material must be consistent.

Isotope Peaks

In nature, most elements of an organic compound can be found as a mixture of isotopes. In the mass spectrum of a substance a group of signals that form an isotopic cluster is always found in the molecular ion.

Fragment Signals

Decomposition reactions of the molecular ion lead to the formation of fragment ions, which are typical for the molecular structure.

Differences (Mass Defects)

Differences between the signals are also typical of molecular structures. For example, they can provide information on functional groups.

Isotope Abundance

The detector registers the frequency (isotope abundance) of the m/z signals and reproduces them in the mass spectrum in the corresponding intensity.

Mass Spectral Databases

Numerous databases are available that contain a large number of mass spectra of various substances as reference material. An evaluation software can be installed in such a database and it ` can be searched for similar reference spectra to identify a substance. However, the results of a search have to be considered cautiously. The plausibility of the results should always be checked, for example, retention time . . . Each software also offers the possibility to create one's own library of reference spectra

Identification and Quantification by Means of MS-data Analysis (see Section 1.1.4)

From the data collected from the GC/MS analysis two essential pieces of information are gained. First, the results provide a certain retention time for every combination under chromatographic conditions, and a mass spectrum characteristic of the fragmentation.

To identify the substances two methods can be used and partially combined.

Comparison of mass spectra obtained with the National Institute of Standards and Technology (NIST) spectral library

Comparison of the mass spectra and retention times obtained with the values of standard substances (see Section 1.1.4)

The raw data of an analysis obtained using MS detection are illustrated in Figure 8.12. By clicking on a certain peak in the MS chromatogram the respective line MS spectrum can be extracted (see Section 8.1.4). Corrections can now be made on the desired spectrum (possibly background corrections).

The spectrum is then compared with the NIST spectral library.

Conducting Library Searches

The library search is performed in two steps, the "presearch" and the "main search".

In the presearch a rough preliminary search is done to filter out suitable candidates for the actual pattern recognition from the extensive database. The presence of eight significant mass signals serves as a criterion. For this purpose, the unknown spectrum is reduced to its eight most intense mass peaks and finally compared with library spectra. Only reference spectra which match all eight mass numbers will be admitted to the main search.

The main search is the decisive step in which the candidates found in the preliminary search are compared with the unknown spectrum and printed as a hit-list of similar spectra. What is essential for accuracy is a process that is referred to as "local normalization." This process compares characteristic fragmentation or isotopic patterns to the unknown mass spectrum. As a result of the local normalization two values are determined for the spectral comparison. The FIT value gives a percentage measure of how well the signals of the reference spectrum are contained in the unknown spectrum. The RFIT (reverse fit) value indicates the opposite view, namely to what extent the unknown spectrum with its lines is contained in the library spectrum. The combination of both values gives information about the degree of purity of the unknown spectrum. For example, if the FIT value is high and the RFIT value is low, it can be assumed that the measured spectrum contains considerably more lines than the library spectrum used for comparison.

For further manual processing, the difference between a library spectrum and the measured unknown spectrum can be generated. With the remaining share of the spectrum a new library search is possible. In individual cases, this helps to clarify the simultaneously identical co-elution of components, as can be frequently observed in the field of environmental analysis of complex samples.

Fragmentation Reactions

As described above, if a molecular ion has enough excess internal energy, which, in addition, at least corresponds to the activation energy of a decomposition reaction, it can disintegrate into fragments. Apart from the energy content of the molecular ion, other factors play a vital role in the occurrence of fragmentation reactions such as steric arrangements, relative binding strengths, and stability of the products formed in the ion source.

A charged and an uncharged fragment ion are normally the result of a decomposition reaction.

The charge of the original molecular ion is, therefore, only transmitted to a fragment. In the mass spectrum, only the charged fragments can be seen, while the uncharged are only visible in the form of mass differences. The charged fragment ions can continue to disintegrate and form smaller fragment ions. All charged fragments that originate are detected and displayed in a mass spectrum. This forms the basis of mass spectrometric interpretation.

Fragmentation Rules in Mass Spectrometry

1. **Allyl radical cleavage:** preparation of particles (allyl, $m/z = 41$), for example, in KW
1. **Elimination of a substituent:** for example, of a halogen
2. **Alpha-cleavage:** resonance stabilization, that is, "stable" intermediate state of the fragment ion is shown in the spectrum.
3. **Beta-cleavage:** onium cleavage, that is, a complex consisting of a central atom, which has undergone a further binding over and above the number of its valence electrons, is split leaving a defined ion.
4. **Elimination of smaller molecules:** results in stable neutral particles, such as water, CO, HCN . . . . . .
5. **McLafferty-rearrangement** [4]: the migration of a hydrogen atom to a six-membered transition state is called the McLafferty rearrangement. The prerequisite for this is the existence of a double-bonded carbon atom, either as C=C, C=N or C=O. The transition state is then formed by migration of the H-atom to the point at which a stable neutral particle arises. Thus, this is comparable with Rule 5.
6. **Benzyl-cleavage:** cleavage of alkyl aromatics, such as $m/z = 91$.
7. **Split that forms a ring:** similar to Rule 7, rings are formed which have the characteristic mass at 91.

Isotope

Most of the elements of an organic compound are found in nature as isotopic mixtures. Exceptions are the pure elements ^{19}F, ^{31}P and ^{127}I. In the mass spectrum not only the molecular ion can be seen, but also always a group of signals. These are a mixture of molecular ions with different elemental composition from the same compound. The relative intensity of the signals for the molecular ion is dependent on the type of the number of elements contained in the compound and their natural mixing ratio of isotopes. The relative frequency of the isotope in nature is so consistent that one could derive the elemental composition of a compound from the isotope cluster if the signals are sufficiently intense and accurate.

With the help of isotopic frequency, a range of useful information can be derived from the spectrum. The elements can be divided into three classes according to their isotopes:

A-Elements are pure elements (elements with only one natural isotope), *A+1 elements* have two natural isotopes, in which the heavier isotope has a heavier mass unit, and *A+2 elements*, in which the heavier isotope is two mass units heavier.

A+1-Elements

This group theoretically includes the three elements hydrogen, carbon and nitrogen. With the increasing number of carbon atoms in an ion the likelihood increases that one of them is a ^{13}C-isotope. In the case of a C_{10}-ion the likelihood for C_1 is ten times as high; [A+1]/[A] therefore corresponds to $10 \times 1.1\% = 11\%$. This allows a number of carbon atoms in a molecule to be channeled away.

A+2-Elements

In the spectrum a second isotope is particularly conspicuous when it is larger than the most frequent isotope by more than one mass unit.

This group includes oxygen, silicon, sulfur, chlorine and bromine. The presence of these elements can often be simply detected by the isotope pattern in the spectrum. You should, therefore, first seek this element. In general, an A-peak cannot contain any of the elements Si, S, Cl or Br when the relative intensity of the (A+2) peaks (in relation to the A-peak) is less than 3% ([A+2]/[A]<3%). The determination of the oxygen atom is very inaccurate at the 100% peak, since the relative intensity of the A+2 peak is only 0.2%.

A-Elements

Once the number of "A+2"- and "A+1"-elements have been determined in the spectrum, the remaining mass difference is based on elements with only one isotope.

Their number can be ascertained by taking the binding rules into consideration and thus the sum formula of the molecule (different options) can be completed.

The reason for the individual A+l and A+2-peaks are possibly not only the individual contributions of different isotopes of an ion, but also overlaps of peaks caused by other fragment ions. To avoid interference from other fragment ions as far as possible, isotopic frequency calculations should, therefore, be made for the intense peaks and high masses of a peak group.

Important Isotopic Distributions for MS-Interpretation

^{13}C: the heavier isotope in carbon occurs in nature with a probability of 1.1%. This value is sufficient in practice to channel away the number of carbon atoms in a compound. For each carbon atom in a compound the signal intensity increases $M^{+}+1$ by 1.1% in relation to M^{+}.

^{34}S: In the case of sulfur, the proportion of the heavy isotope is ^{34}S 4.4%. With the presence of sulfur, the signal intensity increases to $M^{+}+2$ in 4.4% in relation to M^{+}.

^{37}Cl: In the case of chloride, the proportion of heavy isotope is ^{37}Cl 24.2%. In the presence of a chlorine atom, the signal intensity increases at $M^{+}+2$ to 31.9% in relation to M^{+}.

^{81}Br: In the case of bromine the proportion of the heavy isotope is ^{81}Br 49.3%. In the presence of sulfur signal intensity increases at $M^{+}+2$ to 97.3% in relation to M^{+}.

8.1.2.3
Material and Methods

Equipment

GC: 7890 A (Agilent), MSD: quadrupole5975 Series MSD (Agilent), mobile phase: helium, ion source: IS.

Parameters

Flow:	1.2 ml min^{-1}
Injector temperature:	250 °C
Detector temperature:	280 °C
Injection volume:	0.3 µl
Interface temperature:	280 °C
Ion source temperature:	250 °C
Analyzer temperature:	150 °C
Modus:	positive

Stationary Phase

HP-5	(5% Phenyl-/ 95% dimethyl polysiloxane)
Column length:	30 m
Column diameter:	0.25 mm
Layer thickness:	0.25 µm

Software

The electronic data acquisition, graphical representation as a chromatogram and assessment of the measurements were conducted using the ChemStation software (*Agilent*). The mass spectra of the compounds analyzed were compared with spectra from the MS spectra library NIST/EPA (US Environmental Protection Agency)/NIH (National Institute of Health) Mass Spectral Library 1.5a (match factors> 850).

The hits proposed by the library are ranked according to the degree of agreement, thus according to their probability. This serves as an aid in analyzing the spectra.

Chemicals

Ethanol	CAS- Number 64-17-5	F
Limonene	CAS- Number 5989-27-5	Xi, N
Dodecane	CAS- Number 112-40-3	Xn
Benzoate	CAS- Number 93-89-0	Xn
Toluene	CAS- Number 108-88-3	F, Xn
Benzyl chloride	CAS- Number 100-44-7	T

Reagents

Mixed Standard for Mass Spectra Interpretation

For each 10 µl of toluene, dodecane, methyl benzoate and benzyl chloride are pipetted into a 5 ml volumetric flask. 4 ml of ethanol are added to bring the substances into solution. The flask is then filled to the mark.

Sample Solution

50 µl of perfume are pipetted into a 5 ml volumetric flask. 2 ml of ethanol are added to bring the substances into solution. The flask is then filled to the mark.

Stock Solution of Limonene

50 mg limonene are pipetted into a 50 ml volumetric flask. 40 ml of ethanol are added to bring the substances into solution. The flask is then filled to the mark.

Stock Solution of Dodecane (Internal Standard)

250 mg of dodecane are put in a 10 ml volumetric flask. 8 ml of ethanol are added to bring the substances into solution. The flask is then filled to the mark.

Mixed Standard

For the quantitative determination the following standard solutions are prepared in a 5 ml volumetric flask:

Standard 1: 2.50 ml of stock solution of lime and 100 μl stock solution of dodecane
Standard 2: 1.25 ml of stock solution of lime and 100 μl stock solution of dodecane
Standard 3: 0.50 ml of stock solution of lime and 100 μl stock solution of dodecane
Standard 4: 0.25 ml of stock solution of lime and 100 μl stock solution of dodecane

4 ml of ethanol are added to the substances each time to bring them into solution. The flask is then filled to the mark.

The concentrations of the solutions may need to be adapted to the content of the perfume.

Experiment

After optimization of the GC-method the method is to be transferred to a GC-MS system. The oven program starts at 20–30 °C below the boiling point of the solvent. Then, the final temperature is adjusted with a temperature increase of approximately 10 °C min^{-1}. 70 eV is set as the ionization energy.

(A) Optimization of the chromatographic separation of the mixed standards

Inject 0.3 μl of the mixed standards (toluene, benzyl chloride, dodecane, methyl benzoate) and measure with the following temperature program:

60 °C for 2 min, then heat to 220 °C at 10 °C min^{-1} and hold for 5 min.

By testing different temperature programs, a suitable method for the separation of the substances is to be optimized.

Think about the order in which the substances are detected.

How do you decide when a temperature program is suitable?

(B) Identification of the components in the mixture by standard spectral library

Identify the individual components of the mixed standards using the database, which provides the MS software.

(C) Interpretation of mass spectra

C.1 Interpretation of the results of database searches

Now try to confirm the results obtained in the data base search for the individual substances, by interpreting the mass spectra (fragments, isotope clusters, differences, etc.) using the above theory and other literature. In so doing, try to adopt the language used in the literature for fragments and decomposition reactions.

C.2 Search for suitable mass fragments for a SIM (single ion monitoring) analysis

Assume you need to quantify each one of the four compounds in the mixture by means of SIM. For each substance select a specific fragment mass (from the mass spectrum). What should you pay attention to when choosing a mass for a SIM?

(D) Determination of the composition of a perfume

D.1 Inject 0.3 μl of the perfume solution and measure it with the following temperature program: 60 °C for 2 min, then heat to 220 °C at 10 °C min^{-1} and hold for 5 min.

If necessary, a suitable method for the separation of the substances should be developed by testing various temperature programs.

D.2 Identification of the components in the fragrance by database search.

Identify the individual components by using the database which provides MS software. Do the results make sense? Do research on fragrances in perfumes.

D.3 Quantitation of limonene by using SIM.

Inject 0.3 μl of a standard solution (limonene, dodecane) and measure it using the method described under D.1. If necessary, optimize the method. Select from each of the mass spectra of both substances a suitable mass for a SIM.

Now perform the quantitation of limonene in the perfume by using SIM.

Finally, validate the method.

8.1.2.4
Questions

(1) What are the major components of a mass spectrometer?

(2) How are the masses represented in a mass spectrum?

(3) Identify the main differences between EI and CI.

(4) What are the main MSD applications in chemistry?

(5) Why should the temperature of the transfer line always be 10 °C higher than the currently selected final temperature of the oven program and what limitation is this?

(6) Consider which changes in parameter lead to a change (decreasing or increasing signal intensity) of ionization and justify your decision.

(7) What should you pay attention to in a database search? Do the search results always make sense?

(8) List frequent fragment masses from impurities.

References

1 Watson, J.T., and Sparkman, O.D. (2008) *Introduction to Mass Spectrometry*, 4th edn, John Wiley & Sons, Inc.
2 Harrison, A.G. (1992) *Chemical Ionisation Mass Spectrometry*, 2nd edn, CRC Press, Boca Raton, FL.
3 Hillenkamp, F., and Katalinic, J.P. (2007) *A Practical Guide to MALDI MS, Instrumentation, Methods and Applications*, Wiley-VCH Verlag GmbH, ISBN-13: 978-3-2731440-9.
4 MacLafferty, F.W. (2009) *Wiley Registry of Mass Spectral Data*, 9th edn, John Wiley & Sons, Inc., ISBN-13: 978-0-470-52035-2.

8.1.3
High-Performance Liquid Chromatography – HPLC, Project: "Stricter Control of Drugs"

8.1.3.1
Analytical Problem

At Zurich Airport it has been noticed that shipments of drugs from "exotic" countries are increasing. This poses a considerable risk to health. The unpredictable factor is that many capsules and tablets contain an unknown, harmful cocktail. The fakes are mostly externally perfect copies that consumers are unlikely to detect.

Evaluation in Zürich Mülligen in June 2008 (Swiss Agency for Therapeutic Products, Swissmedic)

- *673 shipments: 326 of 347 from India and Thailand (370 kg total)*
- *311 suspicious packages (46%) were opened, including 141 from India and 170 from Thailand*
- *Medicinal products were discovered in 90 of a total of 326 shipments from India (27%) and 55 of 347 shipments from Thailand (16%)*

At least 50 000 illegal drug imports by private persons per year can be expected.

The counterfeits discovered so far may well be only the "tip of the iceberg." In addition to larger seizures of powders, tablets and capsules are increasingly being found and samples of all these need to be analyzed. The airport management would, therefore, like to introduce much more stringent controls, to be carried out by a newly formed analytical team. The aim is to be able to determine the composition and concentration of each component of the products seized. A project team in collaboration with Swissmedic needs to develop and validate a reliable and speedy method for the quantification of ingredients in drugs.

Object of Investigation/Investigation System
Drugs

Samples
Analgesic tablets

Examination Procedure
High-performance liquid chromatography (HPLC) has become one of the most popular chromatographic methods of instrumental analysis for separating and quantifying components in complex liquid mixtures. Chromatographic separation is defined as a technique involving mass transfer between stationary and mobile phases. HPLC technology has developed to the extent that a vast number of organic compounds can be analyzed by existing methods that can be found in the analytical literature, such as professional journals, protocol books like for example, Pharmacopeias or AOAC (Association of Official Analytical Chemists) manuals. Its main areas of application range from pharmaceutical, biochemical, and food chemical analysis to environmental and clinical-chemical analysis.

Analytes
Paracetamol, caffeine, propyphenazone

Keywords
Resolution of chromatographic peaks, liquid chromatography, instrumental components in HPLC, polarity in chromatography, separation efficiency of a column, calibration with an internal and external standard, retention of compounds, height equivalent of a theoretical plate (HETP), selectivity

Bibliographical Research (Specific Topics)

Snyder, L.R., Kirkland, J.J., and Dolan, J.W. (December 9, 2009) *Introduction to Modern Liquid Chromatography*, 3rd edn, Wiley, ISBN-13: 978-0470167540.

Dong, M.W. (June 12, 2006) *Modern HPLC for Practicing Scientists*, 1st edn, Wiley-Interscience, ISBN-13: 978-0471727897.

Meyer, V.R. (May 1, 2006) *Pitfalls and Errors of HPLC in Pictures*, 2nd Revised and Enlarged Edition, Wiley-VCH.

Kromidas, S. (February 21, 2005) *More Practical Problem Solving in HPLC*, Wiley-VCH, ISBN-13: 978-3527311132.

Project Process/Task

(A) HPLC column test

For HPLC columns, which are constantly exposed to changing conditions (in a laboratory class this is permanently the case), it is important to monitor and document the quality of the column continuously. This can be achieved for HPLC systems with the test described by V. Meyer [1]. In this test, a mixture of toluene, phenol, ethylbenzene and benzaldehyde dissolved in eluent is injected onto the column under constant conditions (isocratic flow, constant temperature). Then the plate number and from that the HETP of the column can be determined from the eluting peaks.

(B) Optimization of the isocratic eluent

B.1 By studying the literature find out the molecular structure of the three analytes and consider in what order they will elute from a HPLC column with a methanol–water eluent.

B.2 Perform optimization experiments by varying the methanol content.

(C) Determination of paracetamol, caffeine and propyphenazone in analgesics (e.g., Saridon) and validation of the method

C.1 Perform a calibration curve

C.2 Determine the working (linear) range

C.3 Validate the optimized method

C.4 Determine the concentrations of paracetamol, caffeine, and propyphenazone in for example, Saridon.

For all peaks in the chromatograms, the following chromatographic parameters are to be determined from the data: retention time, net retention time, capacity, plate numbers, resolutions.

(D) Project reporting

The results are to be submitted in the form of a report (or presentation). Comparisons with bibliographical references are to be made.

8.1.3.2
Introduction

High-performance liquid chromatography (HPLC)is a method of liquid chromatography in which separation takes place under high pressure (up to 300 bar at a flow rate of typically 1 ml min^{-1}). An HPLC apparatus consists of four main parts: pump, injection system, column and detector with the processing system. The HPLC-columns use separation particles with sizes between 3 and 10 μm, thereby achieving high plate numbers. However, at the same time a relatively high back pressure must be overcome during transport of the mobile phase through the thin column (2–6 mm diameter). All parts must be connected to each other with the lowest dead volume and pressure stability (Figure 8.14).

Distribution Isotherms and Chromatographic Behavior of Components

The basic idea of chromatography is the separation of substances by their interactions with a stationary phase. The components that interact with the stationary phase in varying degrees are transported by an eluent (mobile phase) at different

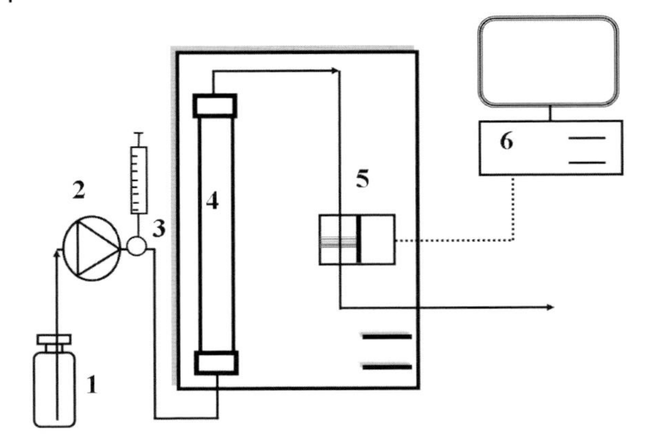

Figure 8.14 Schematic illustration of an HPLC instrument. **1** solvent reservoir(s), **2** pump, **3** injection valve with sampling loop, **4** column, **5** detector, **6** data acquisition and analysis system.

rates, resulting in a separation of these. The most important basis of chromatography is therefore the specific distribution of a component between two phases. The ratio of the concentrations of the substance in the stationary and mobile phases is described by the distribution coefficient, which as a thermodynamic basis following the Nernst distribution law. A chromatographic system suitable for separating certain substances must lead to the circumstance that the substances to be separated have different distribution coefficients. There are three cases of isothermal distribution in the column. Ideally, the distribution between the mobile and stationary phase is independent of concentration and a symmetrical peak appears in the chromatogram.

Normally, under real conditions, an adsorption isotherm deviating from the ideal case is observed. At high concentrations this is expressed by tailing of the peak. Due to oversaturation of the stationary phase large sample concentrations are inadequately withheld.

Chromatographic Parameters (see also Section 8.1.1)

Retention (R)

The retention (R) is an absolute characteristic, which shows the quality of a separation:

$$R = \frac{n_M}{(n_M + n_S)} \tag{8.14}$$

n_M Molar quantity of component X in the mobile phase.
n_S Molar quantity of component X in the stationary phase.

Retention Time, Dead Time (t_0)

The retention time (t_R) is the most important chromatographic property of a component.

The retention time is the time required by component X to migrate through the separation system. A component Y, which does not interact with the mobile and stationary phases, elutes from the separation system at the dead time (t_0).

It is a characteristic of the measuring apparatus at a given flow rate (F) and temperature.

The dead volume (V_0) is the volume of the mobile phase, which runs through the separation system in the dead time. Consequently, the dead time is identical to the residence time of the compound in the mobile phase without interaction with the stationary phase.

Therefore after this time a non-retained component elutes from the system.

$$t_0 = \frac{V_0}{F} \tag{8.15}$$

Net Retention Time (t'_R)

During the net retention time the component does not migrate with the mobile phase. Rather it is interacting on the stationary phase. The net retention time is the difference between total retention time and dead time:

$$t'_R = t_R - t_0 \tag{8.16}$$

Distribution Coefficient (K)

The distribution coefficient (K) is another characteristic parameter of the separation:

$$K = \frac{c_S}{c_M} \tag{8.17}$$

where c_S is the concentration of component X in the stationary phase and c_M the concentration in the mobile phase.

Phase Ratio (β)

Volume ratio of the phases

$$\beta = \frac{V_M}{V_S} \tag{8.18}$$

V_M Volume of the mobile phase
V_S Volume of the stationary phase

Capacity Ratio (k') (Also Called Retention Factor (k) or Mass Distribution Ratio (D_m))

The capacity ratio k' (or retention factor (k)) is a relative characteristic of the separation. It is the ratio of the concentrations of the component X in the stationary and mobile phases (c_S and c_M):

$$k' = \frac{(c_S.V_S)}{(c_M.V_M)} = \frac{K}{\beta} = K.\left(\frac{V_S}{V_M}\right) \tag{8.19}$$

K Distribution coefficient
β Volume ratio of the phases

$$k' = \frac{(t_R - t_0)}{t_0} \tag{8.20}$$

t_R Retention time,
t_0 Dead time

Selectivity or Separation Factor (α)

The separation factor α describes the rate of migration of two substances in relation to each other.

$$\alpha = \frac{k'_2}{k'_1} = \frac{t'_{R2}}{t'_{R1}} = \frac{(t_{R2} - t_0)}{(t_{R1} - t_0)} \tag{8.21}$$

By checking the selectivity of a mixture of known compounds, the quality of the separation column can be determined. If $\alpha = 1$, the two peaks have the same retention time and co-elute.

By convention a separation factors of two adjacent peaks greater than 1.25 indicates a baseline separation of two compounds.

Resolution (R_S)

The selectivity factor does not say anything about a complete separation of two substances. Here peak width still plays a role. Resolution R is calculated as follows:

$$R_S = 2\frac{(t_{R2} - t_{R1})}{(W_1 + W_2)} = 1.18\frac{(t_{R2} - t_{R1})}{(W_{1/2_1} + W_{1/2_2})} \tag{8.22}$$

Eddy or turbulent diffusion leads to an undesirable broadening of the peak width, among other things. When passing through the column, sample molecules traverse slightly different paths. Molecules that have to circumvent large particles in the stationary phase take longer than molecules that bypass the small particles. With inconsistent packing of the column, regions in the column are created in which the sample molecules are transported faster than in other areas of the column. Other causes of peak broadening can be longitudinal diffusion, impurities of the packing material (e.g., metal ions lead to nonspecific interactions), hindrance of mass exchange between the mobile phase and stationary phases.

Tailing (*T*) (for Asymmetric Peaks)

T is defined as the ratio of the α and β distances at 10% of their height (Figure 8.15):

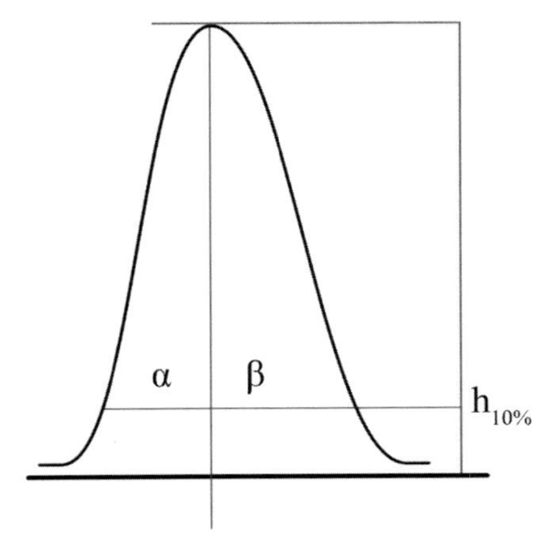

Figure 8.15 Peak symmetry.

$$T = \frac{\alpha}{\beta} \tag{8.23}$$

For Gaussian peaks, the symmetry is $T = 1$. Deviations from larger T-values are described as tailing, and from smaller values as fronting. In practice, symmetry factors of $T = 0.9$ to 1.1 are desirable.

Selection of Mobile and Stationary Phases

The successful application of HPLC requires the correct selection of the chromatographic separation system and accurate work. One must be able to evaluate the properties of the substances that are to be separated, their possible behavior in a separation system and what combination of mobile and stationary phase would be suitable to get a separation of the substances. Further, one should be able to explain separation problems and, at best, one must acquire a wide range of experience based on lots of practice.

Mobile Phase

Solubility: The components must be completely soluble in the mobile phase.

Miscibility: All components of the mobile phase must be completely miscible.

Phase system: The mobile and stationary phases must form a phase system in which separation is possible (selection according to eluotropic series and selectivity triangle, see Section 8.1.6).

Stationary Phase

Properties that strongly determine the function of the stationary phase are:

- Material
- Particle size (as even as possible for a complete package, sizes between 3 and 10 μm).
- Quality of column packing
- Pore structure
- Pressure stability of the packing
- Type of chemical modification
- Column dimensions (length, diameter)

Properties of HPLC Columns [2]

Modes of Separation

Normal-phase (NP) chromatography is the classic form of liquid chromatography using polar stationary phases and nonpolar mobile phases. The analyte is retained by the interaction of its polar functional groups with the polar groups on the surface of the packing. Analytes elute from the column starting with the least polar compound, followed by other compounds in order of their increasing polarity. It is useful in the separation of analytes with low to intermediate polarity and high solubility in low-polarity solvents. Water-soluble analytes are usually not good candidates for NP chromatography.

Reversed-phase (RP) chromatography has become the most common mode of liquid chromatographic separation. In RP the stationary phase is non-polar and the mobile phase is polar. The analytes are attracted to the surface by their non-polar functional groups. The most polar analyte elutes from the RP column first followed by other analytes in order of decreasing polarity. RP chromatography is useful for the separation of compounds having high to intermediate polarity.

Ion-exchange chromatography separates analytes by their ionic functionality. Ion-exchange stationary phases are usually comprised of ionic species attached to the surface of the silica substrate or ionic functional groups evenly distributed throughout a polymeric media. Weak ion-exchange phases are usually pH dependent. Strong ion-exchange phases are always charged and, therefore, independent of typical pH changes.

Affinity chromatography is based on specific interactions in a lock-and-key paradigm between analytes and matrix-bound ligands. Dependent on the secondary structures of biological macromolecultes affinity chromatography is without question the most highly specific, and consequently the most powerful, mode of chromatography.

Substrate Materials of the Packing

Porous silica particles are the most common substrate material used for HPLC column packings. Silica-based columns can withstand high pressures, are compatible with most organic and aqueous mobile-phase solvents, and come in a wide range of bonded phases. Silica-based columns are often used for separations of low molecular weight analytes using mobile phase solvents and samples with a pH range of 2 to 7.5.

Polymeric highly cross-linked styrene-divinylbenzene based packings are compatible with most mobile phase solvents and samples with a pH of 1 to 14. Polymer-based columns tend to have lower efficiencies for small molecules compared to silica-based columns due to their smaller average surface area. In addition, polymer-based columns typically have lower mechanical strength and therefore cannot withstand the highest system back-pressures. A polymer-based packing is often a good alternative if the sample requires a mobile phase pH outside the normal operating range of standard silica-based columns. Polymeric packings are often used for ion-exchange separations, and are also useful in non-aqueous gel permeation chromatography (GPC) size-excluding analyses and ion exclusion analyses of organic acids and carbohydrates.

Particle Properties of the Packing

The first available HPLC columns were packed using irregularly shaped silica particles. Because of this, many standard analytical methods are still based on these materials. Irregular particles are also used in large-scale preparative applications because of their high surface area, capacity and low cost.

The majority of new HPLC methods are performed on spherical shaped or spheroidal (almost spherical) particles. Spherical particles provide higher efficiency, better column stability and lower back-pressures compared to irregularly shaped particles.

Particle size for HPLC column packings refers to the average diameter of the packing particles. Most HPLC packings contain a narrow range of particle diameters. Particle size affects the back-pressure of the column and the separation efficiency. Column back-pressure and column efficiency are inversely proportional to the square of the particle diameter. This means that as the particle size decreases, the column back-pressure and efficiency increase. A well packed column with 3 μm particle size produces almost twice the separation efficiency of a comparable 5 μm column. However, the 3 μm column will have about a three-fold higher back-pressure compared to the 5 μm column when operated with the same mobile phase at the same flow rate. Highly efficient, small-particle (3 μm and 4 μm) columns are ideal for complex mixtures with similar components. Fast, high-resolution separations can be achieved with small particles packed in short (10–50 mm length) columns.

Larger particle (5 and 7 μm) columns are typically used for routine analyses where analytes have greater structural differences. Large 10 μm packings have only moderate column efficiencies. Columns packed with 10 μm particle size are generally used as scout columns for future preparative separations, semi- preparative applications, or routine QA/QC methods where high chromatographic efficiencies are not required. Large particles (15–20 μm) are used for preparative-scale separations.

The *pore size* of a packing material represents the average size of the pores within each particle. The size of the analyte should be considered when choosing the appropriate pore size for the packing material. The molecular weight of an analyte can be used to estimate the size of the molecule. As a general rule, a pore size of

100 or less should be used for analytes below 3000 MW. A pore size of 100–130 is recommended for samples in the range of 3000 MW–10 000 MW. For samples above10 000 MW, including peptides and proteins, a 300 material provides the best efficiency and peak shape.

Pore volume is a measurement of the empty space within a particle. Pore volume is a good indicator of the mechanical strength of a packing. Particles with large pore volumes are typically weaker than particles with small pore volumes. Pore volumes of 1.0 ml g^{-1} or less are recommended for most HPLC separations. Pore volumes of greater than 1.0 ml g^{-1} are preferred for size-exclusion chromatography and are useful for low-pressure methods.

The physical structure of the particle substrate determines the *surface area* of the packing material. Surface area is determined by pore size. Pore size and surface area are inversely related. A packing material with a small pore size will have a large surface area, and vice versa. High surface area materials offer greater capacity and longer analyte retention times. Low surface area packings offer faster equilibration time and are often used for large molecular weight molecules.

Carbon Load

The carbon load is a measure of the amount of bonded phase bound to the surface of the packing. High carbon loads provide greater column capacities and resolution. Low carbon loads produce less retentive packings and faster analysis times.

Surface Coverage

Surface coverage is calculated from the carbon load and the surface area of a packing material. Surface coverage affects the retention, selectivity and stability of bonded phases.

End-Capping

A reversed-phase HPLC column that is end-capped has gone through a secondary bonding step to cover unreacted silanols on the silica surface. End-capped packing materials eliminate unpredictable secondary interactions. Basic analytes tend to produce asymmetric tailed peaks on non-end-capped columns, requiring the addition of modifiers to the mobile phase. Non-end-capped materials exhibit different selectivity than end-capped columns. This selectivity difference can enhance separations of polar analytes by controlling the secondary silanol interactions.

Theoretical Maximum Column Volumes in ml

Column Length, mm	Column Internal Diameter, mm					
	1.0	2.1	3.0	4.6	10.0	22.5
30	0.024	0.104	0.212	0.5	2.35	11.93
50	0.039	0.173	0.353	0.83	3.93	19.88
100	0.079	0.346	1.06	1.66	7.85	39.76

Column Length, mm	Column Internal Diameter, mm					
	1.0	2.1	3.0	4.6	10.0	22.5
150	0.118	0.519	1.06	2.49	11.78	59.64
250	0.196	0.866	1.77	4.15	19.63	99.40

Typical Flow Rates

Column Dimensions ID, mm × Length, mm	Flow Rate, ml min^{-1}
22.5 × 250	24.0
10.0 × 250	5.0
4.6 × 250	1.0
3.0 × 250	0.5
2.1 × 100	0.2
1.0 × 100	0.05

Column Length, Diameter and Volume

Once a packing material has been selected, the physical dimensions of the HPLC column hardware should also be optimized for the desired separation. It is important to understand the relationship between column length, diameter and volume. The column length (L) and internal diameter (d) determine the bed volume (V) of an HPLC column by the following equation: $V = L \times \pi d^2/4$. This equation does not account for the reduction in liquid volume due to the packing material. It is an upper limit on the column volume–the minimum volume of mobile phase required to elute an unretained analyte from the column. Small column diameters provide higher sensitivity than larger column diameters for the same injected mass because the concentration of the analyte in the mobile phase is greater. Smaller diameter columns also use less mobile phase per analysis because a slower flow rate is required to achieve the same linear velocity through the column. HPLC instrumentation may need to be modified for columns with very small internal diameters to eliminate band broadening due to extra-column effects, that is, mixing volumes outside the column. Longer columns often provide increased resolution. Larger-diameter columns provide greater sample loading and lower back-pressure. Column back-pressure for a given flow rate increases as the column length increases and as the internal diameter decreases.

Column Flow Rate

When the column dimensions are changed to optimize a separation or to scale a separation to a preparative application, the mobile phase flow rate should be adjusted proportionally to the cross-sectional area of the column to maintain consistent linear velocity and retention times. The following equation will assist in adjusting the flow rate (F) for columns of different internal diameters (d).

$$F_2 = F_1 \times (d_2/d_1)^2 \quad (8.24)$$

8.1.3.3
Material and Methods

Equipment

HPLC System (Agilent 1200 Series), diode array detector, degasser, multi-position valve with sample loop.

Parameters

Flow: 1 ml min^{-1}
Lower pressure limit: 0 bar
Upper pressure limit: 280 bar
Injection volume: 10 µl
Detection: 254 nm, 240 nm

Stationary Phase

Reversed Phase C18, NUCLEODUR® C8
Column length: 15 cm
Column diameter: 4 mm
Particle size of 5 µm
Pore diameter 90 Å

Chemicals

Caffeine	CAS- Number 58-08-2	Xi 
Paracetamol	CAS- Number 103-90-2	Xn 
Propyphenazone	CAS- Number 479-92-5	Xn 
Methanol	CAS- Number 67-56-1	T, F 
Benzaldehyde	CAS- Number 100-52-7	Xn 
Acetonitrile	CAS- Number 75-05-8	F, Xn 
Ethylbenzene	CAS- Number 100-41-4	F, Xn 
Toluene	CAS- Number 108-88-3	F, Xn 
Phenol	CAS- Number 108-95-2	T, C 

Reagents

Standard Stock Solution

50 mg of caffeine, 100 mg paracetamol and 100 mg propyphenazone are weighed in a 100 ml volumetric flask and dissolved in 5 ml methanol. The flask is then filled to the mark with the intended eluent.

Standard Solutions

Standard S1:	0.5 ml stock solution is filled to the 100 ml mark with eluent, (accordingly 2.5 mg l^{-1} caffeine, 5 mg l^{-1} paracetamol, 5 mg l^{-1} propyphenazone)
Standard S2:	0.75 ml of the stock solution is added to eluent and filled to the 100.0 ml mark (accordingly 3.75 mg l^{-1} caffeine, 7.5 mg l^{-1} paracetamol, 7.5 mg l^{-1} propyphenazone)
Standard S3:	1.0 ml of the stock solution is filled to the 100 ml mark with eluent, (accordingly 5 mg l^{-1} caffeine, 10 mg l^{-1} paracetamol, 10 mg l^{-1} propyphenazone)
Standard S4:	1.5 ml of the stock solution is filled to the 100 ml mark with eluent. (accordingly 7.5 mg l^{-1} caffeine, 15 mg l^{-1} paracetamol, 15 mg l^{-1} propyphenazone)
Standard S5:	2.0 ml of the stock solution is filled to the 100 ml mark with eluent. (accordingly 10 mg l^{-1} caffeine, 20 mg l^{-1} paracetamol, 20 mg l^{-1} propyphenazone)

Sample Solution

25 mg of the analgesic powder is weighed in a 100 ml Erlenmeyer with a stopper, mixed with 50 ml of the eluent and stirred for 30 min. The suspension is filtered through filter into a 100 ml measurement flask.

The Erlenmeyer and the filter are rinsed with approx. 40 ml of eluent (several smaller portions).

Finally, the volumetric flask is filled to the mark with eluent.

This sample solution is diluted 1 : 10 and then measured by HPLC.

Experiment

Important: All samples and standard solutions must be filtered before injection (pore size 0.45 μm).

(A) HPLC column test

- The system is flushed with an eluent of acetonitrile : water = 65 : 35 conditioned (flow rate 1 ml min^{-1}) for at least 30 min. The UV detector is set at $\lambda = 254$ nm and should reach a stable baseline.
- 300 μl are taken from the standard stock solution (benzaldehyde, ethyl benzene, toluene and phenol) and dissolved in 50 ml of eluent. After that, the test solution is injected and the chromatogram recorded.

- The retention times should be determined, plate numbers calculated and the pressure entered in a column logbook. Have there been any significant changes compared with the values of the previous groups?

In general, the theoretical plate numbers calculated from the four peaks vary greatly. Why is that so?

(B) Optimization of the isocratic eluent

- To determine the ideal eluent mixture of methanol/water, the standard S3 is analyzed with two different eluents. First with methanol : water = 40 : 60, then with methanol : water = 60 : 40.
- In both chromatograms the retention times of all peaks should be determined, the dead time and a representative value for the plate number of the column.

How is the retention time affected, when the proportion of methanol is increased in the eluent? Consider whether the different retention times could possibly be influenced by changing the pH of the eluent, for example, by adding acetic acid.

Optimal Eluent: The optimal separation system is characterized, on the one hand, by the fact that all significant analytical peaks display sufficient resolution (R_S at least 1.2) and, on the other hand, by the fact that retention times are as short as possible (capacity factor k' is not too large).

Conduct the analysis with different ratios. Then calculate all *RS* and *k'* values as a function of the methanol : water ratio. Decide with what composition the analysis is performing best.

If an attempt is made to shorten the retention times by changing the isocratic eluent, the resolution of the early eluting peaks will soon be insufficient. This problem can at least be partially solved with a gradient, that is, by continuously changing the methanol : water ratio during analysis.

How should this gradient be chosen (in analogy to temperature programing in gas chromatography)?

If the composition of the isocratic eluent is changed greatly, the system must be conditioned for at least 30 min with the new eluent before injections are made.

(C) Determination of paracetamol, caffeine and propyphenazone in analgesics (e.g., Saridon) and validation of the method

C.1, C.2

- Now all standard solutions are measured in replicate and, for all three analytes, a measurement function with the peak area as an analytical signal should be created.

- Using linear regression determine the analytical sensitivity and the intercept of the three test functions with their standard deviations.
- Calculate the limits of detection and the quantitation limits of the method.

C.3

Validate the method.

C.4

Determine the concentrations of paracetamol, caffeine and propyphenazone in, for example, Saridon.

8.1.3.4 Questions

(1) What data must be available when an HPLC-method is passed on?
(2) What is the elutropic series, how was it determined and what organizing principle does it follow?
(3) Give two reasons why eluents should be degassed.
(4) Name four possible methods of eluent degassing.
(5) Name four other types of HPLC detectors besides UV/VIS.
(6) How can excessive peak broadening be avoided?
(7) Can an HPLC run be stopped during operation without impairing the separation?
(8) List system suitability tests in HPLC.

References

1 Meyer, V. (2010). *Practical High-Performance Liquid Chromatography*, 5th edn, Wiley Online Library, Chapter 8. HPLC Column Tests Published Online doi: 10.1002/9780470688427.ch8.

2 http://www.discoverysciences.com/print.aspx?id=2604

8.1.4 High-Performance Liquid Chromatography Coupled with Mass-Selective Detection – LC/MS, Project "Cocaine Scandal: Hair Sample with Consequences"

8.1.4.1 Analytical Problem

There is a relationship between the element concentrations present in hair and environmental conditions, diet and living conditions. Hair consists largely of keratin, a protein that is rich in the sulfur-containing amino acid cysteine. This leads to permanent storage and accumulation of minerals in the hair. This enrichment is particularly pronounced for trace elements and heavy metals. Their concentrations in the hair are one to two orders of magnitude higher than those in blood or urine. Hair is thus suitable for detecting metallic exposure and showing under-supply of essential elements. Hair analysis is increasingly used in forensics to detect the use of narcotics. In 1961, a sample of hair taken from *Napoléon Bonaparte* was investigated. It was proved that he had gradually ingested arsenic. Whether this was due to deliberate poisoning or the living conditions on St. Helena is still the subject of heated discussion. Ten years ago, a scandal shook the world of German soccer: on 20 October 2000 the coach of Bayer Leverkusen (the premier league soccer team in Leverkusen) and designated national coach was tested positive for drug abuse on the basis of a hair analysis that showed traces of cocaine.

Object of Investigation/Investigation System
Anesthetic Drugs

Samples
Hair

Examination Procedure
The mass spectrometer has not only stood its ground in liquid chromatography for structurally determining organic materials, but it has increasingly been used as a detector. Following chromatographic separation, it enables the specific proof and quantification of a desired compound. In the combination HPLC-MS the mass axis gives a second dimension of analytical information in addition to the chromatographic time axis. This detection method is based on the fact that after the formation of ions through ionization energy, separation or filtering takes place due to the mass-to-charge (m/z) ratio. Tandem mass spectrometry has proven to be of particular importance in the analysis of complex mixtures generating a chemical fingerprint for the compound of interest, ensuring correct peak assignment in the presence of complex matrices. Due to the low detection limit and high selectivity (co-eluting peaks can be isolated by mass selectivity and are not constrained by chromatographic resolution), this method provides excellent results, especially in trace analysis, the determination of molar masses of macromolecules, in origin determination of biological samples, and in isotopic separation.

Analytes
Caffeine, Theobromin

Keywords (see also Section 8.1.2)
Functional principle and parts of the LC-MS system, electrospray ionization (ESI), atmospheric pressure chemical ionization (APCI), molecular ion, MALDI-TOF, vacuum systems in MS-technology, interface LC-MS, flow injection analysis (FIA)

Bibliographical Research (Specific Topics)

Niessen, W.M.A. (2007) *Liquid Chromatography-Mass Spectrometry*, 3rd edn, CRC Taylor & Francis, ISBN-13: 978-0-8247-4082-5.

Watson, J.T., and Sparkman, O.D. (2007) *Introduction to Mass Spectrometry, Instrumentation, Applications and Strategies for Data Interpretation*, 4th edn Wiley, ISBN 978-0-470-51634-8.

Ekman, R., Silberring, J., Westman-Brinkmalm, A., and Kraj, A. (2008) *Mass Spectrometry (Instrumentation, Interpretation and Applications)*, Wiley & Sons Inc., Hoboken, online library doi: 10.1002/9780470395813.ch4.

McMaster, M. (August 10, 2005) *LC/MS: A Practical User's Guide*, Wiley-Interscience, ISBN-13: 978-0-471-65531-2.

Project Process/Task

(A) Optimization of an LC-MS method

Based on the given initial parameters, one should first optimize a chromatographic method for the separation of a standard mixture of caffeine and theobromine.

(B) Determination of the mass spectra of caffeine and theobromine

With the optimized chromatographic conditions (A) the mass spectra of reference compounds are to be obtained and identified.

(C) Optimization of MS parameters by flow injection analysis (FIA)

With the appropriate eluent mixture the optimal MS parameters are to be determined by direct injection of the standards.

(D) Determination of caffeine content in hair

D.1 By calibration with the external standard method the linearity and the linear range are to be determined.

D.2 The caffeine and the theobromine content in the hair are to be determined with the help of the given sample preparation.

D.3 The method is to be validated.

(E) Project reporting

The results are to be submitted in the form of a report (or presentation). Comparisons with bibliographical references are to be made.

8.1.4.2
Introduction

Principle

Section 8.1.3 has already dealt with the separation method that precedes the mass spectrometer. Following chromatographic separation in the column, eluent and analyte are directed into the ionization chamber. More frequently there is a diode array detector installed between the column and the mass spectrometer, where detection is possible without decomposition of the substances. Other detectors could be fluorescence, NMR (nuclear magnetic resonance) or ICP (Induction Coupled Plasma) detectors. The eluent enters now via an interface the MS system. In the ion source the mobile phase is first atomized and the analyte ionized using different ionization techniques. Volatile and neutral components are removed (desolvatization), so that only the target substances can reach the analyzer. Former ionization methods usually started with the removal of the solvent by heating (moving belt interface, thermospray interface) or pulse differences between solvent and analyte molecules (particle beam interface). Today, ionization under atmospheric pressure is most popular.

After ionization the molecular ions are transferred through various elements in the high vacuum region of the mass spectrometer where they are accelerated and focused. The separation takes place in the mass filter, which is timed and only allows the passage of ions of a particular mass : charge ratio (m/z). In the detector the frequency (abundance) of each m/z-ratio is recorded, converted into an electrical signal and displayed as a function of the m/z-ratio as a mass spectrum (a diagram of the LC/MS system can be seen in Figure 8.16).

Interface LC/MS

In contrast to the GC/MS, for the LC/MS coupling the analyte from the liquid phase must be transferred under atmospheric pressure in the gas phase and a high vacuum. Large amounts of solvent must be removed for this purpose. This places high demand on the LC/MS coupling systems ("interfaces") – like teaching a fish how to fly; difficult but not impossible (Figure 8.17).

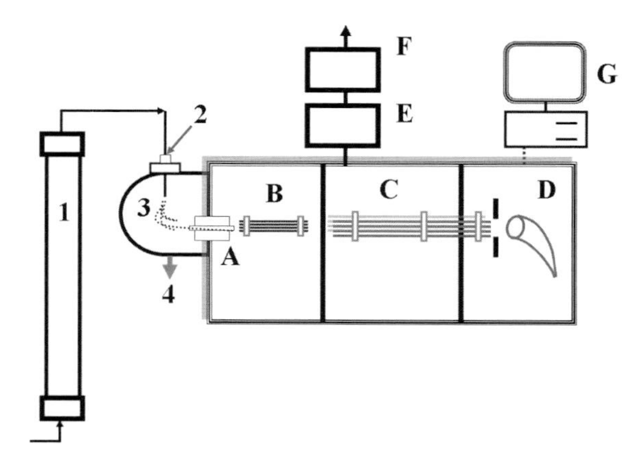

Figure 8.16 Schematic diagram of an HPLC/MS system. **A** interface LC/MS, **B** ion focussing device, **C** mass filter, **D** detector, **E** rough mechanical vacuum pump, **F** high vacuum turbo pump, **G** data acquisition and elaboration, **1** HPLC system, **2** HPLC inlet and nebulizer gas inlet, **3** ion source, **4** waste.

Figure 8.17 The flying fish is enabled to fly by having fins which approach in extent the wings of a bird. (Source: Worthington Hooker, *Natural History*, Harper & Brothers, New York, **1882**, 211).

Major difficulties are:

- The MS vacuum system has to cope with high loads associated with the introduction of up to 1 ml min^{-1} eluent. The vacuum is essentially predetermined by the source geometry and the vacuum pump system, that is, by constructive device parameters, which cannot be changed during the analysis. By contamination of the interface, however, deposits can form on inlet capillaries, thus changing the vacuum in the transport region. The collision energy is determined by the potential that is created at the end of the capillaries ("fragmentor" voltage).
- The incompability with respect to the eluent composition
- The need to produce sufficient heat to achieve complete evaporation of the eluents.

Vacuum System

See Section 8.1.2.

Ion Source

When combining HPLC and MS ionization has already taken place under atmospheric pressure (atmospheric pressure ionization–API) for the following reasons:

- Controlled gentle evaporation of the solvent, that is, minimal thermal stress.
- Suitable for both nonvolatile compounds (electrospray ionization–ESI) as well as partially volatile compounds (atmospheric pressure chemical ionization–APCI).

The following API ionization techniques are now possible:

- Electrospray ionization (ESI)
- Atmospheric pressure chemical ionization (APCI)
- Combined source (DUIS)
- Photoionization (APPI)
- Coordination ion spray ionization
- Electron-capture ionization
- Charge transfer ionization

Electron Spray Ionization

The most common ionization method in LC-MS is (still) ESI [1] (Figure 8.18). In this process the sample enters a cylindrical spray chamber through a capillary following chromatographic separation. Very small charged droplets are generated in a voltage field (3–5 kV) attached to the cylinder, the end plate and the capillary. Through evaporation of the solvent in the heated source, the droplets very rapidly diminish in size, causing the ratio of charge to volume to increase. The "Rayleigh" limit is the maximum charge that a droplet can hold while maintaining its volume. If the load exceeds this limit, Coulomb explosions occur. In the ESI process there is no indirect charge transfer. Here the ionization of the analyte takes place directly through the applied tension. After complete solvent evaporation only charged molecular ions reach the analyzer in the form of $[M+H]^{n+}$ and $[M-H]^{n-}$. An inert, high velocity gas flow (mainly N_2) is used to enhance atomization (nebulizer effect).

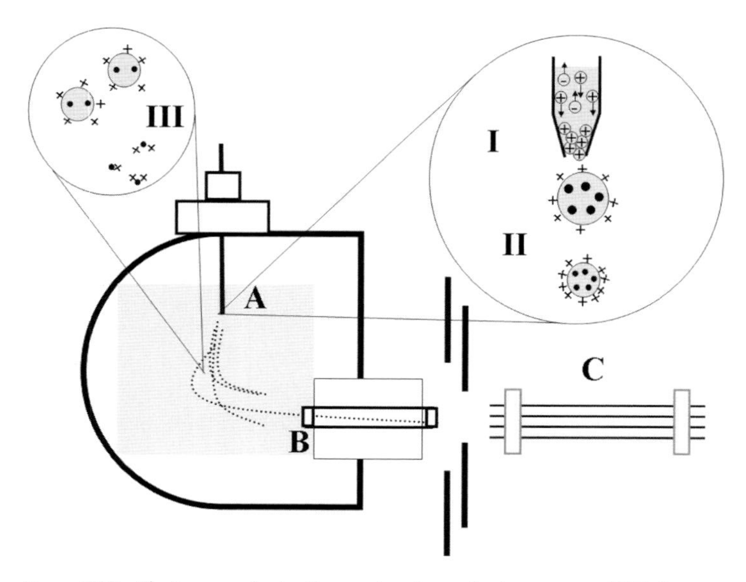

Figure 8.18 Electrospray ionization under atmospheric pressure (ESI). **I** spray needle tip, **II** multiply charged droplet- solvent evaporation-the Rayleigh limit is reached- Coulomb explosion, **III** multiply charged droplet-analyte ion, **A** ion source, **B** capillary, **C** mass filter.

Ionization is affected by several parameters of the mobile phase: buffer in use, surface tension, conductivity, vapor pressure, and pH of the eluent flow rate, as well as field strength at the capillary ends.

Atmospheric Pressure Chemical Ionization (APCI) [2]

In chemical ionization under atmospheric pressure the solvent is atomized at a metal tip (corona discharge needle) with a heated, pneumatically assisted nebulizer (Figure 8.19) by applying high voltage (up to 5 kV). Depending on their polarity, negatively or positively charged primary ions are formed. These can be both gas ions (N_2^+, O_2^+) and also ions of the solvent molecules ($CH_3OH_2^+$, CH_3O, CH_3CNH^+, H_3O^+, OH^-) and buffer salts (N_4^+, CH_3COO^-).

The primary ions transfer charges to a chemical reaction on the analyte molecule (Brønsted–Lowry acid–base theory: a transfer of protons is only possible when the target substance can act as a base). The formation of small clusters of up to four solvent molecules and an analyte ion can be observed. Subsequently, the clusters are destroyed by passing them through a heated capillary. This type of ionization is suitable for substances which contain heteroatoms and which have low to medium polarity, and still have low humidity.

Important Aspects of APCI

- $[M+NH_4]^+$-adducts (NH_3 formed from corona discharge or from ammonium acetate) are observed relatively frequently.
- Ionization losses possible through the formation of ion pairs.
- Flow rates of eluents are in the range 50 μl min (minimal flow) to 1 ml min^{-1}.
- Derivatization to increase the ionization yield is also possible.
- APCI (+): Requires the introduction of O or N without changing the polarity (e.g., acetylation of an OH group) in order to increase proton affinity.

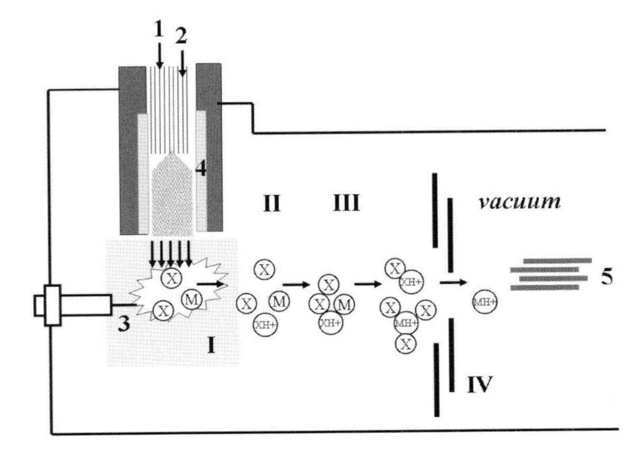

Figure 8.19 Chemical ionization under atmospheric pressure (APCI). **1** nebulizing gas, **2** HPLC inlet, **3** corona discharge, **4** heater/vaporizer, **5** mass analyzer, **I** corona discharge, **II** ionization of solvent molecules, **III** reaction with analyte molecules: formation of clusters, **IV** declustering.

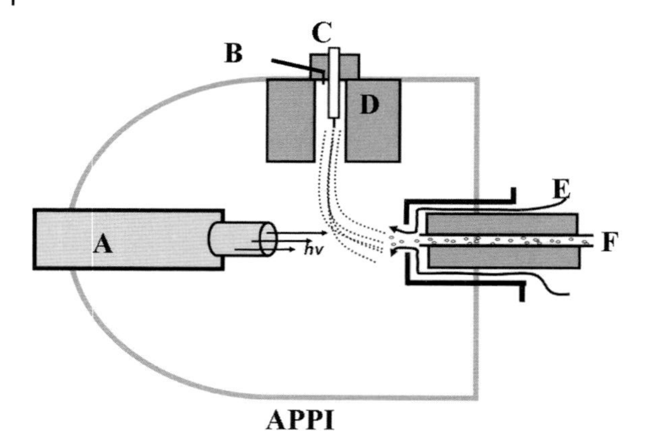

Figure 8.20 Chemical photoionization under atmospheric pressure. **A** UV or krypton lamp, **B** nebulizing gas, **C** HPLC inlet, **D** heater/vaporizer, **E** drying gas, **F** capillary.

Atmospheric Pressure Photoionization (APPI)

This method of ionization is particularly suitable for nonpolar molecules, which are not ionized using ESI or APCI. Photoionization can be improved by an ionization medium ("dopant"). Photon emissions of a krypton lamp are used to ionize the sample (Figure 8.20). Here, as in the ESI and APCI processes, a spray process and evaporation of the solvents occurs. Subsequently, the molecules are irradiated with high-energy photons (10 eV):

$$M + h\nu \rightarrow M^{+} + e^{-} \tag{8.25}$$

The energy is stored in the molecule and leads to ionization if the photon energy is greater than the ionization energy of the molecule.

Electron Capture APCI [3]

In the corona discharge in addition to N_2^{+} radical cations even low-energy electrons are formed, which can form radical anions from molecules of high electron negativity through electron capture (Figure 8.21). Alternatively, fragment anions of high electron affinity can arise by virtue of dissociative electron capture:

- Allows selectivity gains and lower detection limits by appropriate derivatization.
- Enables us to distinguish isomers.

Dopant-Assisted Atmospheric Pressure Photoionization (DA-APPI) [4]

By adding a "dopant" D (usually toluene) a significant increase in the sensitivity of the photoionization is achieved. The dopant, which is available in great excess, is directly ionized by the photons. Charge transfer to the sample molecule is then effected by the proton or electron.

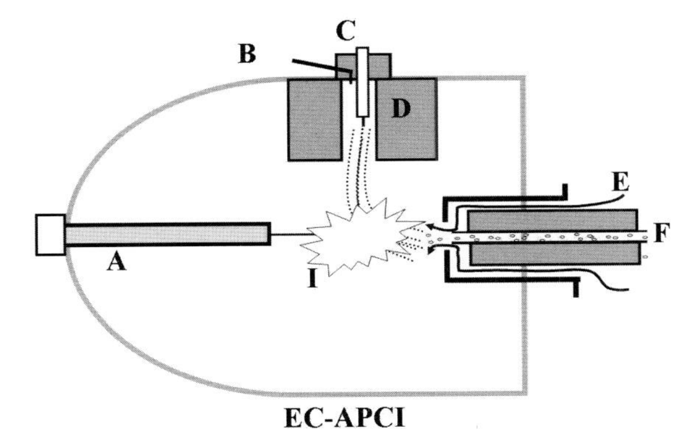

Figure 8.21 Electron capture-APCI (EC-APCI). **A** corona needle, **B** nebulizing gas, **C** HPLC inlet, **D** heater/vaporizer, **E** drying gas, **F** capillary, **I** corona discharge.

$$D + h\nu \rightarrow D^{+} + e^{-} \quad (8.26)$$

$$D^{+} + M \rightarrow M^{+} + D \quad (8.27)$$

The DA-APPI is a photochemically induced chemical ionization.

Matrix-Assisted Laser Desorption and Ionization (MALDI)
See Section 8.1.2.

Focus
A high-frequency eight polar alternating field (without DC component) is often used for ion steering and focussing. Quadrupoles or hexapoles can also be used. The advantage of the octopole lies in the fact that a larger m/z range can be dealt with. This steering of ions permits a better transmittance of all ion masses, thus achieving better resolution, higher sensitivity and lower ion energy dispersion. The ion concentration aims to remove neutral atoms and molecules (air, solvents, non-ionized samples etc.) and to break up adducts between charged analyte molecules and uncharged molecules.

Mass Filters

- Quadrupole mass filters (see Section 8.1.2)
- Ion trap detector (ITD)

Classical Ion Traps
These ion traps are also named after their inventor, that is Pauli ion traps or three-dimensional ion traps. *Wolfgang Pauli* was awarded the Nobel Prize in 1989 “for developing the ion trap, an electrical quadrupole field for the trapping and examination of a few ions or electrons for a long enough period”.

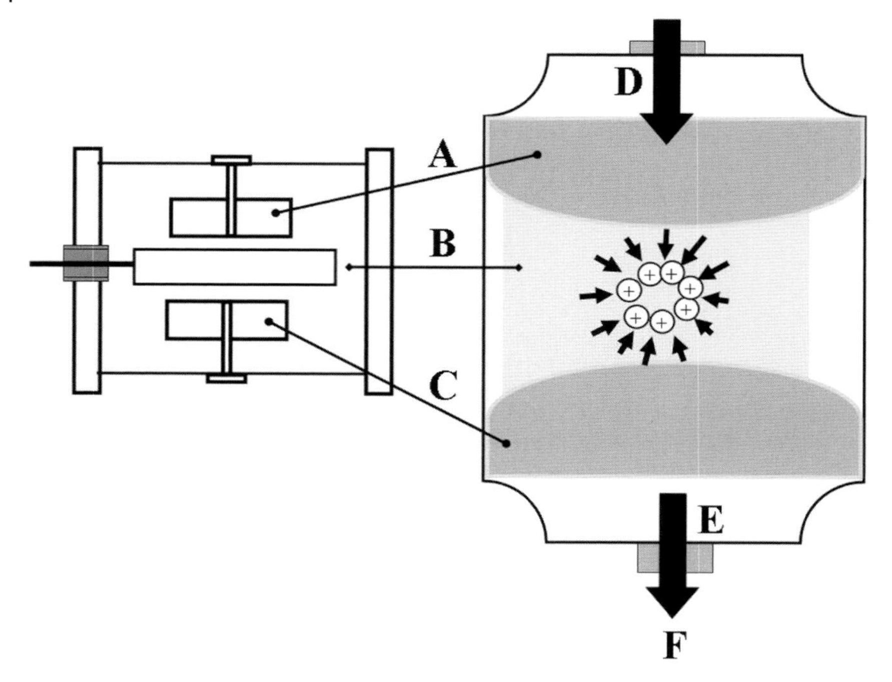

Figure 8.22 Ion trap detector (ITD). **A/C** End Cap Electrode, **B** Ring Electrode, **D** Ion entry, **E** Ion exit, **F** Detector.

As the name suggests, the principle of the ion trap is to keep ions "captive" in an area. The ion trap consists of three electrodes: a ring electrode, an inlet electrode and and outlet electrode (Figure 8.22). The quadrupole field is set up between the ring electrode and the two end electrodes ("end caps"). The amplitude of the alternating voltage (about 100 V) of the electromagnetic field is switched between the two electrode systems with frequency Ω, which leads to stabilization of ions in the trap by oscillation. This structure makes it possible to keep the ions "captive" and to catapult them from the trap towards the detector by changing the potential masses with specific m/z-ratios. The energy is again released by collision with He atoms at a residual pressure of 0.1 Pa, (10^{-3} mbar) and the trajectory is stabilized. If molecular ions are in the ion trap these ions can be excited to fragment. The resulting fragment ions can be further detained and again fragmented. This procedure can be performed several times (MS^n). In this way it is possible to analyze a fragmentation pattern.

Performance of ion traps in the LC-MS:

- For recording complete mass spectra ("full scan") some 100 times less substance is required than for quadrupoles.
- Mass range varies between 50–2000 u, 100–4000 u or 1000–20 000 u.

- u mass accuracy or better at a reduced scan speed and mass range.
- Inclusion of multiple daughter ion spectra (MS^n) possible.

Time of Flight (TOF) [5]

After ionization, the ions are accelerated into an electric field at high speed. Depending on the m/z-ratio, the ions behave differently with respect to their terminal velocity. Ions with different m/z-ratios and the same charge require different flight times (times of flight) to cross a given distance. The resulting time of flight spectrum can be converted into a mass spectrum.

$$t = \sqrt{\frac{m}{2zU}}L \qquad (8.28)$$

m/z Mass to charge ratio of the ion
z Charge of the ion (C)
t Flight time (s)
U Acceleration voltage (V)

The relatively broad ion beam of a TOF-MS is detected by using a "microchannel plate", which consists of a thin layer of glass beads sintered together, which form irregular channels and which are equipped with an electron-emitting layer. When a voltage of 1–3 kV is applied, electrons are emitted and strengthened upon a clash of ions. The thickness of the panel decides when already activated channels can detect new ions. Most often TOF analyzers are coupled with MALDI (MALDI-TOF-MS). The ions are already accelerated at the time of ionization or are given a start pulse, respectively.

Characteristics of Flight Mass Spectrometers

- Sensitivity of ion traps and quadrupoles in the "full scan" mode.
- Sensitivity depends on the–"duty cycle", that is, how large is the percentage of the TOF conducted in the ions (varies between 5% and 50%).
- In combination with quadrupole MS/MS is possible.
- Linear range 2–3 sizes, thus suitable for quantitative analysis only to a limited extent.

MS/MS and MS^n -Techniques

MS/MS techniques can be performed with both quadrupoles and ion traps, as well as with TOFs. Thereby, fragments are further fragmented by additional impact collision. Fragmentation is a process which creates relatively high energy in the collision cell and in which the molecules (parent ions) decay in a specific manner (daughter ions). The pattern of fragmentation is characteristic for each molecule and can be regarded as a fingerprint of a compound. Advantages of MS/MS are both additional information on the structure of fragments and a strong improvement in signal-to-noise due to reduction of the "chemical" noise.

Important: Only MS/MS and MS^n spectra contain enough structurally relevant information to make meaningful comparisons of spectra [6].

Resolution
See Section 8.1.2

Detection
The central task of the detector in mass spectrometry is the measurement of the abundance of ions and conversion of that information into an electronic signal.

High Energy Dynode Detector (HED)
See Section 8.1.2

Data Collection and Processing

- Full scan mode (total ion chromatogram, TIC)
- Single ion monitoring (SIM)

Representation of the Spectra
See Section 8.1.2

Analytical information in the Mass Spectrum
See Section 8.1.2

Adducts or Quasi-Molecules
In mass spectra, it often happens that quasi-molecules can be observed in the masses sought. These are molecular ions, to which an additional ion has been added during the ionization process. $[M+H]^+$ and $[M-H]^-$ are almost always to be found in the LC-MS. In addition to these signals other adduct signals can also occur, see Table 8.1.

The presence of various compounds in the ion source may lead to such deposits of analyte molecules.

In analytics the formation of adducts is often deliberately forced by adding substances (such as Na-salts) to increase the signal intensities.

For example, if the two largest signals in a mass spectrum have a distance of 22 m/z, this makes it easier to identify the molecule ion $[M+H]^{+\cdot}$ As mass 22 cannot be either a fragment or an atom, the first signal must be $[M+Na]^{+\cdot}$

The quasi-molecule in this case is therefore also an identifier.

Table 8.1 Possible adduct molecules or quasi-molecules, which may arise from ionization.

Positive Ions		Negative Ions	
$[M+H]^+$	M+1	$[M-H]^-$	M−1
$[M+NH_4]^+$	M+18	$[M+HCOO]^-$	M+45
$[M+Na]^+$	M+23	$[M+OAc]^-$	M+59
$[M+K]^+$	M+39	$[M+TFA]^-$	M+113
$[M+H+CH_3CN]^+$	M+42	$[M\text{-}Cl]^-$	M+35/37

Direct Infusion or Flow Injection Analysis (FIA)

For a quantitative analysis by LC-MS it is very important to adjust the measurement parameters so as to make it possible to detect a signal for the analyte that is as high as possible. For this there is a possibility to conduct a standard solution of analytes directly into the ion source (without chromatograph separation). So you can avoid chromatography in this step. In this way, analyte is constantly ionized and the change in signal during parameter changes are directly kept track of. Ideally, a chromatographic method is first developed for the analytical problem. First, however, no MS coupling is necessary, as a diode-array detector (DAD) is usually placed before the MSD. The optimized method shows in which eluent mixture and at what flow rate the analyte reaches the ionization chamber.

With this eluent mixture a diluted standard solution is prepared, which is used for the optimization of the MS-parameters by direct infusion or FIA.

For optimal ionization the following parameters must be adjusted:

ESI	Gas flow	APCI	Gas flow
	Cone gas		Cone gas
	Temperature		Temperature
	Nebulizer pressure		Nebulizer pressure
			Corona voltage

As start-up parameters, values proposed by the device manufacturer can also be accepted. These are usually also given as value ranges and can well be used for mass scans. However, they must always be adjusted to the analyte for quantitative analysis. Depending on the device or device manufacturer, the above-mentioned parameters can vary.

Suitable Buffers

The buffering of the eluent facilitates the charge transfer during ionization reactions.

The choice of buffer used depends on the p*K*s of the substances to be analyzed, that is, the analyte should arrive in the spray chamber in ionized form.

Conventional buffers for LC-MS are:

Trifluoroacetic acid	pK_a 0.3	Ammonia	pK_a 9.2
Formic acid	pK_a 4.76	Hydrogen carbonate	pK_a 10.3
Acetic acid	pK_a 3.75	1-Methyl piperidine	pK_a 10.2
Ammonium acetate	pK_a 9.2	Trimethylamine	pK_a 10.7
Ammonium formate	pK_a 9.2	Pyrrolidine	pK_a 11.3

Each buffer area ranges from $pK_a - 2$ to $pK_a + 2$.

The most important feature of the compatible buffer for MS-coupling is its volatility. Other non-volatile buffers such as tris (hydroxymethyl)-aminomethane cannot get into the vapor phase and contaminate the machine or suppress the sensitivity of the measurement.

The use of TFA should be avoided whenever possible. TFA can suppress ionization and lead to insensitive results.

The concentration of the buffer should be kept as low as possible (max. 10 mM).

For formic and acetic acid concentrations of 0.1–1.0% are customary, while a maximum of 0.1% is added to TFA.

Example of a Simplified Identification and Quantification in MSD Data Analysis

Premise

- Only two analytes (Figure 8.23)
- Confirmation of assumed compounds by "plausible" fragmentation (simplest case)
- Use of a single quadrupole mass spectrometer
- Matrix is water
- Quantification by external standard

Procedure

- First, an HPLC method is developed for the separation of two compounds using standards in water (diode-array DAD UV-VIS detector).
- After LC method optimization, the compounds are selectively detected, identified and quantified according to their mass-to-charge ratios (m/z).

1. **HPLC-Separation and UV/VIS-DAD Detection**

   For details about HPLC technique see Section 8.1.3.

   In Figure 8.24 the chromatogram of the separated mixture can be seen (λ 254 nm). For luteolin-7-glycoside the retention time (R_T) is 5.0 minutes and for luteolin it is (R_T) 15.6 minutes.

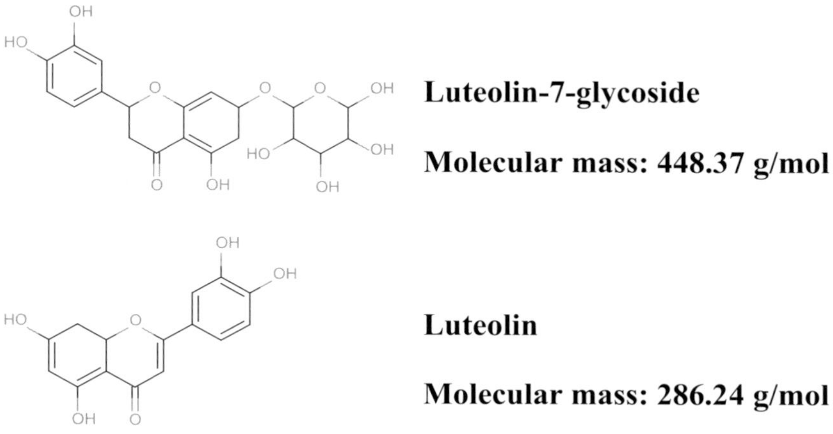

Luteolin-7-glycoside

Molecular mass: 448.37 g/mol

Luteolin

Molecular mass: 286.24 g/mol

Figure 8.23 Characteristics of the examined substances.

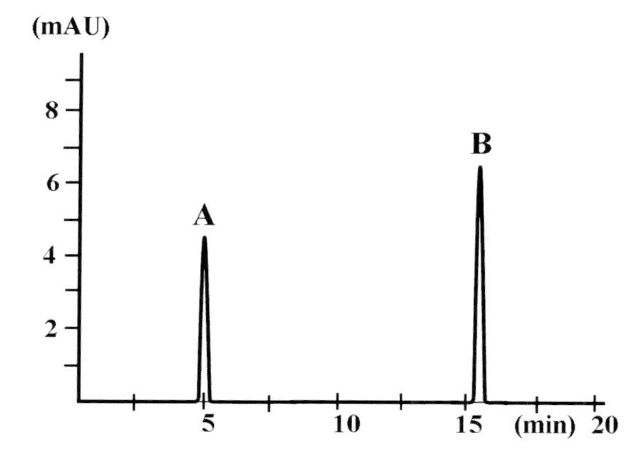

Figure 8.24 Chromatographic separation of the mixture luteolin-7-glycoside (A)/luteolin (B) standards in water.

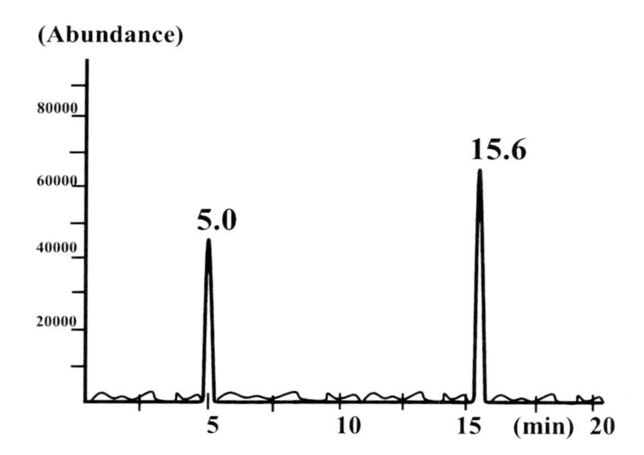

Figure 8.25 Full SCAN signal of the luteolin-7-glycoside/luteolin mixture.

2. LC/MS Measurements

Full SCAN Analysis (Total Ion Monitoring) The mixture A/B (standards in water) is now analyzed with the LC/MS system using the optimized LC chromatographic parameters. Two signals were recorded (DAD and full SCAN). In full SCAN mode (or total ion chromatogram–TIC) the intensities of all fragments are plotted against time (chromatogram over all masses from m/z 200 to 800) as shown in Figure 8.25. Abundance, when viewed as similar to absorbance displayed on a UV detector, the vertical increase in the signal above background indicates an increased occurrence of the total ions present. From the substance-specific retention times, the sample signals can be related to the two substances in the SCAN mode.

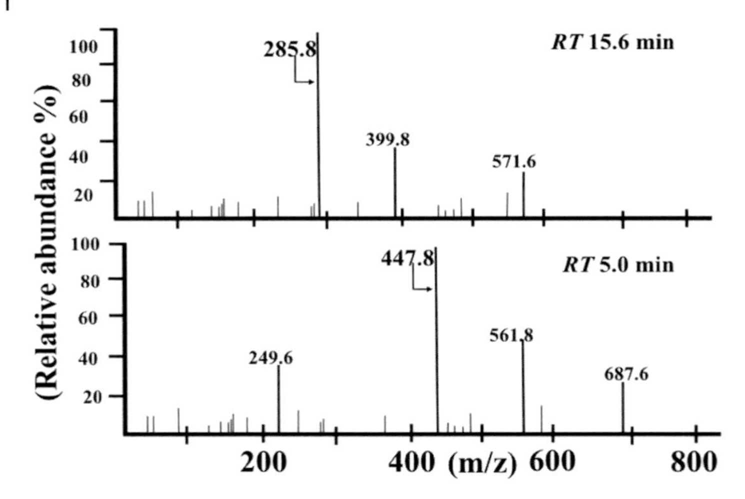

Figure 8.26 Line spectra of the substances extracted from the Full SCAN-chromatogram (R_T 5 min = luteolin-7-glycoside, R_T 15.6 min = luteolin).

Analytical conditions

Ionization:	API-ESI (atmospheric pressure ionization-electrospray-ionisation)
Dry gas:	12 l/min
Dry temperature:	350°
Nebulizer:	60 psig
Full scan:	(*m/z* 200–800)
Fragmentor:	50
Capillary voltage:	4000 V

Substance-Specific Ion Fragments (Diagnostic Ions) From the Full SCAN Mode (or TIC)-chromatograms, the mass spectra of the substances can be extracted and used to determine the characteristic fragments (or molecular ion) of the substances (diagnostic ions). The expected ions [M_1-1] are present in both line spectra and in both cases it is the most intensive ion; for luteolin-7-glycoside = m/z 447.8 and for luteolin = m/z 285.8.

The fragments or molecular ions are weighted as to their plausibility and used for quantification.

Figure 8.26 shows the line spectra from the peak at (R_T) 5.0 minutes and (R_T) 15.6 minutes.

Identification and Confirmation of Luteolin (Luteolin-7-Glycoside Analogously)

In its directive 96/23/EC (*Off. J. C.* Ser. L.221, 8-36, 2002) the EU requires for identification and confirmation purposes, 3-4 identification points (IP) depending on the unknown analyte.

- Matching retention time (R_T 15.6 minutes) results in one point of identification.
- Fragment m/z 285.8 [M_1-1], retention time to be matched results in two points of identification in total.
- Fragment m/z 571.6 [2. (M_1-1)] (see Figure 8.26), retention time to be matched results in three points of identification in total.

Quantification Using the Selected-Ion Monitoring (SIM) Method

For quantification, the simply deprotonated molecular ion [M_1-1], found as the most intensive fragmented ion for both compounds, is used. The peak areas of the SIM signals at m/z 447.8 and m/z 285.8 (simultaneously recorded) are integrated and a calibration line can be calculated using various standard concentrations (Figure 8.27).

Real-Life Example

In contrast to the simplified example illustrated in Figure 8.27, here is an example from real life. It shows the great advantage of the mass selective detector (MSD). Single ion measurements which are highly selective enable quantitative analysis that avoids the effects of matrix even when separation by LC is inadequate (see Figure 8.28).

In terms of providing both broad applicability for a wide range of substances and high selectivity, mass spectrometers offer excellent characteristics as an LC detector.

8.1.4.3 Material and Methods

Equipment

Shimadzu Nexera UHPLC, Shimadzu LCMS-2020 Single Quadrupole

LC Conditions

(SHIMADZU LC-MS Application data Sheet No. 016)

Flow	0.8 ml min^{-1}
Injection volume	20 µl
Column temperature	40 °C
Column	STR ODS-II (4.6 mm I.D. × 150 mm)
Mobile Phase	A: 1.0 ml of formic acid is added to 1 l of water and dissolved. Then, the solution is degassed in an ultrasonic bath for 10 min.
	B: 1.0 ml of formic acid is added to 1 l of methanol and dissolved. Then, the solution is degassed in an ultrasonic bath for 10 min.

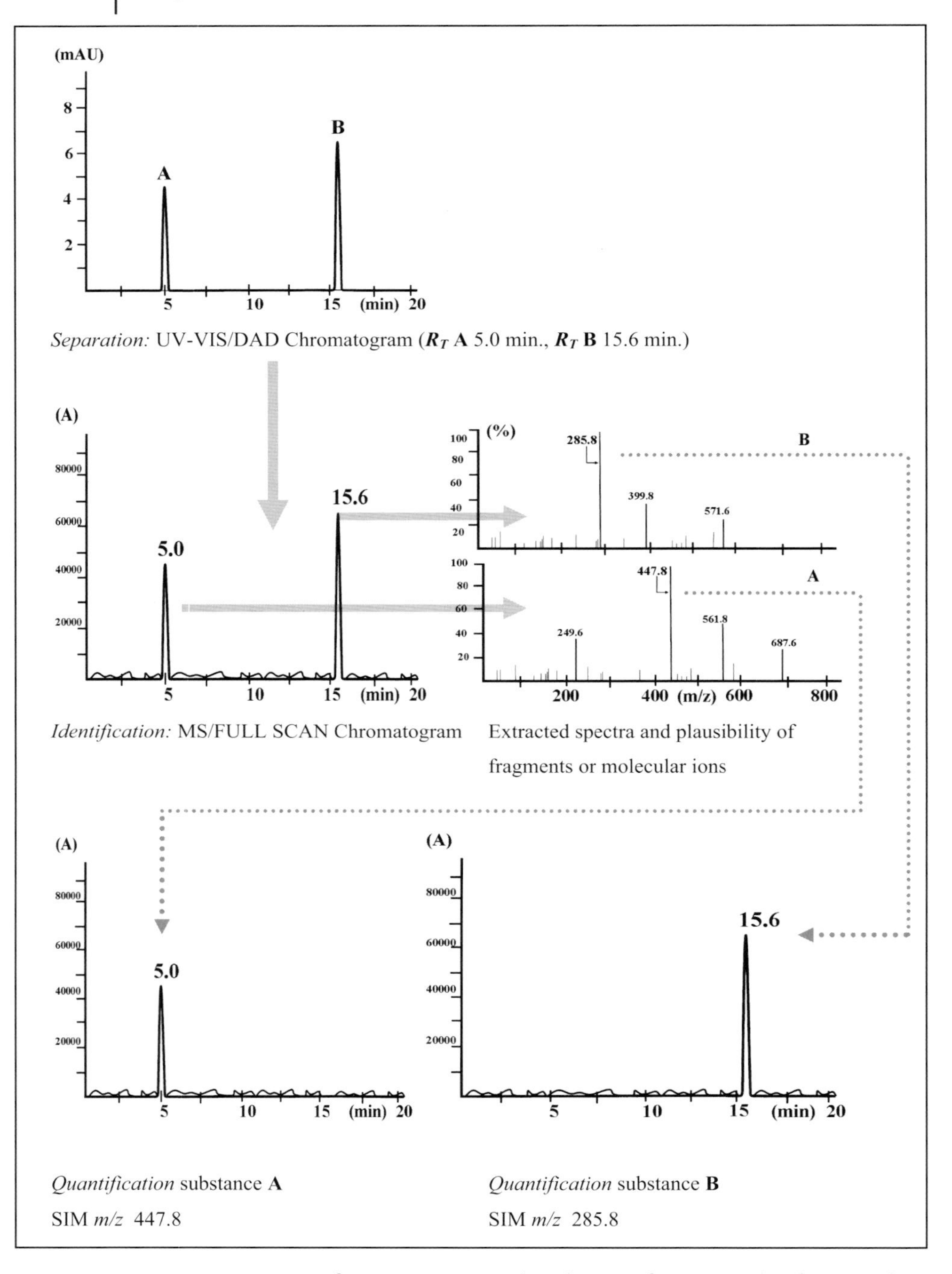

Figure 8.27 Overview of a systematic approach in the case of an LC/MS identification and quantification. (a) Luteolin-7-glycoside, (b) luteolin.

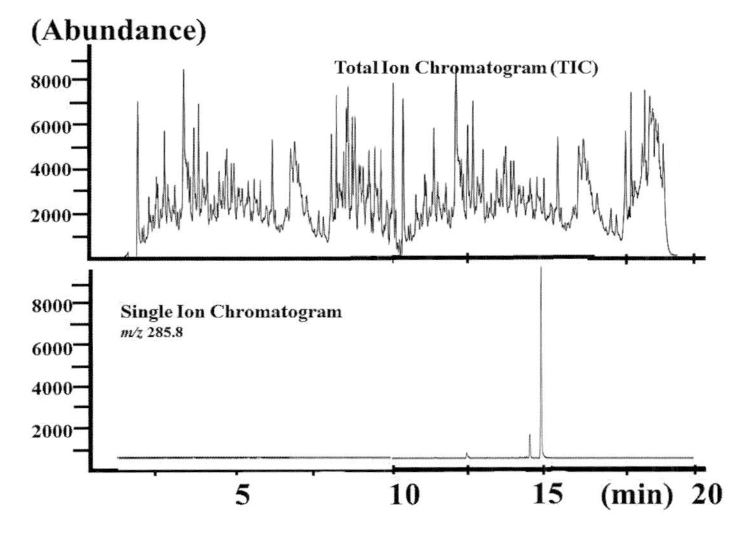

Figure 8.28 MSD chromatogram analyzed in a complex matrix. Note the clean-up from background signal.

MS Conditions

(SHIMADZU LC-MS Application data Sheet No. 016)

Source	APCI
Mode	positive
Sample voltage	+3.5 kV (APCI-Positive mode)
Sample temperature	400 °C
Nebulizing gas flow	$2.5\,l\,min^{-1}$
CDL voltage	−50 V
DEFs voltage	+55 V

Chemicals

Methanol	CAS- Number 67-56-1	T,F
Formic Acid, MS grade	CAS- Number 64-18-6	C
Caffeine	CAS- Number 58-08-2	Xn
Theobromine	CAS- Number 83-67-0	Xn
Hydrochloric acid	CAS- Number 7647-01-0	C

Standard Solutions

Stock Solution

$20\,mg\,l^{-1}$ of caffeine and theobromine in methanol/water 1/1 (v/v)

Working Solutions

20 ppb caffeine and theobromine in methanol/water 1/1 (v/v)
200 ppb caffeine and theobromine in methanol/water 1/1 (v/v)
500 ppb caffeine and theobromine in methanol/water 1/1 (v/v)
700 ppb caffeine and theobromine in methanol/water 1/1 (v/v)
1000 ppb caffeine and theobromine in methanol/water 1/1 (v/v)

Experiment

After optimization of the LC-method, this is to be transferred to the LC-MS system.

(A) Optimization of the LC-MS method

Optimize the chromatographic conditions to separate the standard mixture. As detector use the DAD.

Inject the stock solution and conduct measurements using the following parameters and gradient program:

Flow:	0.8 ml min^{-1}
Injection volume	20 µl
Column temperature	40 °C

Time (min)	A (%)	B (%)
0	70	30
5	50	50
20	30	70

Optimize the gradient program, if necessary. The method should be as short as possible.

(B) Determination of the mass spectra of caffeine and theobromine

Analyze the stock solution with the optimized gradient and the specified MS-start parameters. In this case, you must conduct measurements in the TIC mode.

Determine the molecular ion for each compound. What masses do you also see in the mass spectra?

Knowing the specific molecular ions, the analysis can now be conducted in SIM (single ion monitoring) mode.

(C) Optimization of MS parameters by flow injection analysis (FIA)

After optimization of the chromatographic method, the MS parameters are optimized by FIA of the standard solution of the analyte.

For this purpose, the 200 ppb standard solution is injected and a flow of 0.8 ml min^{-1} is applied.

The optimization is performed for each substance in the corresponding SIM.

Assuming the above start parameters, try to achieve a signal increase by deliberately changing these parameters.

ESI	Gas flow	APCI	Gas flow
	Cone gas		Cone gas
	Temperature		Temperature
	Nebulizer pressure		Nebulizer pressure
			Corona voltage

What do you observe with respect to the ionization of the substances under changed parameters? What are the appropriate parameters?

(D) Determination of caffeine content in hair

D.1 Conduct a calibration by using an external standard and determine a calibration curve of the SIM signal of each substance.

What can be said about linearity?

D.2 Now determine the content of caffeine and theobromine in the hair, by performing the following sample process:

50 mg hair sample is weighed in a test tube and extracted with 5 ml of methanol/5 M hydrochloric acid solution (20: 1). After filtration, the solution is evaporated to dryness and the residue is dissolved in 0.5 ml of water.

A threefold determination is conducted.

D.3 Validate the method.

8.1.4.4 Questions

(1) Discuss the results of the hair sample you have determined.
(2) In what way do temperature, voltage and gas flow influence ionization? Justify your answers for each of the three parameters.
(3) How does the ionization behave when the buffer concentration is changed?
(4) How can the sample matrix be reduced?
(5) What is the difference between MALDI-MS and ESI-MS?
(6) How do the various sources of ionization from various equipment manufacturers differ in their structure?
(7) On what do bond cleavage in collision cells (MS/MS) depend?
(8) What do you understand by signal saturation?
(9) What possible forms of contamination are there in an LC/MS system?

References

1 de Hoffmann, E., and Stroobant, V. (2007) *Mass Spectroscopy: Principles and Applications*, 3rd edn, Wiley-Interscience, ISBN-13: 978-047003311-1.
2 Harrison, G. (1983) *Chemical Ionisation Mass Spectroscopy*, CRC Press, Boca Raton, FL, USA, ISBN 0-8493-5616-4.
3 Karst, U. (2005) *Analytical and Bioanalytical Chemistry*, **382**, 1744.
4 McCulloch, R.D., Robb, D.B., and Blades, M.W. (2008) A dopant introduction device for atmospheric pressure photoionisation with liquid chromatography/mass spectrometry. *Rapid Communications in Mass Spectrometry: RCM*, **22**, 3549–3554.
5 Boyle, J.G., and Whitehouse, C.M. (1992) TOF-MS with an ESI ion beam. *Analytical Chemistry*, **64**, 2084.
6 Berger, U., Kölliker, S., and Oehme, M. (1999) Structure elucidation and quantification by HPLC ion trap multiple mass spectroscopy. *Chimia*, **53**, 492.

8.1.5
Ion Chromatography (IC), Project: "Water Is Life"

8.1.5.1
Analytical Problem

Drinking water is our most precious natural resource. It comes mainly from groundwater and surface water. Surface water consists of lake and reservoir water, bank filtrate, enriched groundwater and river water.

Drinking water must comply with the following DIN 2000 basic requirements:

It should be colorless, clear, cool and free from any foreign odor and taste. It should be as free as possible in its natural state of pathogens and harmful substances. It should not contain too many salts, especially hardening salts, iron, manganese, and organic materials (peat and humic substances). In addition, every precaution should be taken to ensure that it does not cause corrosion. The quantity of drinking water should be adequate to the needs of the population served. Water Technology Ltd in Milan has been commissioned to assess potential impacts on areas where drinking water is extracted. The motivation for this step was primarily an outdated and vulnerable activated charcoal filter system and various renovations to the drinking water network. The water analysis performed at the Water Technology Ltd should show whether there is a change in water quality and whether the statutory limits are being adhered to.

Object of Investigation/Investigation System
Drinking water quality of the city of Milan and its surroundings

Samples
Tap water collected at various sites and commercial mineral water (produced in Milan)

Examination Procedure
The most important application of ion chromatography (IC) today is the routine investigation of aqueous systems, in which the analysis of drinking water plays a

central role. In addition, the IC is used for elemental species analysis of anionic elements that occur as complexes, whereby predominantly environmentally relevant issues are dealt with. The third large area of application of ion chromatography is ultra-trace analysis in high-purity process chemicals, especially in the semiconductor industry. Today, the term ion chromatography comprises all fast liquid chromatographic systems for the separation of inorganic and organic substances that can decompose in water into cations or anions. Since 1979 it has been possible to conduct ion analyses even without a suppressor system using a conductivity meter, whereby ion exchangers with low exchange capacity can be used with low conductivity eluents.

Analytes

Fluoride, nitrate, nitrite, sulfate, fluoride, bromide, phosphate

Keywords

Suppression, suppressor, adsorption, distribution, size exclusion, affinity, ion exchange, ion-pair formation, ion-exclusion, detection in IC

Bibliographical Research (Specific Topics)

Fritz, J.S., and Gjerde, D.T. (April 14, 2009) *Ion Chromatography*, 4th Completely Revised and Enlarged Edition, Wiley-VCH Verlag GmbH, ISBN-13: 978-3527320523.

Weiss, J. (January 3, 2005) *Handbook of Ion Chromatography*, 3rd Revised Updated Edition, Wiley-VCH Verlag GmbH, ISBN-13: 978-3527287017.

Eith, C., Kolb, M., Rumi, A., Seubert, A., and Viehweger, K.H. (eds) *Practical Ion Chromatography, An Introduction, Metrohm Monograph*, 2nd edn (Source of the theoretical part of this project).

Project Process/Task

(A) Selectivity of separation columns

If the strength of the eluent is plotted against the logarithm of the retention times of mono- and divalent ions then a straight line is obtained. The slope of this line is greater for divalent ions than for monovalent ions. The increase in the elution strength of the eluent therefore has a greater influence on the retention time of the divalent ions than on the retention time of monovalent ions. The influence on the sulfate ion can be observed. As the eluent concentration increases the elution of sulfate is accelerated more than that of the monovalent ions. In general divalent eluent ions are stronger eluents than monovalent ones as they can form two bonds with the stationary phase and therefore enter into stronger interactions with it.

At the same molarity, sodium hydroxide has a higher pH than carbonate/hydrogen carbonate. The elution power of the OH^- ion is lower than that of the carbonate as the hydroxide ion does not interact as strongly with the stationary phase. The retention time of the phosphate ion depends strongly on the pH value. At high pH values the equilibrium is displaced from HPO_4^{2-} towards PO_4^{3-}. If sodium hydroxide is added to a sodium carbonate eluent, this results in the retention time of the phosphate being increased while the retention times of all other ions are shortened.

(B) Determination of anions in drinking water and mineral water

B.1 Investigation of food No. 1 = drinking water

Compliance with the limits (fluoride, nitrate, nitrite and sulfate) in tap water should be checked.

B.2 Furthermore, the information on commercially bottled mineral water should be checked.

(C) Method validation

Linear range, detection limit, quantification limit, reproducibility are to be determined.

(D) Project reporting

The results are to be submitted in the form of a report (or presentation). Comparisons with bibliographical references are to be made.

8.1.5.2 Introduction

The beginnings of ion chromatography (IC) or, more exactly, ion exchange chromatography go back to the middle of the previous century. Between 1935 and 1950 knowledge about ion exchangers and their applications was considerably extended by the "Manhattan Project". In the 1950s and 1960s theoretical models for understanding the phenomenon of ion exchange and of ion chromatography, which is based on this, were worked out. Continuous detectors were used in the 1970s; this allowed the jump from low-pressure to high-pressure or high-performance chromatography to be completed. The term "ion chromatography" was coined in 1975 with the introduction of detection by conductivity combined with a chemical reduction in conductivity by Small, Stevens and Baumann; it was subsequently used as a trade name for marketing purposes for a long time. In the meantime the abbreviated term ion chromatography has become established as the superordinate term for the ion exchange, ion exclusion and ion pair chromatography methods included under high-performance liquid chromatography (HPLC) [1]. IC today is dominant in the determination of anions while the atomic spectrometry methods, commonly used for the determination of cations, are hardly useful for determining the electronegative anion formers of the fifth to seventh main groups of the periodic system. The most important field of application today for anion chromatography is the routine investigation of aqueous systems; this is of vital importance in the analysis of drinking water [2–4]. IC is also used for the analysis of the element species in anionic elements or complexes; this is mainly for solving environmentally relevant problems. The third important field of application for anion chromatography is ultratrace analysis in ultrapure process chemicals, required chiefly in the semiconductor industry.

Today the ion exchangers normally used for HPLC consist of spherical polymer particles with a diameter of about 5 to 15 μm. Various methods are used to attach so-called anchor groups to the surface of the polymer; these are used as spacers between the basic polymer and real functional groups. These normally consist of quaternary ammonium ions which are chemically attached to the anchor groups. The total number of functional groups is known as the exchange capacity; this is a basic characteristic of ion exchangers.

Commercial packing materials for anion chromatography are of a low-capacity nature with exchange capacities of 50 to 100 μmol per separating column. The reason for this is the dominant application of conductivity detection, which is all the more sensitive the lower the inherent conductivity of the elution system. Low-capacity anion exchangers allow the use of very dilute aqueous solutions of NaOH or carbonate buffers, whose inherent conductivity can even be additionally reduced by chemical suppression [2, 4].

In anion chromatography it is chiefly functional groups of Type I (trimethylammonium, TMA) and Type II (dimethylethanolammonium, DMEA) which are now used. As the real interaction between the stationary phase and the analyte anions takes place at the functional group this means that its structure has a decisive influence on the selective behavior of packing materials. According to current knowledge the polarity of the functional groups, which can be controlled by the number of hydroxyethyl residues ($-CH_2CH_2OH$) on the quaternary nitrogen, is of particular importance [2, 4].

The term ion chromatography includes all separations of ionic species within HPLC with on-line detection and is, therefore, largely independent of apparatus limitations [5]. IC has developed into the method of choice in anion analysis thanks to the wide range of separating columns, elution systems and detectors which are now available. The reason for this is that only a few separation processes exist for anions; these are hardly suitable for practical use. Gravimetric and volumetric methods are limited by their sensitivity and their selectivity. Even the meteoric development of gas chromatography from 1965 onwards did not have any great advantages for anions, as the non-volatile ions needed to be derivatized first and the sensitivity did not meet the demands placed on trace analysis today [6]. For the analysis of cations powerful atomic spectrometry alternatives to IC exist, e.g., ICP-AES/MS, so that the value of cation chromatography is considerably less than that of anion chromatography. However, cation chromatography has achieved a certain importance in the analysis of alkali and alkaline earth metals and in the determination of ammonium-nitrogen (drinking water analysis). In the speciation of ionic compounds IC in combination with element-specific detectors is indispensable. A good overview of the applications of IC in various sectors is given in the works of Haddad *et al.* and Weiss [2, 4].

Modern Liquid Chromatography (LC)

Liquid chromatography (LC) is to be regarded as the generic term for numerous modern liquid chromatography separation methods. It can be used for a wide range of different substances, is characterized by its excellent analytical performance and is probably the most important separation method used in modern analytical chemistry [3]. Liquid chromatography also includes ion chromatography (IC).

HPLC is a logical further development of the classical liquid chromatography. In classical LC, introduced by Twett in 1906, glass columns with a diameter of 1 to 5 cm and a length of up to 500 cm were used; these were filled with separation phases with particle sizes of 150 to 200 μm. Even separations of simple mixtures of substances could often take several hours with an average separation performance. As a result of the understanding of the chromatography process which was later developed, it became clear that increased performance could only be achieved by a dramatic reduction of the particle diameter in the stationary phase; however, this placed completely new demands on the equipment used for chromatography.

Since about 1970 special and powerful instrument technology has become available which is able to overcome the high counterpressures of 10 to 50 MPa which occur when packing materials with a particle diameter of 3 to 10 μm and separating columns of 125 to 250 mm length × 4 mm ID are used.

As a result of the dramatic miniaturization HPLC has developed into a purely analytical separating method; in contrast classical LC is today practically only used for preparative purposes. The advantages of HPLC in comparison with classical LC are chiefly:

- excellent chromatographic efficiency
- continuous working process
- on-line detection of the separated substances
- high sensitivity and reproducibility
- utilization of the retention time for qualitative identification of substances
- short analysis times

Separation Principles in LC

HPLC can be differentiated according to the different physicochemical interactions between the substances in a sample and the stationary phase. Although in reality there are usually several different mechanisms responsible for a successful separation, a rough classification according to the following separation mechanisms is possible:

- adsorption
- distribution
- size exclusion
- affinity
- ion exchange
- ion pair formation
- ion exclusion

Adsorption chromatography is defined by interfacial reactions, in which liquid or gaseous substances are enriched at a solid phase. Various models are available for providing a qualitative and quantitative description of adsorption processes. Two different techniques are used. In normal phase chromatography the stationary phase is usually silica gel and, therefore, considerably more polar than the mobile phase (hydrocarbons). In reversed phase chromatography (RPC) the conditions are exactly the opposite. For practical reasons, which chiefly concern eluent handling, virtually only RPC is used today [3].

In *distribution chromatography* the stationary phase is a liquid which is immiscible with the mobile phase. Separation is based on the differing solubilities of the analytes in the two phases. In an ideal case the Nernst distribution law applies. This separating mechanism plays an important role, particularly in gas chromatography when capillaries coated with separating liquids are used as the stationary phase. Distribution chromatography may also occur in HPLC if silica gels modified with nonpolar hydrocarbons, e.g. so-called octadecyl phases, are used as the separating material.

Size exclusion chromatography (SEC) allows separation according to the molecular size as a result of sieve effects. Silica gels or organic polymer resins with a defined pore structure are used as the stationary phase. Smaller analytes can diffuse into the pores and are retarded. As the molecule size increases any interaction with the pores becomes less likely, until at a particular size molecules are completely excluded and practically elute in the dead volume. SEC is widely used in polymer analysis and bioanalysis.

Affinity chromatography allows the separation of mixtures of substances by selective or specific interactive forces. Highly specific interactions can be observed between antibodies and antigens (key–keyhole principle), as well as with enzymes and their substrates in particular. In practice, enzymes or antibodies are chemically immobilized on a stationary phase. If there is a corresponding substrate or antigen in the sample then this is retarded with extreme selectivity. This is why bioaffinity chromatography is indispensable in the active substance analysis sector (pharmacology).

Theoretical Concepts for Describing Chromatography

- The model of theoretical plates (see Section 7.1.1)
- The dynamic theory, van Deemter Theory (see Section 8.1.3)

Basic Principles of Ion Chromatography (IC)

Terminology and Classification in LC

Ion exchange chromatography or ion chromatography is a subdivision of HPLC. According to IUPAC ion exchange chromatography is defined as follows [7]:

> In ion exchange chromatography separation is based on differences in the ion exchange affinities of the individual analytes. If inorganic ions are separated and can be detected by conductivity detectors or by indirect UV detection then this is also called ion chromatography.

For several reasons this definition is an unhappy choice. The detection technique should be considered separately from the existing separation mechanism. Apart from this, a limitation of the term "ion chromatography" to inorganic ions is difficult to understand as, in practice, in any given system both organic and inorganic ions can be separated and identified simultaneously.

An older, more general definition is more suitable for defining ion chromatography:

> *Ion chromatography includes all rapid liquid chromatography separations of ions in columns coupled on-line with detection and quantification in a flow-through detector.*

This definition characterizes ion chromatography irrespective of the separation mechanism and detection method while at the same time distinguishing it from classical ion exchange. The following separation principles apply in ion chromatography:

- ion exchange
- ion pair formation
- ion exclusion

Chromatography methods are defined by the chief separation mechanism used. Today ion exchange chromatography is simply known as ion chromatography (IC), while ion pair chromatography (IPC) and ion exclusion chromatography (IEC) are regarded as being more specialized applications.

Ion Exchange

Ion exchange chromatography is based on a stoichiometric chemical reaction between ions in a solution and a normally solid substance carrying functional groups which can fix ions as a result of electrostatic forces. In the simplest case in cation chromatography these are sulfonic acid groups, in anion chromatography quaternary ammonium groups. In theory, ions with the same charge can be exchanged completely reversibly between the two phases. The process of ion exchange leads to a condition of equilibrium. The side towards which the equilibrium lies depends on the affinity of the participating ions to the functional groups of the stationary phase.

Ion exchangers normally consist of solid phases on whose surface ionic groups are fixed. Because of the condition of electroneutrality there is always an oppositely-charged counter-ion in the vicinity of the functional group. The counter-ion usually originates from the mobile phase and is, therefore, also known as the eluent ion.

If a sample is added which contains two analyte ions A^- and B^- then these briefly displace the eluent ions E^- and are retained at the fixed charges before they are in turn exchanged for eluent ions. For anion chromatography this results in the following reversible equilibria:

$$\text{resin}-N^+R_3E^- + A^- \leftrightarrow \text{resin}-N^+R_3A^- + E^- \tag{8.29}$$

$$\text{resin}-N^+R_3E^- + B^- \leftrightarrow \text{resin}-N^+R_3B^- + E^- \tag{8.30}$$

The different affinities of A^- and B^- to the functional groups mean that separation is possible. The equilibrium constant K is also known as the selectivity coefficient.

If it can be assumed that the concentration of the eluent ions is normally higher than that of the analyte ions by several powers of ten then $[E^-]$ can be regarded as being a constant in the mobile and stationary phases. This means that the distribution coefficient D_A (Eq. 8.31) and the retention factor k'_A (Eq. 8.32) can be calculated.

$$D_A = \frac{[A]_S}{[A]_M} \tag{8.31}$$

D is defined as the ratio of the concentrations of a substance A between the stationary (index $_S$) and the mobile phases (index$_M$)

$$k' = D \cdot \frac{V_S}{V_M} = \frac{t_S}{t_M} = \frac{u \cdot t_R}{L} - 1 \tag{8.32}$$

The retention factor is mathematically defined as the product of the distribution coefficient D and the phase volume ratio V_S/V_M between the stationary and mobile phase, or as the ratio of the net retention time t_S to the dead time t_M. A calculation over the length L of the migration path and the speed u of the mobile phase is also possible (Eq. 8.32).

Strictly speaking, such calculations are only permissible if the concentrations correspond to the activities; however, this is only the case for an infinite dilution. In principle the activities of the ions in the stationary phase are inaccessible [4]. For the most frequently used ion exchangers of low capacity, which can only be used as the mobile phase with very dilute electrolytes, the activities are simply disregarded. Such very coarse approximations are no longer valid for high-capacity (>200 mmol g^{-1}) packing materials and concentrated eluents; these show clear variations from the "ideal" behavior.

Ion Pair Formation

With the aid of ion pair chromatography it is possible to separate the same analytes as in ion exclusion chromatography, but the separation mechanism is completely different. The stationary phases used are completely polar reversed phase materials, such as are used in distribution chromatography. A so-called ion pair regent is added to the eluents; this consists of anionic or cationic surfactants such as tetraalkylammonium salts or n-alkylsulfonic acids. Together with the oppositely charged analyte ions the ion pair reagents form an uncharged ion pair, which can be retarded at the stationary phase by hydrophobic interactions. Separation is possible because of the formation constants of the ion pairs and their different degrees of adsorption.

Ion Exclusion

Ion exclusion chromatography is mainly used for the separation of weak acids or bases [2, 4]. The greatest importance of IEC is for the determination of weak acids, such as carboxylic acids, carbohydrates, phenols or amino acids. In IEC a completely sulfonated cation exchanger whose sulfonic acids groups are electrically neutral with protons as counter-ions is frequently used as packing material. In aqueous eluents the functional groups are hydrated. The hydrate shell is limited by an (imaginary) negatively charged membrane (Donnan membrane). It is only passable by uncharged, non-dissociated molecules such as water. Organic carboxylic acids can be separated if strong mineral acids such as sulfuric acid are used as the mobile phase. Due to their low acid constants (pK_A values) the carboxylic acids are present almost completely in a non-dissociated form in strongly acidic eluents. They can pass through the Donnan membrane and be adsorbed at the stationary phase, whereas the sulfate ions of the completely dissociated sulfuric acid are excluded.

Electrochemical Detection Methods

Conductivity Detection

Conductivity detection, also known as conductometric detection, has a market share of about 55% in the ion chromatography sector [4]. If the number of ion chromatographs which have been sold is taken into account then this share is probably much higher today. Conductivity detection is a non-selective detection principle; in this case both direct and indirect detection determinations are possible. As aqueous electrolytes are frequently used as the mobile phase in ion chromatography, the detector must be able to respond to the relatively small changes in the total conductivity of the eluent caused by the analyte ions. By the use of so-called suppression techniques the inherent conductivity of certain eluents can be dramatically reduced; in the case of strong acid anions it is thus possible to considerably improve the sensitivity.

In ion chromatography the conductivity of an eluate can be determined either directly or after passage through a suppressor. These versions are known as IC without or with chemical suppression. The version which is to be preferred can be determined by carrying out a rough calculation.

Suppressors for ion chromatography can be arranged either as discontinuously working packed-bed suppressors or as continuously working membrane suppressors (Figure 8.29).

The packed-bed suppressor in the revolver version used by Metrohm has three identical suppressor units; while one unit acts as the suppressor the second is regenerated and the third is rinsed with ultrapure water. After an analysis has been carried out the revolver is rotated by 120 degrees and the column which has just been rinsed is used as the suppressor. This makes quasi-continuous work possible.

Amperometric Detection

In principle, voltammetric detectors can be used for all compounds which have functional groups which are easily reduced or oxidized. The amperometric detector is the most important version. In this detector a certain potential is applied

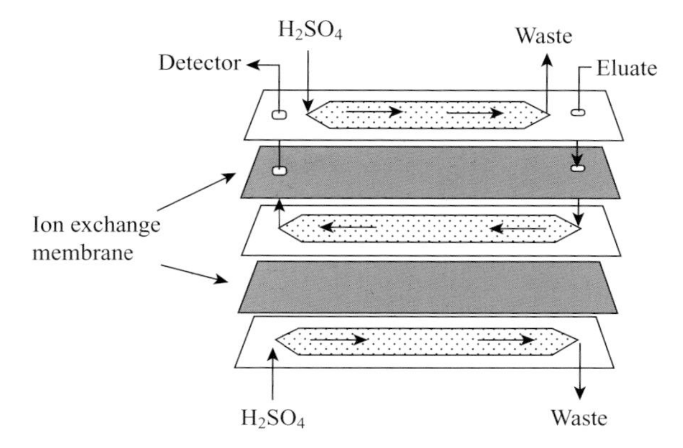

Figure 8.29 Schematic construction of a continuously working membrane suppressor.

between a working electrode and a reference electrode. If an electrochemically active analyte, whose half-wave potential is such that the applied potential causes reduction or oxidation, now passes between the electrodes then a current will flow; this represents the measuring signal. Amperometry is very sensitive; the conversion rate is only about 10%. Apart from a few cations (Fe^{3+}, Co^{2+}) it is chiefly anions such as nitrite, nitrate, thiosulfate, as well as halides and pseudohalides which can be determined in the ion analysis sector. The most important applications lie, however, in the analysis of sugars by anion chromatography and in clinical analysis. Owing to its different working principle, the coulometric detector provides a quantitative turnover without, however, there being any increased sensitivity.

Potentiometric Detection

In potentiometric detection ion-sensitive electrodes are used, some of which have a very high selectivity. Despite their necessary high degree of miniaturization the sensors must function reliably; this still causes problems in practice. This is the reason why up to now potentiometric detection, in the ion chromatography sector, is limited to a few special applications.

Spectroscopic Detection Methods

Photometric Detection

Because of its extremely wide range of application photometric or UV/VIS detection is the most important detection method used in HPLC, as virtually all organic molecules contain chromophore groups which are able to absorb in the UV or VIS spectrum. A requirement is that the eluent used does not absorb in the range of wavelengths used. With direct detection at the maximum absorption of an analyte UV/VIS detection is practically selective. Substances which either only have a limited absorption or no absorption at all in the particular wavelength range can be determined indirectly by measuring the maximum absorption of the elution system. In the field of inorganic ion analysis UV/VIS detection plays a smaller role. While of the simple anions only analytes such as nitrate, bromide or iodide absorb, important analytes such as fluoride, sulfate or phosphate can only be measured indirectly [4]. Many cations do not absorb at all, but multivalent and transition metals, in particular, can be converted in a post-column derivatization with chelate formers such as 4-(2-pyridylazo)-resorcinol (PAR) or Tiron to form colored complexes. Redox-active analytes such as bromate and other oxohalides can be analyzed by UV/VIS detection after undergoing a post-column reaction with an electrochemically active indicator.

Fluorescence Detection

Fluorescence detection is very sensitive and is always possible whenever analytes can be excited to fluoresce; this is mainly the case for organic compounds with extended π-electron systems. This means that typical applications are found in the fields of clinical and organic analysis. In connection with ion chromatography, fluorescence detection is used in a few special cases, as only particular ions such as Ce^{3+} are directly accessible and non-fluorescing ions can only be detected after derivatization. It is extremely difficult to develop elution systems for this detection method because of its great susceptibility to interference by contaminants. Furthermore, the linear range of the method is relatively small (often less than two powers of ten) owing to self-absorption effects.

Coupling Techniques

So-called *coupling techniques* represent the link-up of a chromatography system with an independent analytical method, usually spectrometry [3]. In recent years these methods have greatly increased in importance. Although the coupling of a gas chromatograph with a mass spectrometer (GC-MS) is well established, coupling HPLC with spectrometric methods causes great technical problems. In classical HPLC, that is, the analysis of organic compounds, couplings with a mass spectrometer (LC-MS), IR-spectrometer (LC-FTIR) and nuclear magnetic resonance spectrometer (LC-NMR) are available [3]. In particular, powerful atomic spectrometric detectors are used in ion chromatography (IC). Examples are atomic emission and mass spectrometry with inductively coupled plasma (IC–ICP-AES, MS); as a result of their element specificity and sensitivity these provide excellent performance data. This is the reason why, despite their relatively high costs, such systems are used for the analysis of species and in the ultratrace analysis of elements.

Refractometry

Differential refractometry is a further detection method based on an optical method. This detector is also called an RI detector (from refractive index). It is wholly nonspecific and is suitable for universal use as the quantity measured is the change in the refractive index of the pure eluent caused by the analyte. However, the great temperature sensitivity of the refractive index means that the method is very susceptible to interference. Provided that the temperature is absolutely stable the method has a linear range of three powers of ten. As simple inorganic ions have an extremely low refractive index they can only be determined indirectly by using eluents to which strongly refracting compounds have been added.

Stationary Phases in Ion Chromatography

Efficient ion chromatography requires packing materials which are made of very small particles which are as spherical as possible and have a narrow particle size distribution range. Particles with diameters from 2 to 10 μm are used. In addition the packing material should display ion exchange kinetics which are as rapid as possible. Apart from the particle this also determines the performance of ion exchangers.

Overview of Common Stationary Phases

A wide range of different organic and inorganic materials is suitable for use in ion chromatography. What they all have in common is that their surfaces bear functional groups which are able to exchange ions. The following classes of substances can be used [4]:

- modified organic polymer resins
- modified silica gels
- inorganic salts (e.g., polyphosphates)
- glasses
- zeolites
- metal oxides (e.g., Al_2O_3)
- cellulose derivatives

Apart from these, systems with a very complex constitution are also possible, such as anion exchangers with functional groups consisting of alkali metal ions bound to a stationary phase by crown ethers. In practice, the most important representatives are based on modified organic polymer resins and silica gels. Figure 8.30 provides an overview of separation materials used in IC.

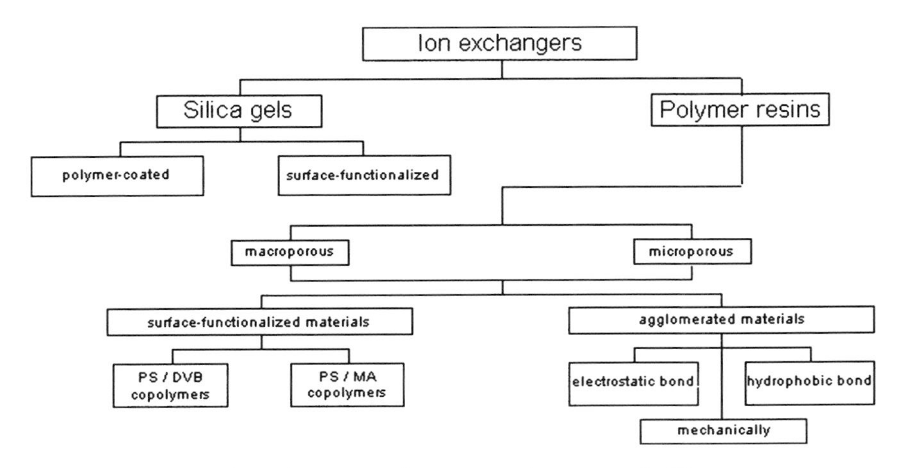

Figure 8.30 Commonly used stationary phases in ion chromatography [4].

All stationary phases can be further differentiated according to their type of application (anion or cation chromatography) or the structure of the functional group. The packing materials based on silica gel were originally used for ion chromatography. Although they have a very good separating performance and are mechanically extremely stable, their chemical instability means that they can only be used in the pH range 2 to 7.

From about 1980 onwards ion exchangers became available which were based on organic polymers and could be used in ion chromatography; these were manufactured by modifying commercially available adsorber resins. Today the packing materials used are normally based on either polystyrene-divinylbenzene copolymers (PS-DVB) or methacrylate polymers (MMA).

These two basic copolymer types differ mainly in their polarity. The PS-DVB copolymers are completely nonpolar and represent RP phases whereas the MMA polymers are relatively polar. This situation is an advantage in IC as the more polar separation phases have a lesser tendency to secondary interactions such as adsorption.

The greatest advantage of the organic polymer resins is their great chemical stability throughout the whole pH range. After initial problems had been overcome their chromatographic efficiency is similar to that of the silica gels. However, MMA phases may have a limited mechanical stability, this could limit the length of the separating column used or the maximum possible eluent flow rate.

In ion chromatography today two types of stationary phases are used which differ in principle; surface-functionalized ion exchangers and pellicular ion exchangers. In the first type the functional groups are located directly on the polymer surface or in the pores; the pellicular materials have very small particles (also surface-functionalized) which are bound to larger central particles [4]. The bonding can either be mechanical or result from hydrophobic or electrostatic interactions.

The pellicular packing materials have a higher chromatographic efficiency as the diffusion paths are kept very short owing to the greater distance of the functional groups from the base material; this results in excellent mass transfer. However, the chemical stability of these separation phases is considerably less than that of the surface-functionalized materials.

Stationary Phases for Anion Chromatography

The functional group used in anion chromatography is normally produced by converting an anchor group with a suitable amine. This produces ammonium ions which are fixed on the polymer surface. Functional groups based on nitrogen are virtually the only ones used for the separation of anions by IC. This is mainly due to their unusually good chemical stability and the almost unlimited number of possible substituents at the nitrogen atom.

The alkyl residues at the positively charged nitrogen can be varied over a wide range. In the simplest case R = H and a primary ammonium ion is formed.

However, this can be deprotonated at higher pH values and lose its charge. With this type of packing material the exchange capacity depends on the pH of the eluent, which is why they are also known as weakly basic. If the hydrogen atoms are now substituted successively by alkyl groups then first secondary and then tertiary ammonium groups are formed; these can also be deprotonated. Only when all the residues R consist of alkyl groups is the capacity or charge irrespective of the pH of the eluent and a strongly basic quaternary anion exchanger is obtained. In chromatography a pH-independent capacity is normally the aim so that only completely alkylated materials are used. For special applications such as protein analysis or preconcentration techniques weakly basic materials are also used.

The two most important functional groups in anion chromatography are derived from trimethylamine (TMA) and dimethylethanolamine (DMEA, 2-dimethylaminoethanol). Practically all commercially available separation materials use one of these two groups. TMA groups are also known in the literature as Type I groups, DMEA groups as Type II.

Stationary Phases in Cation Chromatography

In cation chromatography materials based on both silica gel and polymer materials are in use. In contrast to anion chromatography with its usual alkaline eluents the conditions for cation chromatography also allow the use of silica gels.

Cation Exchangers Based on Silica Gel

With cation exchangers based on silica gel a differentiation is made between directly functionalized materials and those which are coated by a polymer.

Practically the only reports concerning directly functionalized materials are about strongly acidic exchangers with sulfonic acid groups [2, 4]. These have a good chromatographic efficiency, but are unsuitable for the simultaneous determination of alkali and alkaline earth metals due to the large differences in affinity. With polymer-coated silica gels, the so-called Schomburg phases, the silicate surface is coated with a "prepolymer" which is then immobilized by cross-linking. By subsequent functionalization various different types of exchanger can be realized. In order to obtain weakly acidic cation exchangers polybutadiene-maleic acid (PBDMA) is used, this is then radically cross-linked *in situ*. As a result of the thin polymer layer of about 1 to 5 nm [8] the analyte diffusion paths are short so that a high degree of chromatographic efficiency is the result. A large number of applications for silica-gel-based ion exchangers exist. The simultaneous separation of alkali and alkaline earth metals is one of the most frequent applications; the separation of transition metal and heavy metal ions is also possible. However, several disadvantages prevent the universal use of ion exchangers based on silica gels:

- At pH <2 the bond between the silicon matrix and the functional group becomes increasingly weaker; this leads to a gradual erosion of the functional groups and a loss in capacity.

- At pH >7 the solubility of the silica gel increases considerably; this results in a reduction in the mechanical stability of the packing, particle break-up and a dead volume at the start of the column.

Cation Exchangers Based on Organic Polymers

Resins produced by styrene-divinylbenzene copolymerization are mainly used as the matrix. The limitations mentioned for silica gel exchangers do not exist here; cation exchangers based on organic polymers can be used over the whole pH range from 0 to 14 and are also inert to fluoride. Although their pressure resistance is less than that of silica gels it is generally still adequate, with the exception of a few methacrylate resins.

The majority of commercially available cation exchangers use sulfonic acid groups as the functional groups. These can be bound to the aromatic system either directly or via a spacer whose length is variable. Sulfonic acid exchangers with spacers have a considerably better chromatographic efficiency; the spacer length is of secondary importance. They are not suitable for the simultaneous determination of alkali and alkaline earth metals due to the large differences in affinity.

Pellicular Cation Exchangers

As well as the directly functionalized resins, pellicular exchangers are also available. They have a double layer arrangement as it is not possible to present a completely aminated substrate particle. This is why a fully sulfonated substrate particle is first surrounded by a layer of aminated and then by a layer of sulfonated latex particles. Fixing takes place via electrostatic and van der Waals interactions. The relatively large diameter of the substrate particles (10–30 μm) results in a comparatively low counterpressure. The small diameter (20–250 nm) of the latex particles allows short diffusion paths with correspondingly rapid exchange processes and, therefore, a high separating column efficiency. The disadvantage of pellicular materials is their sensitivity to organic solvents [9] and to mobile phases with a high ionic strength as in this case the latex particles are gradually displaced.

The simultaneous determination of alkali and alkaline earth metals is possible by using a modified version in which the aminated layer is covalently bound. However, the differences in selectivity between potassium, magnesium and calcium are unnecessarily large so that relatively long analysis times are required. This can be traced back to the strongly acidic sulfonic acid groups which are used.

Stationary Phases in Ion Exclusion Chromatography

The choice of stationary phases for IEC is very limited. The formation of the Donnan membrane by dissociated functional groups is important for the separation process. Their number should be relatively high. In addition, the packing material should avoid analyte adsorption on the stationary phase as much as possible. As the analytes are mainly from the carboxylic acid and sugar groups, as polar a surface as possible is the aim. Practically no demands are placed on the kinetics of the ion exchange reaction as the separation is based on an exclusion mechanism. Low cross-linked PS/DVB polymers are almost the only ones used; these are completely sulfonated.

The Meaning of the Capacity of Ion Exchangers

Apart from the sort of matrix and the type of functional group, the exchange capacity Q is the decisive quantity for characterizing ion exchangers. It provides information about the number of places which are available for ion exchange.

The exchange capacity is normally given in grams of dry packing material, either in microequivalents ($\mu eq\,g^{-1}$) or micromols ($\mu mol\,g^{-1}$) [2]. For analytical purposes ion exchangers can be divided roughly according to their capacity into:

low-capacity materials:	$Q < 100\,\mu mol\,g^{-1}$
medium-capacity materials:	$100 < Q < 200\,\mu mol\,g^{-1}$
high-capacity materials:	$Q > 200\,\mu mol\,g^{-1}$

The separation phases used in classical ion exchangers are all high-capacity types with 3 to 5 $mmol\,g^{-1}$ [4].

The above definition of the capacity is based on a complete equilibrium being attained between the stationary and mobile phases; this is why it is also known as the static capacity. In contrast, the dynamic (effective) capacity is understood to be the number of functional groups which are actually available during a chromatographic process. It is always smaller than the static capacity [2, 4].

The exchange capacity can be determined in different ways:

- by titration
- by elemental analysis
- by determining retention times

All the methods provide different values for Q for the same material. Volumetric methods are most frequently used in practice. For anion exchangers a packed separating column or a defined amount of resin is loaded, for example, with chloride solution. After the residual chloride has been washed out the column or resin can be eluted with nitrate. The amount of chloride eluted, which corresponds to the static exchange capacity under equilibrium conditions, can be titrated against $AgNO_3$ solution and quantified.

The pH dependence of ion exchangers can usually be neglected. With cation exchangers the capacity of the weakly acidic carboxylic acid groups (R–COOH) is only given at higher pH values (deprotonation), while the strongly acidic sulfonic acid groups ($R–SO_3H$) are always completely deprotonated and provide a pH-independent capacity.

In combination with the retention models the importance of Q for the selection of the elution and detection systems becomes obvious. Universal conductivity detection was practically impossible to realize with eluents with a high ionic strength, no matter which special technique was used. This is the reason why, since the introduction of ion chromatography in 1975, virtually only low-capacity separating columns have been used. Up to now high-capacity ion exchangers have only been described in IC applications in combination with detection techniques which are not so strongly limited by the elution system, such as UV/VIS detection [2, 4].

While low-capacity separating materials are very suitable for the analysis of samples with a low ionic strength and therefore a low matrix load, at higher ionic strengths they rapidly reach their limits. Overload effects occur because of their low number of functional groups; these result in deformed peaks and a dramatic drop in efficiency. Problems also occur if the analytes are present in very different concentrations; this is almost always the case in ultratrace analysis.

Eluents in Ion Chromatography

As in all separations by liquid chromatography, also in ion chromatography the mobile phase is the easiest parameter to alter if one wishes to influence the separation. In contrast, the separating column and the detection system are in most cases predefined.

The choice of a suitable elution system can be made by using a wide range of criteria. In anion chromatography the following parameters must be taken into account, among others [4]:

- compatibility with the detection method
- chemical nature and concentration of the eluent ion
- pH
- buffer capacity
- organic solvent content (modifiers)

In the relevant literature [2, 4, 5] the discussion about the mobile phases mainly takes place from the point of view of whether they are suitable for IC without or with chemical suppression. Even though the focal point of the subject of this work is the exchange capacity Q of the separating column, the two techniques will also be considered. As in the previous section, the observations are mainly limited to anion chromatography.

Anion Chromatography

Ion Chromatography without Chemical Suppression

Ion chromatography without chemical suppression was established in ion chromatography by Gjerde and colleagues in 1979 [10]. The outlet of a separating column is directly connected to a detector; this corresponds to the classical arrangement of an HPLC system. This technique is also called "non-suppressed ion chromatography". The eluent containing the analytes leaves the separating column and is not altered chemically. The number of possible elution systems with different chemical characteristics is almost unlimited. Apart from the separation problem itself, only the compatibility of the eluent with the detector needs to be taken into account. For example, if direct UV detection is to be used then the eluent must not absorb in the relevant spectral range. In non-suppressed IC no limitations are placed on the detection system so that, in principle, all the usual HPLC detectors could be used.

For low-capacity anion exchangers ($Q < 100\,\mu$mol/separating column) a whole range of eluents with different chemical characteristics are available for non-suppressed IC. The concentration of the eluents is normally in the lower $\text{mmol}\,l^{-1}$ range or even below. Among others, the following substance classes can be used [4]:

- aromatic carboxylic acids
- aliphatic carboxylic acids
- sulfonic acids
- alkali hydroxides
- inorganic acids such as H_2SO_4, HCl or H_3PO_4

In special cases complexing agents such as EDTA or borate-mannitol complexes can also be used.

The most important class of compounds is the aromatic carboxylic acids and their salts. The representatives which are most frequently used are shown in Figure 8.31.

This class of substances is used so frequently because the solutions of the acids and their salts have a very high elution power combined with a relatively small inherent conductivity, so that they can be used for direct conductivity detection. With polybasic acids the charge and, therefore, the elution power can be controlled via the pH which, however, must be maintained very accurately. Aromatic carboxylic acids have a high absorption in the UV spectrum so that they can be used advantageously in indirect UV/VIS detection.

For aromatic sulfonic acids such as *p*-toluene sulfonic acid the same statements apply in principle; these are always present in a deprotonated state and, therefore, possess no buffer capacity. Their elution power can only be controlled via the concentration.

Aliphatic carboxylic acids, such as oxalic acid and citric acid, have a high inherent conductivity but are UV-transparent, so that they can be used for direct UV/VIS detection. This also applies to aliphatic sulfonic acids, of which methane

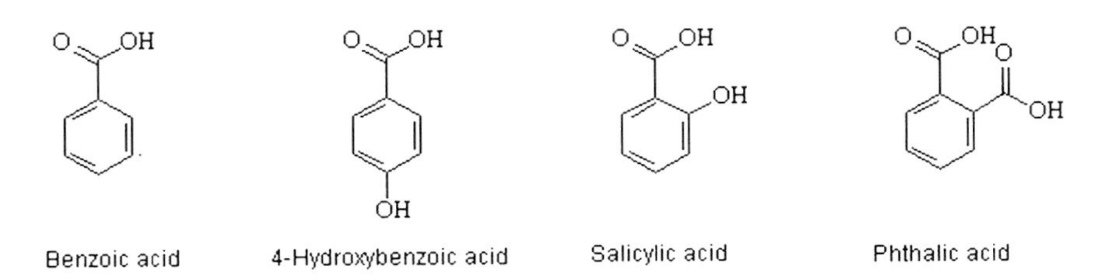

Figure 8.31 Structures of the most important aromatic carboxylic acids used in non-suppressed IC.

sulfonic acid is the one which is most frequently used. The higher homologues of both classes of compounds with their long hydrocarbon chains and the low equivalent conductivities with which this is associated offer the possibility of use for direct conductivity detection [2, 4].

Alkali hydroxides have only a very limited use in non-suppressed IC as the OH^- ion has a very low affinity to quaternary ammonium groups. As a result high eluent concentrations are required, even at low capacities, so that at best only indirect conductivity detection can be used. In contrast, UV/VIS detection is easily possible, although the basically UV-transparent hydroxide ion has a distinct absorption below 220 nm at higher concentrations.

If inorganic acids or their salts are used then their virtually complete dissociation and associated high conductivity means that only photometric detectors can be used. If phosphoric acid or phosphates are used then the buffer capacity and the elution power can be controlled via the pH of the eluent.

Most of the eluents mentioned here also allow the use of other detection systems such as amperometry, fluorimetry or coupling techniques. If you wish to use the elution systems mentioned above with these detectors please refer to the relevant literature [2, 4, 5].

Suppressor Technique

The suppressor technique is the detection technique originally used when IC was introduced [1]. In contrast to non-suppressed IC, this method uses only conductivity detection. In the suppressor technique a so-called suppressor is inserted between the separating column and the detector, which is why this technique is also known as "suppressed ion chromatography" or "ion chromatography with chemical suppression" [2, 4]. Both the eluent and the analytes are chemically modified in the suppressor; this is of particular importance with respect to the subsequent conductivity detection. The suppressor has the task of reducing the inherent conductivity of the eluent and, if possible, of increasing the detectability of the analytes.

The principle of chemical suppression is shown in Equations (8.33) and (8.34) for an anion chromatography application. An $NaHCO_3$ eluent is used, the chloride ion is the analyte. Suppression is carried out with a strongly acidic H^+-form cation exchanger.

$$\text{R-SO}_3^-\text{H}^+ + \text{Na}^+ + \text{HCO}_3^- \leftrightarrow \text{R-SO}_3^-\text{Na}^+ + \text{H}_2\text{O} + \text{CO}_2 \tag{8.33}$$

$$\text{R-SO}_3^-\text{H}^+ + \text{Na}^+ + \text{Cl}^- \leftrightarrow \text{R-SO}_3^-\text{Na}^+ + \text{H}^+ + \text{Cl}^- \tag{8.34}$$

In the simplest case the suppressor unit consists of a column inserted after the separating column, which is where the older term "two-column technique" originates (Figure 8.32). The eluent sodium hydrogen carbonate is neutralized according to Eq. (8.33), as the sodium ions are replaced by protons. This dramatically lowers the inherent conductivity of the eluent. The analyte Cl^- itself is not altered (Eq. 8.34), but its counter-ion Na^+ is exchanged for H^+, which has a considerably larger equivalent conductivity. As the detector records the sum of the conductivi-

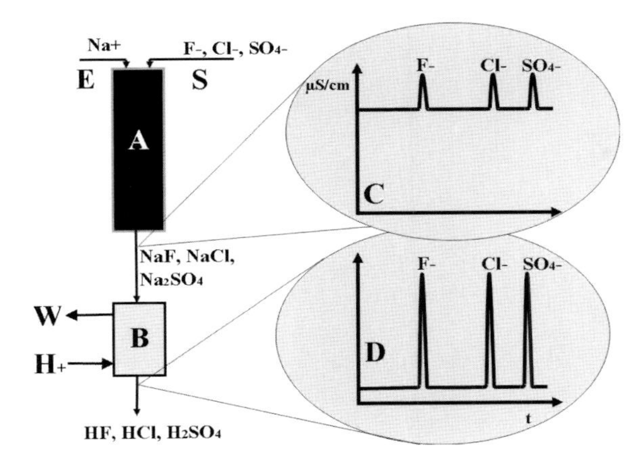

Figure 8.32 Influence of the chemical suppression. **A** analytical column, **B** suppressor, **C** signal without suppression, **D** signal with chemical suppression, **E** eluent, **S** sample, **W** waste.

ties of the analyte and counter-ion as the signal, the two reactions which take place result in a clearly improved sensitivity.

The columns with H^+-form cation exchangers originally used as suppressors made a considerable contribution to broadening the peaks due to their large dead volume. In addition they could only be operated discontinuously, as the cation exchanger had to be regenerated periodically. In contrast to this, modern packed-bed suppressors, such as the Metrohm Suppressor Module, guarantee excellent performance and allow quasi-continuous work. Continuously operating membrane suppressors are also used; in these the regenerant, usually dilute sulfuric acid, and the eluent are led past each other in a counterflow principle. Compared to packed-bed suppressors, membrane suppressors are, however, considerably more susceptible to interferences and less resistant (pressure, organic solvents).

Despite its advantages the suppressor technique also has several crucial disadvantages. In practice, only those eluents which are based on alkali hydroxides or carbonates can be successfully suppressed in anion chromatography. Anions of weaker acids, for example, acetate or fluoride, are almost completely protonated after the suppression reaction, so that their detectability is much lower when compared with non-suppressed IC. High-valency cations must be removed before the analysis as they form insoluble hydroxides which precipitate out on the separating column and can block it.

As has already been made clear, the suppressor technique is synonymous with chemical suppression of the eluent and the use of direct conductivity detection [2, 4]. This technique is unusually widespread in anion chromatography, chiefly because in many cases it is more sensitive than the direct conductivity detection used in non-suppressed IC. The inherent conductivity of the classical eluents for non-suppressed IC (e.g. $2\,\text{mmol}\,\text{l}^{-1}$ phthalate, pH 8) is about $200\,\mu\text{S}\,\text{cm}^{-1}$. For suppressible eluents it is usually lower by at least one power of ten.

Chemical suppression of the inherent conductivity is only possible for a few eluents. These are solutions of [4]:

- alkali hydroxides
- alkali carbonates and hydrogen carbonates
- borates (e.g. $B_4O_7^{2-}$)
- amino acids

In practice, of the eluents mentioned above only the alkali hydroxides and carbonate buffers are of any great importance, which means that the choice of potential mobile phases is very limited.

The hydroxide ion is an extremely weak eluent ion, so that even with low-capacity separating materials work must be carried out at high concentrations, above $50\,mmol\,l^{-1}$. If very polar functional groups are used then the relative elution power of the OH^- ion can be considerably increased by utilizing the hydroxide selectivity. With OH^- eluents manipulation of the retention times or selectivities can only be carried out via the concentration.

The use of alkali carbonates and hydrogen carbonates permits considerably more flexible elution systems. Both species, HCO_3^- and CO_3^{2-}, are present as carbonic acid H_2CO_3 after suppression; this is only dissociated to a very small extent. Hydrogen carbonate has an even lower elution power than hydroxide, whereas carbonate is a relatively strong eluent. Both anions are normally used together; this results in an eluent with a buffer effect whose elution power can be easily controlled via the concentrations of the two components and the ratio of their concentrations. Because of the charges on the eluent species the selectivity for monovalent and multivalent analytes can be influenced very selectively. The concentration ratio of the two eluent ions can also be adjusted very accurately via the pH, this is why the pH range used for HCO_3^-/CO_3^{2-} eluents is between 8 and 11. Just as for OH^- eluents, elution can be accelerated by the use of stationary phases with polar functional groups.

Both elution systems can only be used successfully with surface-functionalized anion exchangers when either the matrix (methacrylate copolymers) or the functional group has a high polarity. For separating columns based on PS-DVB very poor peak symmetries and long retention times have been observed for weak analytes (nitrate, bromide), even when polar functional groups are used. With pellicular materials these effects are not so marked, which explains their exceptionally widespread use in the suppressor technique [2, 4].

Cation Chromatography

Cation Chromatography of Alkali, Alkaline Earth and Ammonium Ions with Conductivity Detection

In the separation of alkali metal and ammonium ions, as well as short-chain aliphatic amines, on sulfonated separation phases by ion chromatography, mineral acids such as HCl or HNO_3 are mainly used as eluents [4]. The acid concentration of the eluent depends on the type and capacity of the cation exchanger used and is a few mmol l^{-1}. Divalent cations, such as the alkaline earth metals, cannot be eluted with mineral acids as they have a distinctly higher affinity to the stationary phase, and a dramatic increase in the acid concentration would make detection too insensitive or mean that suppression could no longer be guaranteed. As an alternative an organic base such as ethylenediamine can be used for the separation of alkaline earth ions. At low pH values this is protonated and is present as a divalent cation.

The simultaneous determination of alkali and alkaline earth cations on strongly acidic cation exchangers is mainly carried out with eluents which contain hydrochloric acid and 2,3-diaminopropionic acid [2]. By varying the pH it is possible to alter the degree of protonation of the amino groups and, therefore, the elution power of the 2,3-diaminopropionic acid.

In ion chromatography systems without suppression, weak organic acids can also be used as eluents in addition to the mineral acids. Organic acids used are, for example, oxalic acid, citric acid and tartaric acid. Complexing agents such as 2,6-pyridinedicarboxylic acid (dipicolinic acid) and the crown ether 18 crown-6 are used to selectively influence the analysis times of individual cations. There are two different types of conductivity detection. If the background conductivity of the eluents is high, e.g. with dilute mineral acids because of the high conductivity of the H^+ ions, then direct conductivity detection is possible with the analytes having a clearly lower conductivity than the eluent. This means that negative peaks are produced which, however, can be converted into a normal chromatogram by reversing the polarity of the detector or by simple inversion. If the quality of the conductivity detector does not permit sensitive conductivity measurements at a high level then suppression is also used for cation chromatography. The construction of suppressors for cation chromatography is far more difficult than for anion chromatography and they also have a much shorter working life.

8.1.5.3
Material and Methods

Equipment
Ion chromatography-system with a suppressor module (e.g. 883 Basic IC plus, Metrohm AG, Switzerland), filtration plant, membrane filter 0.45 μm.

Parameters
See Table 8.2.

Chemicals

Sodium carbonate	CAS- Number 497-19-8	Xi
Sodium hydrogen carbonate	CAS- Number 144-55-8	–

Table 8.2 Parameter experiment A1–A4.

Columns	6.1006.020 Metrosep Anion Dual 1 (3×150 mm)
Eluents	1) 1 mmol l^{-1} Na2CO3/4 mmol l^{-1} NaHCO3
	2) 4 mmol l^{-1} Na2CO3/1 mmol l^{-1} NaHCO3
	3) 4 mmol l^{-1} Na2CO3/1 mmol l^{-1} NaOH (pH = 11–12)
	4) 5 mmol l^{-1} Na2CO3/5 mmol l^{-1} NaOH (pH = 11–12)
	Conductivity after chemical suppression approx. 17 $\mu S\, cm^{-1}$
Sample	Standard (fluoride, chloride, nitrite, bromide, nitrate, phosphate, sulfate)
Flow 0.5 ml min^{-1}	
Pressure	3 MPa
Duration of analysis	a) 30 min
	b) 12 min
	c) 10 min
	d) 8 min
Loop	20 µl
Suppressor	Regeneration medium: 50 mmol l^{-1} H_2SO_4, ultrapure water
	Autostep with fill
Polarity	+

Sodium hydroxide — CAS- Number 1310-73-2 — C

H_2SO_4 — CAS- Number 7664-93-9 — C

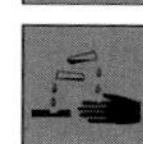

Solutions

Standard Solutions for Experiments A1–A4 (Table 8.2)

Fluoride	2 mg l^{-1}
Chloride	5 mg l^{-1}
Nitrite	5 mg l^{-1}
Bromide	10 mg l^{-1}
Nitrate	10 mg l^{-1}
Phosphate	10 mg l^{-1}
Sulfate	10 mg l^{-1}

Eluent Solutions

A1 1 mmol l^{-1} Na2CO3/4 mmol l^{-1} NaHCO3
Dissolve 106 mg of sodium carbonate (anhydrous) and 336 mg of sodium hydrogen carbonate in 1 l of ultrapure water.

A2 4 mmol l^{-1} Na2CO3/1 mmol l^{-1} NaHCO3
Dissolve 424 mg of sodium carbonate (anhydrous) and 84 mg of sodium hydrogen carbonate in 1 l of ultrapure water.

A3 4 mmol l^{-1} Na2CO3/1 mmol l^{-1} NaOH
Dissolve 424 mg of sodium carbonate (anhydrous) and 40 mg of sodium hydroxide in 1 l of ultrapure water.

A4 5 mmol l^{-1} Na2CO3/5 mmol l^{-1} NaOH
Dissolve 424 mg of sodium carbonate (anhydrous) and 200 mg of sodium hydroxide in 1 l of ultrapure water.

B Pipe Water/Mineral Water
Dissolve 265 mg of sodium carbonate (anhydrous) and 201.5 mg of sodium hydrogen carbonate in 980 ml of ultrapure water and add 20 ml of acetone.

Information about the Practical Work

Bacterial Growth

In order to prevent bacterial growth, the eluent, rinsing and regenerating solutions should always be made up freshly and not used over a longer period of time. If bacterial or algal growth should nevertheless occur then 5% methanol or acetone can be added to the eluent. If membrane suppressors are used this is not possible as these could be destroyed by organic solvents. However, the Metrohm Suppressor Module (MSM) is 100% solvent-resistant.

Cation Analysis

In each analysis carried out the sample must be acidified (pH 2.5–3.5) with nitric acid (approx. 100 μl 2 mol l^{-1} HNO_3 to 100 ml sample), as otherwise no reproducible results will be obtained for divalent cations.

Chemical Grade

All the chemicals used should have at least the grades p.a. (analytical grade) or puriss (extra pure). Any standards used must be specially suitable for ion chromatography; for example, sodium salts should be dissolved in water, not in acid.

Sources of Contamination

All solutions, samples, regenerating solutions, the water and the eluents used should be free of particles as these will block the separating columns in the course of time (rise of column pressure). This is especially important when preparing the eluents as these flow through the column continuously (500 to 1000 ml per working day, whereas the sample throughput is approximately 0.5 ml).

Eluent Degassing

In order to avoid bubble formation it is recommended that the water used for preparing the eluent is degassed prior to adding the reagents. This can be done by applying a vacuum from a water-jet or vacuum pump for about 10 min or by using an ultrasonic bath.

Environmental Protection

One of the great advantages of ion chromatography is that usually aqueous media are used. This is why the chemicals used in ion chromatography are to a large extent non-toxic and do not pollute the environment. However, if acids, bases, organic solvents or heavy metal standards are used then care should be taken that these are disposed of properly after use.

Instrument Shut-Down

If the ion chromatograph is not to be used for a longer period of time (>1 week) then the separating column should be removed and the ion chromatograph rinsed with methanol/water (1:4). Care should be taken that all three chambers of the suppressor are rinsed.

Storing the Separating Columns

Most of the columns can be stored in the eluent in a refrigerator (4 °C). It is essential that the column does not dry out. For long-term storage (weeks) storage should be in solvent/water.

Water Quality

In ion chromatography mainly aqueous media are used. This is why the water quality is crucial for good chromatography results. If the water quality is unsatisfactory then it is certain that the results will also be unsatisfactory. In addition there is the danger of damaging the instrument and separating column. The demineralized water used (aq. demin.) should therefore have a resistivity greater than 18 MΩ cm and be particle free. It is recommended that the water is filtered through a 0.45 μm filter.

Experiment

(A) Selectivity of separation columns

Generally divalent ions elute in the eluent more strongly than monovalent ions, since they can form two bonds with the stationary phase, thus interacting more with this phase. Caustic soda at the same molarity has a greater influence on the pH-value of the eluent than carbonate/bicarbonate. Compared with that of bicarbonate, the eluting power of the OH-ion is lower, as the hydroxy ion has less interaction with the stationary phase.

The retention time of the phosphate is strongly dependent on pH.

With a high pH-value, the balance of PO4 shifts in the direction of PO_4^{3-}

– Compare sodium and carbonate/hydrogen carbonate eluents

– Measure the influence of the eluent pH-value on the retention of phosphate

Describe the influences and try to explain the effect.

See experiment (above) for eluent composition (eluent solutions A1–A4), concentrations of the standard solution and analysis parameters.

(B.1) Determination of anions in drinking water

Take a calibration curve using 10 standard solutions (adjust concentrations of the standards to the limits).

See experiment (above) for eluent composition (eluent solution B), concentrations of the standard solution and analysis parameters.

Different tap water samples are to be determined and quantified and the values must be compared with Table 8.3.

The parameters for experiment B are given in Table 8.4.

Table 8.3 Limits for drinking water ($mg\,l^{-1}$) (D).

Fluoride	1.5	as F^-
Nitrate	50	as NO_3^-
Nitrite	0.1	as NO_2^-
Chloride	250	
Sulfate	250	

Table 8.4 Parameters for experiment B.

Column	6.1006.020 Metrosep Anion Dual 1 (3 × 150 mm)
Eluent	2.4 $mmol\,l^{-1}$ NaHCO 3/2.5 $mmol\,l^{-1}$ $Na_2\,CO_3$ + 2% acetone
Conductivity after chemical suppression approximately	16 µS cm
Sample of tap water	
Flow	0.5 $ml\,min^{-1}$
Pressure	3 MPa
Duration of analysis	about 16 min
Loop	20 µl
Suppressor regenerant: 50 $mmol\,l^{-1}$ $H_2\,SO_4$, ultrapure	Autostep with fill
Polarity	+

(B.2) Determination of anions in mineral water

Determine fluoride, nitrate, nitrite, chloride and sulfate in various mineral waters (with and without gas) and compare the results with the data on the label.

Note that mineral water with carbon dioxide must be degassed before measurement.

(C) Method validation

The method B is to be validated. Essential parameters for analytical determination methods are linear range, detection limit, quantification limit, and reproducibility

8.1.5.4 Questions

(1) Count the advantages/disadvantages of the ion chromatography method.

(2) For which analytical applications is IC predestined?

(3) Evaluate the use of ion chromatography for the present task. Consider, among other things, the detection limit, the amount of work and the costs of the proceedings.

(4) Which chromatographic principles come into play in IC? What similarities/differences are there to other chromatographic methods such as HPLC or GC?

(5) Find out about statutory limits of these ions in mineral water, thermal spring and spring water. How is compliance monitored (by whom, at what intervals, methods)? What happens if the limits are exceeded?

References

1 Small, H., Stevens, T.S., and Bauman, W.C. (1975) *Analytical Chemistry*, **47**, 1801.

2 Weiss, J. (2005) *Handbook of Ion Chromatography*, 3rd Revised Updated Edition, Wiley-VCH Verlag GmbH, ISBN-13: 978-3527287017.

3 Mermet, J.M., Otto, M., and Widmer, H.M. (1998) *Analytical Chemistry*, Wiley-VCH Verlag GmbH, Weinheim.

4 Haddad, P.R., and Jackson, P.E. (1990) *Ion Chromatography – Principles and Applications*, 1st edn, vol. 46, Elsevier, Amsterdam.

5 Fritz, J.S., and Gjerde, D.T. (2000) *Ion Chromatography*, 3rd edn, Wiley-VCH, Weinheim.

6 Skoog, D.A., and Leary, J.J. (2006) *Principles of Instrumental Analysis*, 6th edn, Brooks Cole, ISBN-10: 0495012017, ISBN-13: 978-0495012016.

7 International Union of Pure and Applied Chemistry (IUPAC) (1993) *Pure and Applied Chemistry*, **65**, 819.

8 Pohl, C.A., Stillian, J.R., and Jackson, P.E. (1997) *Journal of Chromatography. A*, **789**, 29.

9 Dorfner, K. (ed.) (1991) *Ion Exchangers*, 4th edn, Walter de Gruyter, Berlin.

10 Gjerde, D.T., Fritz, J.S., and Schmuckler, G. (1979) *Journal of Chromatography. A*, **186**, 509.

8.1.6
High-Performance Thin-Layer Chromatography (HPTLC), Project: "Ensuring Regulatory Compliance by Quantification of Lead Compounds (Markers) in Herbal Combination Products"

8.1.6.1
Analytical Problem

The task of the analytical department of the Cashmere Extract Ltd. is to determine by quantification of three lead compounds (markers) in the end and intermediate products, the amount of individual plant extracts. This regulatory requirement is designed to ensure that the composition at the various processing stages and in the combination product corresponds to the declaration. To cope with high throughput of samples HPTLC technology can be used.

Extract with Salicylic Acid as a Marker
The Greek physician *Hippocrates* (460–377 BC) prescribed infusions made from poplar and willow bark to his patients against pain, fever and rheumatic complaints. The willow bark is recognized to fight the fever with great success and alleviate the symptoms. By far the easiest way to ensure salicylate intake, however, is tea. One cup of black tea delivers 3 mg, five cups a day even the supposedly life-prolonging (no guarantee by the author) dose of 15 mg of salicylate.

Extract with Acetylsalicylic Acid as a Marker
Meadowsweet (Filipendula ulmaria) is one of our traditional medicinal plants. For centuries, it has been grown in the gardens of Central Europe. As a gentle form of medicine for headaches and migraines, it is still known and used in phytotherapy. In 1838, for the first time, it was possible to obtain acetylsalicylic acid from the plant used as a headache drug. The meadowsweet plant, which at that time was attributed botanically to the Spier shrubs (Spirea), contributed to the development of the brand name Aspirin. While the "A" stands for acetyl, "Spirin," is derived from the term "Spirea".

Extract with Phthalic Acid as a Marker
"Ajuga bracteosa" is a plant that grows in Nepal, at an altitude of 1300 meters. This plant is a source of phthalic acid and its esters are used in popular medicine to treat rheumatism and gout.

Object of Investigation/Investigation System
Phytopharmaka

Samples
Plant extracts

Examination Procedure

The classical thin-layer chromatography (TLC) as a simple technique in the field of chromatography, especially for qualitative analysis, has evolved in recent decades to become an independent instrumental method of analysis that, despite progress in column (liquid) chromatography, continues to be of significance. Especially, separations of substances with very different polarity can be performed better with the techniques of TLC. In addition, several samples can be separated simultaneously using TLC. The main applications of instrumental thin-layer chromatography (HPTLC) are now pharmaceutical quality control, food and chemical analysis, and water analysis. In all these areas large numbers of samples can be analyzed with cost-effective processing, even in normal discontinuous operations. The latest developments make the possibilities of extensive automation clear, whereby coupling processes, such as mass spectrometry are possible.

Analytes

Salicylic acid, acetylsalicylic acid, phthalic acid

Keywords

Comparison of DC-HPLC, R_f-value, polarity in chromatography, detection in TLC, eluotropic series, densitometry, instrumental components in the HPTLC (sample dosing, TLC chamber, scanner), derivatization, documentation in the HPTLC

Bibliographical Research (Specific Topics)

Wall, P.E. (2005) *Thin-Layer Chromatography: A Modern Practical Approach*, Royal Society of Chemistry.

Hahn-Deinstrop, E. (2007) *Applied Thin-Layer Chromatography: Best Practice and Avoidance of Mistakes*, Wiley-VCH Verlag GmbH, ISBN 978-3-527-31553-6.

Srivasta, M.M. (2011) *High Performance Thin-Layer Chromatography (HPTLC)*, Springer Science+Business Media, ISBN-13: 978-3-642-14024-2.

Project Process/Task

(A) Purification of the TLC plates

Although today's commercially available TLC plates are of a high quality, there may be surface contamination after prolonged storage. For this reason it is recommended to clean the TLC plates before use.

(B) Optimization of the eluent

B.1 Find out the molecular structure of the three analytes by studying the literature and consider the sequence in which these could be separated on a silica plate (very polar stationary phase).

B.2 Perform optimization experiments by varying the eluent (n-hexane, diethyl ether, concentrated formic acid).

(C) Determination of salicylic acid, acetylsalicylic acid, phthalic acid and validation of the method

C.1 Create a calibration curve

C.2 Determine the working range

C.3 Validate the optimized method

C.4 Determine the concentrations of salicylic acid, acetylsalicylic acid, and phthalic acid in the combination product (to be delivered by assistants)

(D) Investigate the influences of the sample amount and its traveled distance

(E) Project reporting

The results are to be submitted in the form of a report (or presentation). Comparisons with bibliographical references are to be made.

8.1.6.2 Introduction

Definitions

Thin-layer chromatography (TLC), also known as planar chromatography refers to a specific manner of executing solid–liquid or liquid–liquid chromatography in which a support separation bed with thin layers of about 0.1–0.25 mm thickness, and thus an essentially two-dimensional separation process is observed. This very simple analysis technique as it was until about two decades ago has now developed to become an instrumental method of high-performance thin-layer chromatography (HPTLC) which can be automated from the sample application stage to the quantitative analysis. Due to capillary flow motion, a particular flow law applies here (as opposed to column chromatography, in which there is usually a constant flow rate).

During application of a mixture of substances to the starting point of a TLC plate, the substances penetrate into the layer. During "plate development" ("the chromatography process") solvent flow takes place, caused by the evaporation of the solvent from the plate surface (particularly in non-saturated chambers), and pushes the spots to the surface. This fact is of particular significance in view of separation efficiency as well as quantitative evaluations. The basics of HPTLC originate from the van Deemter function (see Section 8.1.4), mainly due to the dependence of plate height H on particle size. In thin-layer chromatography, adsorption chromatography plays a much bigger role than in column chromatography. As a result of the unique use of a stationary phase, changes in the activity of the adsorbent play a much less significant role than in the continuous through-flow in a separation column.

The solvent strength parameter has been defined in relation to the solvent n-pentane with $e_0 = 0$. e_0 is defined as the adsorption energy of the solvent per surface unit in relation to n-pentane. The polarity of the eluent increases with increasing e_0-value (Table 8.5).

Table 8.5 e_0 values of some solvents (when using Al_2O_3 as a stationary phase).

Eluents	e_o -Value
n-Pentane	0.00
Cyclohexane	0.04
Tetrachloromethane	0.18
Xylene	0.26
Toluene	0.29
Benzene	0.35
Diethyl ether	0.38
Trichloromethane	0.40
Tetrahydrofuran	0.45
Acetone	0.56
Ethyl acetate	0.58
Acetonitrile	0.65
Pyridine	0.71
Isopropanol	0.82
Ethanol	0.88
Methanol	0.95
Water	$\gg 1$
Acetic acid	$\gg 1$

Pure solvents and solvent mixtures are used as an eluent. By mixing several solvents, virtually any desired solvent strength parameter can be adjusted for a separation problem. However, in so doing, the e_0-values cannot readily be determined by normal calculation. The solvent mixtures used must be miscible and must not dissolve the sorption layer of the carrier plate. Snyder [1] also determined the relationship between the adsorption coefficients and the activity parameters, whereby the substance-specific variables can often be calculated additively from the increments of molecular structural groups. Polarity scales have resulted from these theoretical considerations, which agree well with the empirically determined sequences.

R_f -Value

The R_f-value describes the behavior of a substance during thin-layer chromatographic separation. This retention factor is calculated by dividing the migration distance of the mobile phase front and the migration distance of the substance zone (Figure 8.33). R_f-values are, by definition, always less than 1. Useful R_f-values are between 0.2 and 0.8 (front runners and those that remain stationary are not usable). Qualitative evaluation is done by comparing the R_f-values of the samples with a standard substance.

Equipment Technology

The apparatus or instrumentation equipment in analytical thin-layer chromatography ranges today from equipment for the automatic application of defined

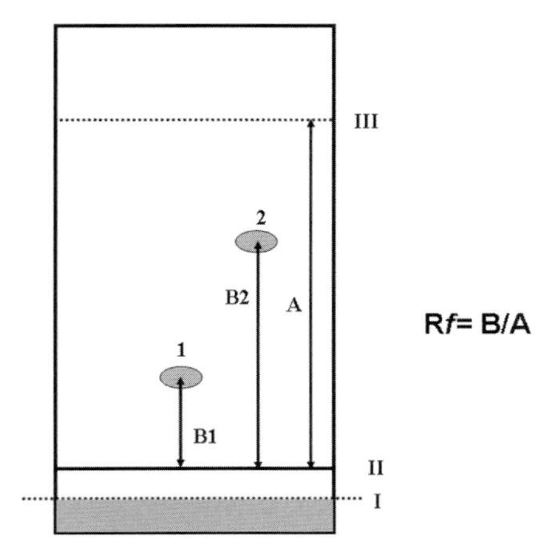

Figure 8.33 Calculation of the R_f-Value. **1 and 2** components, **A** distance line, **III** solvent front. traveled by solvent, **B** distance traveled by component, **I** origin level of the developing solvent, **II** starting.

sample volumes, to various kinds of development chambers, spraying devices and/or dipping chambers for performing chemical reactions on the layer (conversion of not detectable compounds to detectable compounds), to detection systems such as thin-film scanners with spectrophotometric or fluorometric analysis and combination procedures in connection with, for example, FTIR measurements.

Using a microliter syringe, applications of between 5 and 20 μl (maximum 100 μl) can be done very evenly as bands. Such devices for sample dosage could be microprocessor controlled and programmable. So-called nano-dosing devices are also available for HPTLC with volumes of 100 or 200 nl.

A particular advantage of planar chromatography is its enormous flexibility, which is based on the interplay of many parameters.

Chromatogram Development

Chamber Shape, Chamber Saturation

Planar chromatography differs from all other chromatographic techniques by the fact that apart from stationary and mobile phases there is a gas phase, which can significantly affect the separation results.

Processes in the Development Chamber

The classical type of development consists of putting the plate into a development chamber containing a sufficient amount of solvent, so that the lower end of the plate dips in several millimeters. The solvent rises to the layer through capillary action until the desired height is reached and the chromatography is stopped by removing the plate.

The developing eluent is allowed to evaporate and the plate is now ready for detection.

In most cases, planar chromatography runs in non-equilibrium between the stationary, mobile, and gas phases. For this reason, the conditions in the development chamber cannot be recorded in a mathematically correct manner. Reproducible chromatographic results can only be obtained when all factors are kept constant. Chamber geometry and saturation play very important roles in this context.

Parameters of Planar Chromatography

The chromatogram development requires special TLC-chambers. Only one TLC plate fits in a flat-bottomed chamber. In a double-trough chamber, however, two plates can be inserted. The R_f-values are decisively influenced by the degree of saturation with solvent in the chamber. If the chamber atmosphere is not saturated, the solvent evaporates from the layer during the chromatographic process causing change in solvent velocity.

Quantitative thin-layer chromatography uses TLC scanners to evaluate thin-layer chromatograms. They consist of a spectrophotometer, a mechanical, programmable device for transporting a TLC plate through the light beam of the photometer, and a recorder for recording the so-called local-reflectance curves. Depending on the structure of the compound (chromophore, see Section 8.2.1) and concentration in the layer, substance spots absorb part of the light energy. At a certain angle the monochromatic light is radiated from the light source to the TLC plate. After that, the reflected light is measured at the receiver. Since the TLC plate continues to move during the measurement in the direction of the solvent, a local reflectance curve is registered.

Process

Selection of the Stationary and Mobile Phase

Stahl [2] developed a diagram (Figure 8.34) for the selection of the most favorable mobile and stationary phase for a given mixture to be separated that is very well suited as an orientation.

Application of the Stahl diagram: One corner of the movable triangle is set to the polarity of the mixture to be separated. At the two remaining corners the required polarity of the stationary (the most important adsorbents in TLC are silica gel and aluminum oxide) and mobile phases can then be read.

In addition to changing the solvent, elution power can also be changed by using mixtures.

Conventional TLC

To carry out a TLC separation the solution to be tested is applied with the help of a glass capillary, in the conventional TLC approximately 1 cm from the bottom of the plate, for example, in the form of a point. The TLC plate is then placed in a development chamber filled with solvent to less than 1 cm high and the back wall of which is lined with paper (to saturate the atmosphere with the vapor of the eluent). The eluent begins to wander upwards.

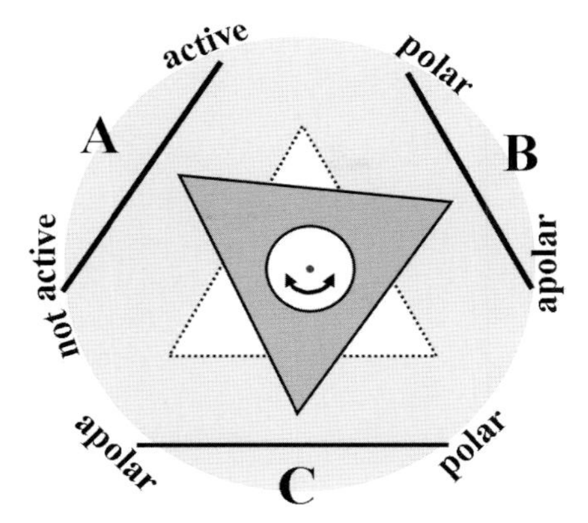

Figure 8.34 *Stahl* diagram for TLC. **A** stationary phase, **B** mobile phase, **C** mixture to be separated.

Without chemical reaction, substances can be made visible by their own color, UV-absorption or fluorescence. Furthermore, substances or groups of substances, can be located after being treated with spray reagents to make them visible. Instead of spraying it is often advisable to use a dipping chamber, resulting in a more uniform distribution of the reagent (which is important for quantitative analysis). The commonly used development technique is the ascending TLC, with point-or line-shaped application of the substance mixtures.

The wedge-strip technique results in improved separation compared to the conventional technology; the same applies to the circular technique. In both cases, the solvent flows not only in the direction of movement, but also vertically. The substances are drawn apart perpendicularly to the direction of flow and do not overlap even with slight differences in the R_f-values. In the circular technique the solvent is led through a bore over a wick in the middle of the TLC plate, whereby the substance mixtures are applied to a concentric ring. With the anti-circular development, on the other hand, the substances are on the outer circle, and the solvent flows inwards from the outside circle. In this way, the enlarging of the spot which arises by cross-diffusion with substances having high R_f-values is minimized. In two-dimensional TLC the substance is applied in the first run at a corner of the plate and dried after development in the first direction. Then the plate is rotated by 90°, so that the partially separated mixture is at the bottom of the plate, and development occurs with a second eluent in the second direction.

Instrumental HPTLC

Densitometry

The fundamental benefit of densitometry is to be able to "measure" and quantify thin-layer chromatograms. This was the basic prerequisite for the acceptance of instrumental HPTLC as being reliable and comparable to the HPLC and GC quantitative analysis methods.

The measurements are basically relative, that is, the properties of unknown substances are compared with those of known substances. With the help of known standards, the measuring system can be calibrated on a plate. For this reason, standards and samples must always be chromatographed together on the plate. Densitometry is the most accurate evaluation method of HPTLC. Compared to visual observation and video-densitometry it has the great advantage of spectral selectivity. The use of monochromatic light in the range 190 to 800 nm and the resulting potential adjustment to the absorption and/or fluorescence excitation maximum of the individual substances, make the measurement very sensitive. Typical detection limits are in the low nanogram range (absorption) or medium picogram range (fluorescence). In most cases, densitometry is performed prior to derivatization. Only compounds without chromophoric groups need to be chemically altered to be detectable.

Sample Application

During sample application, the following basic requirements must be met: the point of application must be exact and the application volume must – especially for quantitative analysis – be correct and precise. Of course, the layer must not be damaged during application. Samples can be either transferred or sprayed in direct contact to the layer. In order to maximize the separation performance of a given chromatographic system, the application zones should be as narrow as possible in the direction of the chromatography. On conventional layers 2 to 4 mm should not be exceeded. On HPTLC layers 1.5 mm is the upper limit. This requirement limits the volume that can be applied point-like at once. In general, TLC plates allow sample volumes of 0.5 to 5 μl, while on HPTLC layers 0.1 to 1 μl are customary. The resolution and detection limit for a given LTC system can be improved if the samples are plotted as narrow lines with the help of the spraying technique. This facilitates a homogeneous distribution of the sample over the entire length of the application zone. As the solvent of the sample evaporates completely during application, there is virtually no limit to the volume that can be applied. If the sample is applied to the layer by contact application, "circular chromatography" takes place at the point of application. This can lead to irregular distribution of the sample components at the starting spot. Broad and asymmetric spots or stains are often a consequence of chromatography. The rule of thumb applies that the solvent of the sample should always have the minimum possible elutropic strength, in order to produce small and compact starting spots. Simplicity and very low time requirements are the advantages of point-like contact application. A very large number of samples can be analyzed in parallel. For example, 72 samples (36 each on opposite sides) fit on a 20 × 10 cm HPTLC plate.

Practical Tips

Contrary to common practice, it is not advisable to mark the point of application on the layer with a pencil. This can damage the surface of the layer and impede the transfer of the sample. Pencil marks can also affect the densitometric evaluation of the chromatogram. Samples should not be applied too close to the lateral edge of the plate, as there are often irregularities in the layer thickness. For quantitative determinations on HPTLC plates, the left and right margins should be at least 15 mm wide and 25 mm on TLC plates. When chromatogram development starts, the point of application should be significantly above the solvent level. If, for example, the solvent level is adjusted to 5 mm (HPTLC) or 10 mm (TLC), the application should be 8 mm or 15 mm above the lower edge of the plate. The distance between (the edges of) zones must be at least 2 mm or 4 mm. As the distance between the centers of the point-like samples 5 mm or 10 mm, respectively, are chosen as a rule.

Samples can also be sprayed as rectangles and focused before the actual chromatography to narrow bands. This option is particularly useful when large sample volumes with a low concentration are to be examined.

Measurement Methods

Densitometry can be conducted both in absorption and in fluorescence mode. During the absorption measurement the large amount of light reflected from the plate base (without sample) forms the baseline signal. This is reduced when the measuring beam meets an absorbing zone in the chromatogram. The resulting negative peak is electronically reversed and positively represented in the densitogram. In contrast to the absorption measurement a very low baseline signal is thereby generated thus facilitating a higher signal. The fluorescent light of a higher wavelength generated by the sample passes through the filter, touches the photomultiplier and generates a positive signal. The measuring beam is positioned on the selected zone in order to directly measure the UV spectra on the plate. During the measurement, the wavelength is continuously altered over a certain wavelength range. UV spectra are usually measured at the peak maximum. If one also records spectra at the peak sides and correlates them with those at the peak maximum, peak purity can be assessed. This is an easy way to track substances that have not been separated.

The UV spectrum provides two main types of information: the position of the absorption maximum of a given substance, and the general shape of the spectrum. Both provide information on the identity. Thus, unknown substances can be identified by comparing the spectra with spectra of reference compounds produced under identical conditions.

Selection of Measurement Parameters

Before a panel is measured by densitometry, it should be visually examined for possible scratches or chromatographic irregularities. Optimum results are achieved when the measurement gap is about 60 to 80% of the length of an applied band-shaped zone. For a point-like applied sample the measurement gap should be adjusted to 120% of the largest zone. To optimize the amount of light and the resolution of the measurement a "micro"-gap of 0.3–0.4 mm in height should be chosen and the measurement velocity should be low, that is, 20 mm s^{-1}. Qualitative measurements can be performed at considerably higher velocities. The measurement usually starts below the point of application and ends above the front. Thus, the total information of the chromatogram path is accessible. The integration that follows the measurement can be limited to the chromatogram zones, provided that the plate base on the track or the base line permits this.

Derivatization

The possibility of easy specific or non-specific derivatization, often connected to the development of heat, is an essential strength of planar chromatography. Reasons for choosing derivatization as a step in the HPTLC process could be any of the following:

- the conversion of non-absorbing substances in detectable derivates
- the improvement of detectability (reduction of the detection limit)
- proof of all sample components
- the selective detection of individual substances
- the induction of fluorescence
- a number of derivatization reagents are available

From a technical point of view it must above all be decided how to transfer the reagent onto the plate. The derivatization can occur through the gas phase or by spraying or dipping.

Derivatization via the Gas Phase

Derivatization via the gas phase allows a rapid and uniform transfer of the reagent. Unfortunately, however, only a few reagents are suitable, for example, iodine, bromine, chlorine, volatile acids and bases, as well as other gases such as H_2S and NO. In practice, derivatization takes place most easily via the gas phase in the twin trough chamber, whereby the reagent is located in the rear tray, or is generated there (e.g., chlorine from HCI and permanganate) and the plate is placed in the front trough with the layer facing inwards.

Spraying

Spraying is the most common way of transferring a reagent easily and quickly to the plate. There is no need for expensive equipment and only small volumes of reagent are needed. Spraying is very flexible, which is particularly advantageous when multiple reagents have to be applied sequentially. Even when one is searching for a suitable reagent during method development, spraying is the method chosen. On the other hand, non-negligible amounts of unpleasant and hazardous fumes arise when spraying, which must be carefully removed using a spray cabinet. Another problem of spraying is the difficulty of achieving a uniform and defined derivatization of the entire plate. Especially for a quantitative analysis spraying requires great skill to generate reproducible results. The spraying device must generate a very fine mist and should not "slosh". With the use of compressed air the best performance is often influenced by the air pressure and also by the distance between the nozzle and the plate. The amount of reagent is difficult to control, but with practice it is possible to achieve uniform results.

Dipping

A very homogeneous derivatization can be achieved by dipping the plate into the reagent. The up and down movement must be even and smooth to avoid the formation of dipping lines. The use of an automatic immersion device can considerably improve the reproducibility of the derivatization step as opposed to the spraying technique. The concentration of the reagent can easily be kept constant, and by changing the vertical speed and immersion time, the amount of the transferred reagent and/or the reaction time can be varied. In this technique, no vapors are generated, and exposure to toxic chemicals is minimal. After dipping, the plate is often covered with excess fluid. In order to avoid an unnecessary broadening of the separated zones, the plate should be dried as quickly as possible. In most cases, the same reagents can be used for spraying and dipping. However, dipping reagents are usually less concentrated (up to ten times, usually four times less) and should be dissolved in less polar solvents.

Heating

Most chemical derivatization reactions require heat. Ovens and plate heaters serve as sources of heat. Routinely, and for quantitative analysis, it is vitally important for the plate to be heated homogeneously and in a reproducible manner. Ovens have two major drawbacks. Due to the sometimes aggressive vapors produced during reactions, the problem of corrosion as well as cross-contamination arises. The other disadvantage is that the derivatization process usually cannot be observed. Plate heaters are, therefore, a frequently used alternative. These produce a constant temperature on their surface. When a plate is placed on the hot surface (temperatures higher than 120 °C), their edges bend due to the poor thermal conductivity of glass. As a result, the middle of the plate is more strongly heated.

Within minutes, these irregularities are, however, usually balanced out. For reproducible results, the temperature of derivatization should be adjusted so that the plate is heated for at least 5 min.

In practice, care should be taken to heed the following:

- Derivatization is usually performed to make a chromatogram visible. Depending on the analytical target and type of subsequent evaluation, reproducibility is either important or it is not.
- Normally, the result of a chemical derivatization – especially the color and/or fluorescence of the derivatized zones – is influenced by the time/temperature of heating and the concentration of the reagent. However, these factors can be easily kept under control. Other factors – such as, for example, drying (temperature and time) of the developed plate prior to derivatization – are often overlooked, but can also affect the outcome. In order to obtain reproducible results, the derivatization procedure should be meticulously documented.
- If it is not necessary to make the zones visible, derivatization can also be disadvantageous. For example, progesterone has a chromophore and absorbs UV light at 254 nm. Thus it can be detected without derivatization. If necessary, the substance can be visualized with sulfuric acid.

8.1.6.3
Material and Methods

Equipment

Automatic TLC Sampler 4 (CAMAG, Muttenz, CH), fully automated sample application
CAMAG ADC 2 development chamber diode array detector
TLC scanner 3 (CAMAG)
Densitometric analysis software winCATS (CAMAG)

Chromatographic Parameters

Application of the Sample (Guidelines)

For point-like contact application the sample is to be dissolved in the solvent with the least possible eluent strength, to minimize the effect of zone broadening during application. For spray application, the feed rate must be adjusted the particularities of the chosen solvent. To avoid volumetric errors, readily volatile solvents should not be used in the manufacture of standard and sample solutions for quantitative analysis. Samples are applied in lines or dotted with the following parameters:

Bottom edge distance in mm for TTC: 8
Minimum edge distance to the left and right in mm: 10
Minimum spacing between bands/zones in mm: 4
Belt length in mm: 8
Maximum diameter of the application zone during dotted application: 5 mm

Quantitative Determination

Basically, the quantitative analysis is performed using the TLC scanner 3 and winCATS software.

Stationary Phase

$20 \times 10\,cm^2$ silica gel 60 F254.

Eluent

n-Hexane, diethyl ether and formic acid (various compositions).

Chemicals

Salicylic acid	CAS Number 69-72-7	Xn
Acetylsalicylic acid	CAS Number 50-78-2	Xn
Phthalic acid	CAS Number 88-99-3	Xi

Absolute ethanol	CAS Number 64-17-5	F
n-Hexane	CAS Number 110-54-3	F, Xn, N
Diethyl ether	CAS Number 60-29-7	F, Xn
Formic acid	CAS Number 64-18-6	C

Reagents

Standard Stock Solutions

0.3 g of phthalic acid, 0.3 g salicylic acid and 0.3 g of acetylsalicylic acid are weighed together in a 100 ml volumetric flask and dissolved in absolute ethanol (approx. 60 ml). After the three substances have been dissolved, the flask is filled with ethanol to the mark.

Mix-Standard Solutions

Standard 1 (S1):	0.15 ml of stock solution are pipetted into a 10 ml volumetric flask and filled up to the mark with absolute ethanol
Standard 2 (S2):	0.35 ml of stock solution are pipetted into a 10 ml volumetric flask and filled up to the mark with absolute ethanol
Standard 3 (S3):	0.65 ml of stock solution are pipetted into a 10 ml volumetric flask and filled up to the mark with absolute ethanol
Standard 4 (S4):	1.00 ml of stock solution are pipetted into a 10 ml volumetric flask and filled up to the mark with absolute ethanol
Standard 5 (S5):	0.65 ml of stock solution are pipetted into a 5 ml volumetric flask and filled up to the mark with absolute ethanol.
Standard 6 (S6):	0.90 ml of stock solution are pipetted into a 5 ml volumetric flask and filled up to the mark with absolute ethanol

Sample Solutions

Dilute the unknown samples with ethanol 1:10. Prepare two sample flasks (P1 + P2).

Reference Solutions

Stock Solutions

Weigh 0.3 g of salicylic acid in a 100 ml volumetric flask and fill up to the mark with absolute ethanol.

Weigh 0.3 g of acetylsalicylic acid in a 100 ml volumetric flask and fill up to the mark with absolute ethanol.

Weigh 0.3 g of phthalic acid in a 100 ml volumetric flask and fill up to the mark with absolute ethanol.

Working Solutions
SR Pipette 1 ml of salicylic acid stock solution into a 10 ml volumetric flask and fill up to the mark with absolute ethanol.

AR Pipette 1 ml of acetylsalicylic acid stock solution into a 10 ml volumetric flask and fill up to the mark with absolute ethanol.

PR Pipette 1 ml of the phthalic acid stock solution into a 10 ml volumetric flask and fill up to the mark with absolute ethanol.

Diluents for LOD

V1	S6 undiluted
V2	Place 1 ml of S6 in a 5 ml volumetric flask and fill up to the mark with absolute ethanol.
V3	Place 0.5 ml of S6 in a 5 ml volumetric flask and fill up to the mark with absolute ethanol.
V4	Place 0.1 ml of S6 in a 5 ml volumetric flask and fill up to the mark with absolute ethanol.
V5	Place 0.05 ml of S6 in a 5 ml volumetric flask and fill up to the mark with absolute ethanol.

Experiment

(A) Purification of the TLC plates

In a development chamber the TLC plates are developed, depending on their degree of contamination, once to twice at right angles to the planned direction of movement (the actual separation). Methanol purum is a suitable eluent.

Between developments and prior to use, the plate must be completely dried.

To speed up drying a hair dryer can be used.

Under the UV lamp (at $\lambda = 254$ nm) check whether there are still any disturbing contaminants visible on the TLC plate.

(B.2) Optimization of the eluent

To save on TLC plates, it is recommended to use the $5 \times 10\,cm^2$ silica gel plates for optimizing the eluent instead of a $10 \times 20\,cm^2$ plate. These plates should also be cleaned in accordance with (A).

When searching for a suitable eluent the following mixture can be started with: 90% n-hexane / diethyl ether / 2% formic acid conc.

Develop LTC in a small LTC-chamber design and observe the result under the UV lamp at $\lambda = 254$ nm. Next, the separation is assessed based on the

R_f-values of the three substances. If an optimal separation of the substances is achieved, the eluent found can be used for separation by HPTLC.

(C) Determination of salicylic acid, acetylsalicylic acid, phthalic acid and validation of the method

C.1, C.2

Now measure all standard solutions five times each and for all three analytes create the measurement function with the peak area as an analytical signal.

Using a linear regression, determine the analytical sensitivity and the intercept of the three measurement functions with their standard deviations.

Calculate the detection limits of the method by means of the signal/noise-ratio method.

Application, development, scanning

To determine the R_f -values of the analytes both the mix-standard (S4) and the solutions (AR, SR and PR) of the individual analytes are applied (sample volume 5 µl). In addition, diluted standard solutions (V1–V5) are applied to the TLC plate at the same time to determine the LOD. In addition, at least one track should remain free (blank) on which to be able to determine the fluctuations in the base line.

The TLC is developed with the optimum eluent mixture, which was determined under item B.2.

C.3

Validate the method.

C.4

Determine the concentrations of salicylic acid, acetylsalicylic acid, phthalic acid in the combination product (to be delivered by the assistant).

Results:

Standard deviation	(p = 95%)
Standard deviation	(p = 68%)
Confidence level	(p = 68%)
Confidence level	(p = 99%)

(D) Investigate the influences of the sample amount and its traveled distance

Examine the influences of the sample volumes 10 µl (instead of 5 µl) and the path 8.5 cm (instead of 5 cm) on separation.

8.1.6.4 Questions

(1) Densitometry: Explain the principle of measurement.
(2) What data do you need to describe when you pass on a HPTLC method?
(3) What variables affect selectivity in the HPTLC?
(4) What is an R_f-value? How is the R_f-value calculated? What is the significance of the R_f-value?
(5) What problems can arise during application?
(6) What problems can arise during development?
(7) How does separation take place in the HPTLC?
(8) What is the hR_f-range?
(9) What is meant by the robustness of a method? Specifically pertaining to the HPTLC?
(10) When is a derivatization carried out in TLC?

References

1 Snyder, L.R., Kirkland, J.J., and Dolan, J.W. (2010) *Introduction to Modern Liquid Chromatography*, John Wiley & Sons, Inc., Hoboken, ISBN: 978-0-470-16754-0.

2 Stahl, E. (1965) *Thin Layer Chromatography: A Laboratory Handbook*, Springer-Verlag.

8.2 Spectroscopy

8.2.1 UV–VIS Spectroscopy, Project: "Evaluation of Potential Saving through Use of Optimized Alloys"

8.2.1.1 Analytical Problem

Ferrit Ltd produces stainless steels with good resistance to both stress corrosion cracking, and pitting. The steels are used in heat exchangers with seawater as the cooling water, in tanks and pipes in desalination plants, in flue gas desulfurization equipment and in pipelines in refineries and petrochemical plants. However, the company's alloys and nickel-based superalloys have limitations, not only in terms of production capacity but also in their high price compared to typical stainless steels. In this context there are on-going studies to develop an improved corrosion-resistant stainless steel by controlling the composition of the alloying elements of the steel.

The Board is promoting the development of steel containing reduced quantities of nickel, which is expensive. For this purpose the development leader of the analytical group has been tasked to develop and validate a rapid analytical method for the quantification of nickel, chromium and manganese in steel samples.

Object of Investigation/Investigation System
Alloys

Samples
Steel plates

Examination Procedure
Ultraviolet and visible spectrometers have been in general use for the last 35 years and over this period have become the most important analytical instrument in the modern day laboratory. In many applications other techniques could be employed but none rival UV–Visible spectrometry for its simplicity, versatility, speed, accuracy and cost-effectiveness.

Due also to the numerous feasibilities to convert non-absorbent substances by a chemical reaction into photometrically detectable derivatives, this method has to this day retained high priority, especially in trace analytics. In environmental analytics photometric determination methods are applied particularly frequently. Photometry can be used in laboratories, for example, with registered, microprocessor-controlled devices, as well as at the site of sampling in rivers, landfills, and so on with simple battery-operated filter photometers.

Analytes
Nickel, chromium, manganese

Keywords
Beer's law, generation and detection of electromagnetic radiation, absorption of electromagnetic radiation, chromophores, multicomponent analysis, determinants-method, digestion as sample preparation

Bibliographical Research (Specific Topics)

Pretsch, E., Buehlmann, P., and Badertscher, M. (April 23, 2009) *Structure Determination of Organic Compounds: Tables of Spectral Data*, 4th edn, Springer, ISBN-13: 978-3540938095.

Hesse, M., Meier, H., Zeeh, B., and Dunmur, R. (2007) *Spectroscopic Methods in Organic Chemistry*, 2nd edn, Thieme Medical Pub, ISBN-13: 978-3131060426.

Perkampus, H.H. (May 27, 1994) *UV-VIS Spectroscopy and Its Applications (Springer Lab Manuals)*, 1st edn, Springer, ISBN-13: 978-3540554219.

Project Process/Task

(A) Photometric determination of manganese in steel (standard addition method)
Determine the manganese content of a steel alloy after a suitable digestion step.

(B) Multi-component analysis of chromium (Cr^{3+}) and nickel (Ni^{2+})

The Cr^{3+} + and Ni^{2+} ion concentration of a decomposition solution of a stainless steel is to be determined by multicomponent analysis.

B.1 Determine the chromium and nickel concentrations (including measurement uncertainty) in the digested steel sample.

B.2 Estimate the measurement uncertainty of the method.

B.3 Validate the method.

(C) Project reporting

The results are to be submitted in the form of a report (or presentation). Comparisons with bibliographical references are to be made.

8.2.1.2
Introduction

Spectrophotometric analysis is based on the measurement of light absorption (visible and ultraviolet radiation) by substances across a range of wavelength (190–400 nm UV, 400–780 nm VIS). The light absorption is caused by electron transitions between different states in the molecule. UV–VIS spectroscopy is also known as electron spectroscopy, since the absorption of electromagnetic radiation in the UV and visible range for the excitation of electrons takes place exclusively in the outer orbitals. Electron spectroscopy deals with the interaction of electromagnetic radiation, in the wavelength range of approximately 100 to 800 nm, with matter. Light absorbance measurements are used for the quantitative analysis of inorganic and organic compounds, and to a lesser extent also for the qualitative analysis of molecules, radicals and ions. The term colorimetry is presently the umbrella term for those physico-chemical methods of measurement that use light absorption of colored solutions for quantitative analysis of dissolved substances. In a narrower sense colorimetry is based on a direct color comparison without a meter. The visible color represents the complementary color of the absorbed light in each case. Photometry is the term used for absorption or transmittance investigations. When monochromatic light is used, the process is termed spectrophotometry. Upon passage of electromagnetic radiation through a solution, its intensity is reduced by absorption. *Bougouer* and *Lambert* found that the absorption band or absorbance (A) of the layer thickness (d) depends on the distance the light travels through a material (path length). *Beer* discovered the proportionality of absorption in which there is constant layer thickness only from the standpoint of the molar concentration (c) of a light-absorbing substance. The combination of both laws is the *Bouguer–Lambert–Beer* law, usually called *Lambert–Beer's* law [1–3].

$$A = \log\frac{I_0}{I} = \varepsilon * c * d \tag{8.35}$$

A Absorbance of the analyte

I_0 Intensity of the incident light (100%)

I Intensity of the emitted light (% of I0)

ε Molar extinction coefficient ($l\ mol^{-1}\,cm^{-1}$)

c Molar concentration ($mol\,l^{-1}$)
d Path length (cm)

Definitions

The dimensionless unit $A_\lambda = \varepsilon_\lambda cd$ is absorption (outdated: extinction, optical density). The absorption coefficient $A = 1 - T$ is often cited instead of the transmittance degree *T*.

Chromophores

In comparison to molecules with only s-electrons, molecules with n-electrons, show a shift in the absorption bands to the visible area of the spectrum. Groups that can bring about this effect are called chromophores. The accumulation of chromophores in a molecule results in a bathochromism, a deepening of color affected by a shift of the absorption maxima to longer wavelengths. A shift to shorter wavelengths is called a hypsochromic effect. A chromophore is a system (atom, ion, functional group, molecule), whose electrons are excited by photons, causing an absorption. During absorption the electron goes into a higher state of energy which can be left again by giving off energy in the form of radiation (light quanta). These excitations take place depending on the degree of affinity and type of affinity of the different photon energies. Therefore, important conclusions can be drawn about the binding energies and the nature of bound atoms from the position of the absorption bands in a UV–VIS spectrum.

Simple, non-conjugated chromophores are generally not very valuable from a diagnostic standpoint for the organic-chemical structural analysis. Conjugated chromophores are essentially more valuable for structural analysis than the simple chromophores.

The Photometric Process

After conversion of non-absorbing compounds to detectable derivatives, non-absorbing compounds such as metal ions can be photometrically determined as a colored metal complex. This area is also the focal point of application. A prerequisite for the applicability of metal reagents apart from the highest possible selectivity in complex formation is the overlapping of a sufficient difference in the absorption spectra of reagent and complex. In this process it is necessary to perform a measurement against a reagent blank solution. Because of the additivity of absorptions it is possible to perform a photometric two-component analysis when there are sufficient differences in the absorption spectra of the individual substances. The measurements are performed here at the positions of two maxima. If the molar extinction coefficients for both compounds at λ_1 and λ_2 are known, the concentrations c_1 and c_2 can be determined with only two measurements according to Eqs. (8.36) and (8.38).

The superposition of signals, that is, of absorption bands, can also be detected with the help of derivative spectrophotometry. In general, during derivative methods the mathematical derivations of a curve are formed. This means that even without separation individual substances in mixtures can be determine without any disturbance.

Multicomponent Analysis

It is assumed that the number of components and their molar extinction coefficients are known. In this case, only the concentrations are unknown. While it is easy to determine the concentration of several components in an analytical mixture together, if for each of these components there is a range of wavelengths acting as the only absorbant, unfortunately, this most favorable case is very rare in practice. Rather, mostly all spectra overlap. Under the assumption that the absorption bands of two components in a sample do not overlap too much, (maximum about 10% overlap), they can be quantified without prior separation.

The absorptions of two components (1 and 2) are additive:

$$A(\lambda)_{\text{mixture}} = A_1 + A_2 = d(c_1\varepsilon_1 + c_2\varepsilon_2) \tag{8.36}$$

Spectrometer/Photometer

A spectrometer or photometer is employed to measure the amount of light that a sample absorbs. The instrument operates by passing a beam of light through a sample and measuring the intensity of light reaching a detector.

An instrument with light sources that enables the recording of the complete spectra in the UV–Vis range is called a spectrometer (in contrast to an apparatus having interference filters that is called a photometer. Filters of colored glass or gelatine are the simplest form of selection, but they are severely limited in usefulness because they are restricted to the visible region, and they have wide spectral bandwidths. Typical bandwidths are rarely better than 30–40 nm.) As the light source a deuterium lamp is used for the UV region, and a tungsten halogen lamp for the visible region.

To generate monochromatic light (light of a single wavelength), polychromatic light of the continuum source is dispersed into its component wavelengths by prisms (dispersive prisms are used to break up light into its constituent spectral colors because the refractive index depends on frequency) or gratings (a fundamental property of gratings is that the angle of deviation of all but one of the diffracted beams depends on the wavelength of the incident light. Therefore, a grating separates an incident polychromatic beam into its constituent wavelength components, i.e., it is dispersive) in a monochromator.

The polychromatic light enters the monochromator through the entrance slit. The beam is collimated, and then strikes the dispersing element at an angle. By moving the dispersing element or the exit slit, light of only a particular wavelength leaves the monochromator through the exit slit.

Photometers, however, differ in both the light source and the wavelength selection. Instead of a continuum light source, a spectral lamp, usually a mercury vapor lamp with emission maxima of 254, 366 and 560 nm, is used. Recently, xenon lamp sources have been introduced, and these cover the UV and visible range. Apart from monochromators, filters can also be used for wavelength selection. A cuvette containing the sample solution is irradiated by the monochromatic light. The cuvettes are either rectangular or round in shape and are made of glass or, in the field of UV measurements, of quartz. In the case of lower accuracy requirements, plastic cuvettes are also suitable. The layer thickness must be strictly adhered to. There should be no dispersion on the surfaces of the cuvettes. The detector of a photometer consists of a so-called quantum detector, electrochemical components, in which the electrical properties change due to the incident radiation or in dependence on the radiation intensity. Photovoltaic cells have an outer photoelectric effect and photoelements an inner one. Furthermore, a distinction is made between single-and dual-beam photometers. In the case of the single-beam process, first the zero point is set with a closed beam path and pre-selected wavelength. Next the cuvette containing the blank solution is placed in the beam path and the absorption is adjusted to zero through the size of the aperture. After that, the absorption of the sample solution can be determined. In the dual-beam method, the light beam is split into two coherent beams using a mirror combination. One of the beams is measured after passing through the analysis solution and the other unhindered. Through a slit of the monochromator light of a defined wavelength range λ enters via a filter, prism or grating. A narrower slit is required to permit, at best, monochromatic light to enter, and thereby to attain the validity of *Lambert–Beer's* law. The spectral bandwidth also affects the resolution of an absorption spectrum. In this connection, particular attention is drawn to the sources of error when radiation passes through a measurement cuvette. Photometric cuvettes must have great optical quality and precise dimensions. Erroneous absorbance measurements can be caused, for example, by light scattering on soiled cuvette windows and by dust particles and/or turbidity in the solution, by light reflection on the wall of the cuvette and self-absorption of the solvent used. Other sources of error are temperature influences, a wrong choice of the wavelength setting, and an insufficient resolution capacity of the device.

Such an arrangement is suitable for measuring absorption at a fixed wavelength. In practice it is, however, often necessary to measure different samples at varying wavelengths or to record spectra of samples. This is accomplished by diode arrays (Figure 8.35). Diode arrays make it possible to record the entire spectrum at once. No wavelength change mechanism is required and output presentation is virtually instantaneous.

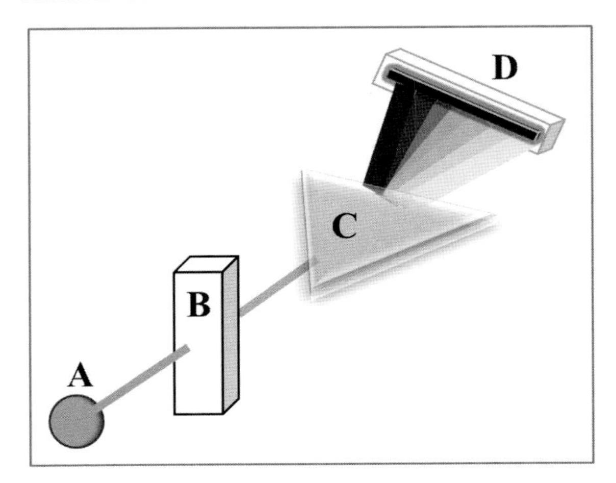

Figure 8.35 Schematic design of a diode array spectrophotometer. **A** light source, **B** sample, **C** polychromator, **D** diode array.

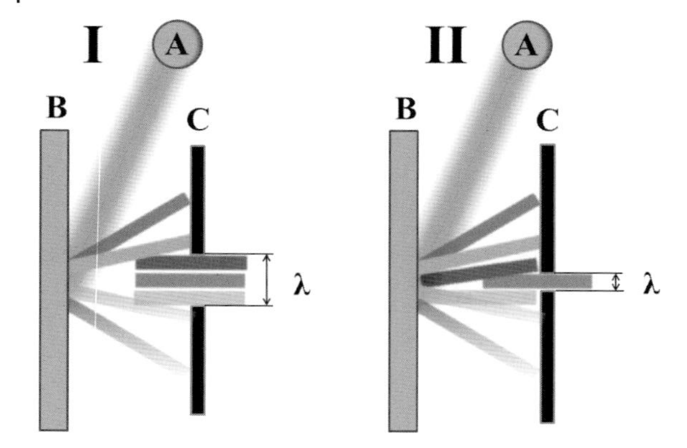

Figure 8.36 Influence of slit width on the spectral bandwidth. **A** light source, **B** grating or prisma, **C** aperture, **λ** spectral bandwidth.

8.2.1.3 Bandwidth

In UV-spectroscopy, the sample is irradiated with monochromatic light. The monochromator, usually an optical grating, fans out the polychromatic radiation of the light source according to the wavelength. A small part of the spectrum is now selected with a slit and used for further measurements. The remaining radiation is lost. The spectral width of the sub-spectrum can be adjusted through the width of the aperture (Figure 8.36). For this one wants as much light as possible, that is, a wide aperture to keep a high signal to noise ratio and a short measurement time. The width of the slit is thus to be chosen wisely. The smaller the spectral bandwidth, the finer the resolution. The spectral bandwidth depends on the physical slit width and the dispersive power of the monochromator. For diode array instruments, the resolution also depends on the number of diodes in the array.

Solvent Interactions, Solvatochromic Effect

The position and structure of a band may depend on the solvent. The choice of the solvent depends mainly on the solubility of the compound to be examined and the spectral range chosen. Due to the strong interaction with the sample molecules, polar solvents such as ethanol often lead to a blurring of the fine-structure of the bands. Since the position of an absorption maximum may depend on the solvent, it is absolutely imperative to specify the solvent. In low-polar compounds, the shift is only a few nanometers. Nevertheless, there are compounds which react particularly sensitively to solvent changes. The pronounced dependence of a compound on the polarity of the environment is called solvatochromism.

The betaine dye shows an absorption maximum in water at $\lambda = 453\,\text{nm}$, and in diphenyl ether at $\lambda = 810\,\text{nm}$. It changes color well according to the polarity (Table 8.6). It was even used to characterize the environment, for example, in a micelle or on a surface, with respect to polarity. With a wingspan of 357 nm, the dye displays one of the largest solvatochromic effects ever observed. The reason lies in

Table 8.6 Influence of polarity on absorption maximum (polarity increasing from top to bottom) of betaine dye.

Solvent	λ max (nm)	Color
Diphenyl ether	810	Colorless
Anisole	771	Yellow
Acetone	677	Green
Isopentanol	588	Blue
Ethanol	550	Violet
Methanol	516	Red
Water	453	Colorless

Polarity

the charge separation in the ground electronic state, which is especially well stabilized by polar solvents. The excited state is less polar. To accomplish the transition, a relatively large energy in a polar solvent is required.

Stainless Steel

Stainless steel is characterized by a high percentage of chromium, which must be dissolved in the ferrite mixed crystal. Other alloying elements such as nickel, molybdenum, manganese, lead to an even better corrosion resistance and more favorable mechanical properties.

In the refinement of iron to steel, manganese is added at a concentration of 0.1 to 1 wt%. It prevents oxidation and increases hardening capabilities. Furthermore, the yield strength, the stability and malleability are improved. An important aspect of manganese in the manufacture of steel is the fact that it is one of the ferrite formers. If, for example, liquid iron is cooled slowly, mixed crystals are separated from the melt as soon as the temperature falls below the melting point. The stability range of this so-called ferrite mixed crystal (also called gamma-iron or austenite) is between 1392 and 911 °C. By means of suitable additives to alloying elements (such as manganese), the stability range of the austenite can be extended to room temperature. Accordingly, these steels are called austenitic (or ferrite) steels. These steels are not magnetic and cannot be magnetized and are characterized by high strengths. Normally, for analysis of the chemical composition, the colored ions of the transition elements Cu, Fe, Ni, Co, Cr, and so on can only be directly used and only in larger concentrations for photometric measurements due to the relatively small values of the extinction coefficients. Manganese is, however, already suitable for photometry, even without complexing agents, due to the intense purple color of the MnO4 ions. Even small amounts can be very well determined after full oxidation (e.g., with KIO4).

8.2.1.4
Material and Methods

Equipment

UV/VIS spectrometer (UV-1800, Shimadzu), microwave decomposition equipment, quartz glass cuvette, PC for data recording.

Chemicals

Chromium salt/ chromium (III) chloride hexahydrate	CAS Number 10060-12-5	Xn	
Nickel(II) chloride	CAS Number 7791-20-0	T, N	
Manganese(II) chloride tetrahydrate	CAS-Number 13446-34-9	Xn	
Hydrochloric acid $1.5 \cdot 10^{-3}$ mol/l	CAS Number 7647-01-0	C	
Sulfuric acid	CAS Number 7664-93-9	C	
Phosphoric acid	CAS Number 7664-38-2	C	
Nitric acid	CAS Number 7697-37-2	O, C	
Sodium nitrite	CAS Number 7632-00-0	O, T, N	
Potassium periodate	CAS Number 7790-21-8	Xn, O	

Ultrapure water
Steel chips

Reagents

Mixed acid
Fill 150 ml of H2SO4 conc. and 150 ml of H3PO4 conc. up to 1000 ml
KIO4 (potassium periodate) solution: 5 g KIO4/100 ml HNO3 (20 vol%)
NaNO2 solution: 2 g NaNO2/100 ml (2 wt% in H2O)
HNO3 (20%ig)
Mn(II) standard solution: 0.1 g of Mn (II) dissolved in 1000 ml of water

Production of Manganese Stock Solution by Digestion

The digestion is carried out to transform the solid sample into a dissolved (aqueous) form and to destroy the matrix, or to chemically modify it in such a way that the analyte is dissolved. The steel chips produced are placed in a 100 ml Erlenmeyer flask with 35 ml of mixed acid and dissolved by heating. If, after the reaction, insoluble carbides are still present, these are dissolved by adding 1–2 ml of nitric acid (20%) (Caution! Allow the solution to cool slightly, otherwise foaming could occur!).

The cooled solution is then mixed with 20 ml of distilled water and 10 ml of nitric acid (20%) and again brought to the boil.

After cooling the sample, it is transferred into a 100 ml volumetric flask and filled to the mark with water.

This solution serves as an analyte stock solution for the manganese analysis.

> The usual laboratory safety rules apply but additional care should be taken in the following instances:
>
> In carrying out digestion, protective clothing must be worn (lab coat, safety goggles, latex gloves).
>
> *Caution!* Nitrous gases are formed when using conc. nitric acid.
>
> The proper functionality of the vapor exhaust hood is to be ensured!
>
> **Only work under the hood!**

"Digested solution" chromium (Cr^{3+}) and nickel (Ni^{2+})

Since the intrinsic color of nickel and chromium as a solution can only be detected photometrically in large concentrations, a synthetic decomposition solution is resorted to.

Guideline: The concentrations are 8 g of Cr^{3+}/l and 6 g of Ni^{2+}/l. The solution is to be acidified.

Experiment

(A) Photometric determination of manganese in steel (standard addition method)

1. Manganese content is analyzed using the standard addition method (see Section 4.3.3.3). First, a test sample is prepared to estimate the manganese content and adherence to the linear range. For this, 5 ml of analyte stock solution is mixed with 10 ml of Mn standard solution and heated to boiling point. This is followed by the addition of 10 ml of potassium periodate solution and after the first appearance of a violet coloration it is heated for another 4 min. The purple sample solution is transferred to a 100 ml volumetric flask after cooling and filled to the mark with water.

2. Next, NaNO2 solution is poured into a test tube drop by drop (!) until the solution is completely decolorized. With this solution, the photometer is set to zero.

3. The photometric measurement is performed at a wavelength of 525 nm.

4. Steps 1 and 3 are repeated with aliquots (5 ml, or possibly less) of the analyte stock solution with various additions of Mn standard solution (2, 5 and possibly 10 ml).

Using a plot of absorbance values measured against the added amount of manganese the content of the stock solution is determined.

(B) Multicomponent analysis of chromium (Cr^{3+}) and nickel (Ni^{2+})

B.1 Prepare a chromium solution with about $1\,g\,l^{-1}$ of Cr^{3+} and a nickel solution with about $3.5\,g\,l^{-1}$ of Ni^{2+} and measure the VIS spectrum of both solutions between 400 and 800 nm. In the spectra look for two different wavelengths suitable for multi-component analysis and determine the respective molar extinction coefficients of each analyte. Prepare several standard solutions and plot their absorbances against the concentrations.

Now measure the spectrum of the unknown sample in the same wavelength range and calculate the absorbance of the two previously selected wavelengths.

Then run through the multi-component analysis according to Eqs. (8.37) and (8.38).

If the molar extinction coefficients of compound X and compound Y at λ_1 are described as $\varepsilon_1(X)$ and $\varepsilon_1(Y)$ and at λ_2 as $\varepsilon_2(X)$ and $\varepsilon_2(Y)$ the following applies:

$$A_1 = d[c_X\varepsilon_1(X) + c_Y\varepsilon_1(Y)] \text{ and } A_2 = d[c_X\varepsilon_2(X) + c_Y\varepsilon_2(Y)] \tag{8.37}$$

By solving the linear system of equations, the desired concentrations $c_A + c_B$ are obtained:

$$c_A = \frac{[A_1\varepsilon_2(Y) - A_2\varepsilon_1(Y)]}{d[\varepsilon_1(X)\varepsilon_2(Y) - \varepsilon_1(Y)\varepsilon_2(X)]} \text{ and } c_B = \frac{[A_1\varepsilon_2(X) - A_2\varepsilon_1(X)]}{d[\varepsilon_1(Y)\varepsilon_2(X) - \varepsilon_1(X)\varepsilon_2(Y)]} \tag{8.38}$$

Several software packages or even mathematical tools on scientific calculators are available for solving the linear system of equations and these can be tested during the internship.

B.2 One way to estimate the uncertainty of the measurement method is to repeatedly combine different pairs of wavelengths. This provides several varying results, from which the mean and its uncertainty can be determined. Do this calculation for six different pairs of wavelengths.

B.3 Validate the method.

8.2.1.5 Questions

(1) What is the difference between a single-beam and a two-beam photometer?
(2) According to what criteria are the two wavelengths of (B.1) chosen?
(3) List the major instrumental components of a photometer.
(4) How can the resolution of a spectrometer be improved?
(5) Comment on the differences in a spectrum, which occur at temperatures of 20 and 50 °C.

(6) List the important applications of UV–VIS spectrometry.
(7) Explain the influence of the pH-value on UV spectra.

References

1 Bouguer, K.P. (1729) *Essai d'Optique sur la gradation de la Lumiere*, Claude Jombert, Paris.
2 Lambert, J.H. (1760) *Photometria, sive de mensura et gradibus luminis colorum et umbrae. Sumptibus Vidae Eberhardi Klet*, Sumptibus Vidae Eberhardi Klett.
3 Beer, A. (1852) Bestimmung der Absorption des rothen Lichts in farbigen Flüssigkeiten. *Annalen der Physik und Chemie*, **86**, S.78–S.87.

8.2.2
Fourier-Transform Infrared Spectroscopy (FTIR), Project: "Benchmarking with a New Competitive Japanese Product"

8.2.2.1
Analytical Problem

The advent of plastics cannot be overlooked, especially in the packaging industry. Here the focus of attention is, in particular, their use in the food industry, to preserve the quality of fresh products for a long time. They are mainly used for conventional packaging foils of polyethylene and polystyrene. The disadvantage is the relatively high oxygen permeability. For this reason, oxygen barriers in the form of multilayer foils are prepared at great expense. A less expensive way to protect the package contents from oxygen is to trap the oxygen in the foil. This can be achieved by mixing additives into the synthetic materials before producing them. This manufacturing process is known as additivation. Such substances can be regarded as oxygen absorbers or O_2 scavengers.

A new product of a Japanese competitor is advertised with the advantage of having reduced oxygen diffusion to a minimum thanks to the increase in film thickness and appropriate additives. These new films should be analyzed and their formulation should serve as the basis for our own product. The R&D department has the task of developing an appropriate testing method to qualitatively determine the layer thickness of the foils, the additive characterization and the composition of the original product.

Object of Investigation/Investigation System
Packaging material

Samples
Foils and granulates

Examination Procedure

IR spectrometry is used primarily in the analysis of organic materials. Depending on the task, we distinguish between the applications of qualitative analysis, structural analysis and quantitative analysis. In qualitative analysis, for example, the identity of a compound is proven by a spectral comparison, whereby the substance to be tested must be in the purest possible form. Contributions to the structural analysis are obtained by interpretation of an IR spectrum, whereby, in contrast to UV–VIS spectrometry, the spectra are not recorded in their dependence on the wavelength, but on the number of waves. The interaction between the infrared radiation and the sample is subject to Beer's law, that is, the absorbance values have a linear relationship to the concentration or layer thickness of a sample, (this applies to diluted samples). FTIR spectroscopy is widely used in the incoming inspection of goods (identity checks).

Analytes

Polystyrene, polyethylene

Keywords

Dipole moment, IR-light sources, harmonic/anharmonic oscillator, ATR-technique, Michelson interferometer, Fourier transformation, molecular vibrations, IR-frequency ranges, pyrolysis, polymer analysis

Bibliographical Research (Specific Topics)

Stuart, B.H. (2004) *Infrared Spectroscopy: Fundamentals and Applications (Analytical Techniques in the Sciences (AnTS) *)*, 1st edn, Wiley, ISBN-13: 978-0470854280.

Griffiths, P., and De Haseth, J.A. (April 25, 2007) *Fourier Transformation Infrared Spectrometry (Chemical Analysis: A Series of Monographs on Analytical Chemistry and Its Applications)*, 2nd edn, Wiley-Interscience, ISBN-13: 978-0471194040.

Hollas, J.M. (January 19, 2004) *Modern Spectroscopy*, 4th edn, Wiley, ISBN-13: 978-0470844168.

Project Process/Task

(A) Determination of the layer thickness of polystyrene/polyethylene foils

A.1 Calculate the thickness of various plastic foils.

A.2 Calculate the combined measurement uncertainty of the layer thickness (d). Does the calculated value deviate significantly from the manufacturer's specifications? If so, what could be the cause of this systematic deviation?

(B) Identification of a plastic foil

The polymer basis of a plastic foil is to be identified.

(C) Identification of a plastic additive

In production plastics are often mixed with additives (plasticizers, stabilizers, dyes, etc.) to improve their properties.

The task is now to clarify the structure of the additive in the present sample.

The following two samples are to be inspected:

Foil No. 1 Plastic without additives

Foil No. 2 Plastic with additives

(D) Specific sample preparation technique – pyrolysis

Identify (by pyrolysis) the polymer basis of a test sample (granulates).

(E) Project reporting

The results are to be submitted in the form of a report (or presentation). Comparisons with bibliographical references are to be made.

8.2.2.2 Introduction

Some Physical Relations

Wavelengths
The measurement dimension for electromagnetic radiation is the wavelength (λ). The unit is μm.

Frequency
Measurement dimension for vibration frequency is (ν). It is defined as the number of oscillations performed by the oscillating electrical (or magnetic) vector of radiation executed per time unit.

$$1\,\text{Hz} = 1\,\text{s}^{-1} \tag{8.39}$$

Frequency and wavelength are linked together by their propagation speed, the speed of light (c_{air}= 3.10^{10} cm s^{-1}) as follows:

$$\nu\,(\text{s}^{-1}) = \frac{c}{\lambda}\,[\text{m s}^{-1}\ \text{cm} - 1] \tag{8.40}$$

Wavenumber
Especially in IR spectroscopy, a third unit is used: the wavenumber $\overline{\nu}$, the reciprocal value of the wavelength:

$$\overline{\nu} = \frac{1}{\lambda}\,[\text{cm}^{-1}] = \frac{10^4}{\lambda}\,[\mu\text{m}^{-1}] \tag{8.41}$$

Radiation Energy

In order to generate electromagnetic radiation, energy must be expended in some way. Electromagnetic radiation is an energy carrier. The energy and frequency of radiation are in the following ratio:

$$E = h\nu = h\frac{c}{\lambda} \qquad (8.42)$$

E Radiation energy (J)
h Planck's constant ($6.626–10^{-34}$ J s)
ν Frequency (s^{-1})

If an electromagnetic wave interacts with matter, it can lead to an energy exchange, whereby the radiation energy E is absorbed in a material sample. This intake of radiation energy is called absorbance. In spectroscopy the term transmittance (permeability) is often used, and refers to that portion of the intaken radiation not absorbed by the sample. Absorbance A and transmittance T are related: $T = 1 - A$.

The absorbance or transmittance is a very characteristic, frequency-dependent property of each chemical. The measurement and interpretation are the subject of optical spectroscopy. What all spectroscopic methods have in common is that they use the interaction of electromagnetic radiation with atoms/molecules to extract information from them. In so doing, chemists use the entire electromagnetic spectrum to obtain a variety of information.

Infrared Spectroscopy

Infrared spectroscopy is one of the oldest and most widely used analytical spectroscopic methods. It enables chemists to make qualitative statements about partial structures of molecules.

IR spectroscopy is a vibrational spectroscopic method, in which molecular vibrations (translations, rotations, vibrations) are excited and detected.

- **Translation:** The movement of the whole molecule in the three spatial directions.
- **Rotation:** Rotation of the whole molecule around the three axes of rotation.
- **Vibration:** Periodic motion of individual atoms or groups of atoms of a molecule in relation to each other. These vibrations are characteristic of certain functional groups within the molecules.

Types of vibrations (see also Section 8.2.3)

Localized Vibrations Localized vibrations are vibrations of individual atoms or functional groups.

These vibrations either occur in the direction of the bond (stretching) or they deform the bond angle (bending vibrations).

Stretching vibrations can be symmetrical or asymmetrical. A distinction is made between spreading, bending, rocking twist and wagging vibrations.

Structure Vibrations Structure vibrations have relatively low frequencies (fingerprint region below 1500 cm^{-1}).

Overtone or Harmonic Vibrations These oscillations are usually a multiple of the fundamental oscillations.

Combination of Vibrations The major types of molecular vibrations are stretching and bending. Infrared radiation is absorbed and the associated energy is converted into these types of motions. The absorption involves discrete, quantized energy levels. However, the individual vibrational motion is usually accompanied by other rotational motions. These combinations lead to the absorption bands, not discrete lines, commonly observed in the mid-IR region.

Frequency Range of Infrared Spectroscopy

By definition, three areas can be distinguished in infrared spectroscopy:

- Near-IR (NIR) Range: 12 800–4000 cm^{-1}, 0.78–2.5 μm

  In the NIR only overtone and combination vibrations of the fundamental vibrations observed in the mid-IR region are registered.

- Average IR (mid-IR): 4000–400 cm^{-1}, 2.5–25 μm

  The analytically most important area of IR-spectroscopy is in the mid-IR region.

  The mid-IR includes the interactions of the valence, deformation and skeletal vibrations. In this way information on functional groups, sub-structures and isomers of a molecule are provided.

- Far-IR range: 400–33 cm^{-1}, 25–300 μm

  In the far-IR primarily torsional and skeletal vibrations are recorded. Measurements in this area are not instrumentally simple, and are mainly used only in special cases.

Dipole Moments

A vibration is IR-active when during vibrational excitation a dipole moment (alternating electromagnetic field) is induced, the basic condition of which differs from the ground state. The oscillation must, therefore, have a transitory dipole moment. The stronger this dipole moment, the more intense the bands in the IR spectrum. Homonuclear diatomic molecules (H_2, N_2, Cl_2, etc.) do not have any IR-active bands. In general, it can be said that in a molecule with high symmetry, all vibrations that occur symmetrically to the center of symmetry, are IR-inactive, as the dipole moment remains unchanged during the process. These vibrations are, however, Raman-active. Raman spectroscopy is also a vibrational spectroscopy, but with different selection criteria.

Both Raman and IR spectra often provide complementary information. In Raman spectroscopy polarizability and not the dipole moment is responsible for absorbance.

Quantitative Analyses

As in UV–VIS spectrometry, Beer's law is valid as a basis for quantitative analyses conducted by IR spectrometry. Different IR-absorbance spectra can also help to analyze multiple-component mixtures, whereby the quantitative evaluations are performed at several wavenumbers. Sensitivity, precision and detection limits in IR spectroscopy are, however, less favorable than in spectrophotometry. For a quantitative analysis aimed at determining the content of a substance, it is first necessary to define the zero and baselines (Figure 8.37) after having selected an absorption band.

In accordance with the baseline method, which allows the elimination of possible background influences, the zero line is drawn through the absorbance maximum by the intersection of the baseline and the vertical line The spectral absorption mass is obtained from $A = \log I_0/I = \log T_0/T_S$ [1].

Instrumentation

IR Spectrometry The basic structures of UV–VIS spectophotometers and IR spectrometers differ only slightly. However, unlike spectrophotometers, the common light sources of an IR spectrometer are Nernst glowers (constructed of rare-earth oxides), which provide IR radiation when electrically heated to about 1600 °C, "globars" (silicon carbide rod), or Nichrome coils. They all produce continuous radiation, but with different radiation energy profiles. The widely used Nernst glower is in turn divided into two light beams of equal intensity in two-beam devices. In conjunction with a rotating mirror, measuring and reference beams are alternately switched to the monochromator behind the cuvette. A thermoelement serves as the detector. Thermal detectors can be used over a wide range of wavelengths and they operate at room temperature. However, this type of spectrometer has considerable disadvantages as opposed to modern Fourier transformation spectrometers with regard to measurement time, resolution capacity and S/N ratio. The reason is that only an approximate isolation of the narrowest possible wavelength intervals by the dispersing element is possible.

The development of modern computer technology has made it possible to automate the Fourier transformation and accelerate it. This in turn has made it possible to render FTIR spectrometers accessible to the whole range of use. In this process, wavelengths are not measured separately. Rather, all wavelengths simultaneously record an interferogram, from which the spectrum is calculated using the Fourier transformation.

The FT technique has prevailed as a standard method and managed to drive the traditional IR-devices from the market completely.

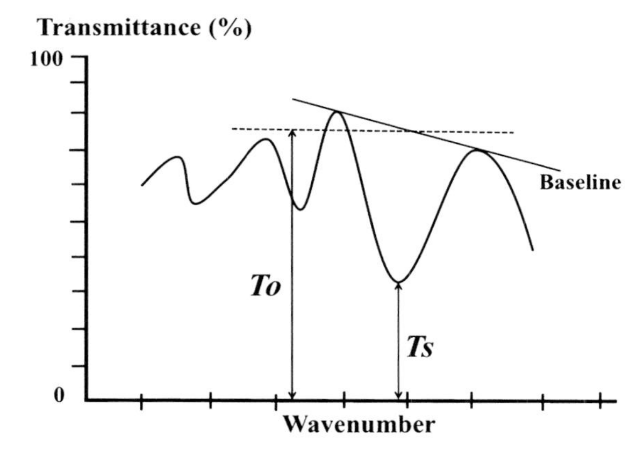

Figure 8.37 Quantitative IR-analysis.

Interferometer An interferometer is an optical design used to separate electromagnetic waves (light), and which causes them to move against each other and overlap again after this shift. The simplest interferometer known is the Michelson interferometer (Figure 8.38).

Based on the principle of the Michelson interferometer, radiation first reaches a beam splitter (a semitransparent optical drive). One part is reflected at the beam splitter to a fixed mirror (located in the reference arm of the interferometer). The other part is transmitted to a moving mirror (in the measurement arm). According to the position of the movable mirror both sub-bundles have a phase shift (run-time or path difference). In addition to the infrared light used for measuring, the beam of a HeNe laser (red, $\lambda = 633\,\text{nm}$) is reflected into the beam path and from this an interferogram is recorded by means of a photodiode. The fact that the exact wavelength of the laser is known and is constant helps to accurately calibrate the position of the interferometer. Both reflected partial beams are returned to the beam splitter and overlap (interfere) again. The infrared light beam first passes through the interferometer and then through the sample. The characteristic signals are recorded by the detector as interferograms and converted into IR spectra by Fourier transformation.

Advantages of the FTIR Spectrometer over the Dispersive IR Spectrometer The FTIR spectrometer requires no slit in the traditional sense for resolution. By eliminating the discharge gap the total energy reaches the detector and thus contributes to the signal. As a result, the S/N ratio is significantly improved (this advantage is called the *Jacquinot* or throughput advantage). While all frequencies are excited and detected, in contrast to the sequential analysis of wavenumbers in a dispersive spectrometer the measurement time will shorten (multiplex or *Fellgett* advantage). By using the HeNe laser, a wavelength accuracy of the spectrum of

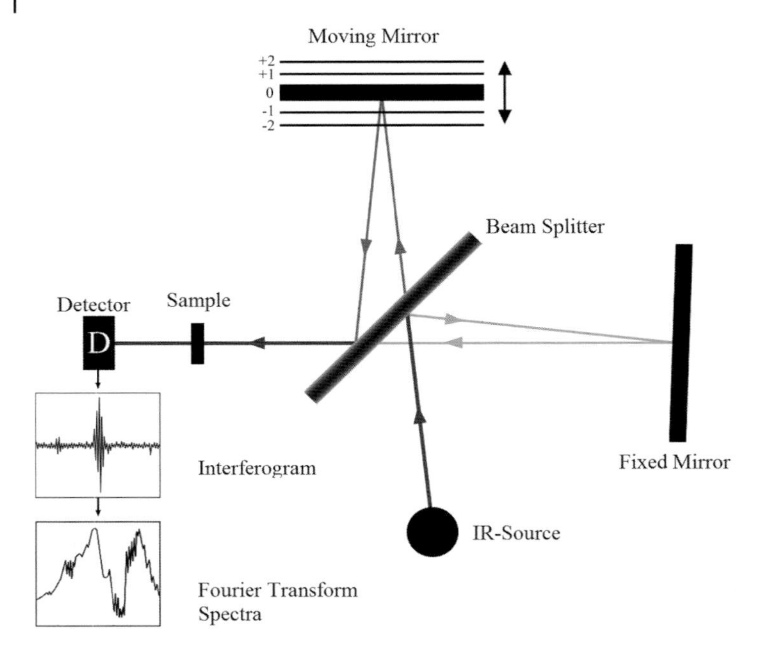

Figure 8.38 Diagram of a Michelson interferometer.

more than 0.01 cm^{-1} (*Connes* advantage) is achieved. Improved resolution capacity and higher energy flow are the advantages of this technique.

The IR Spectrum

An IR spectrum consists of two large areas. Above approx. 1500 cm^{-1} there are absorption bands, which can be allotted to the individual functional groups, while the region below 1500 cm^{-1} contains many bands that are characteristic of the molecule. This area is, therefore, also referred to as a "fingerprint" area and is used, among other things, to establish the identity of a substance with an authentic reference. The identification of the bands within the "fingerprint range" is often difficult and should only be considered to be an additional help. However, it is by no means conclusive. Indeed, the presence of a band is still no evidence of the fact that this functional group is also present in the molecule. It could also be a so-called artifact. In contrast, the absence of a characteristic band is indicative of the absence of this functional group. In IR spectroscopy the transmittance (in %) is usually recorded, as opposed to the wavenumber (cm^{-1}). A usable spectrum has a transmittance of about 10% maximum at absorption bands. The concentration of the sample is to be set at approximately this value. The position of the bands depends significantly on the masses involved in the vibration and the affinity between the two atoms.

Traces of water or other stabilizers in the solvents can lead to bands in the vicinity of 3700 and 3600 cm^{-1} and to an additional signal at about 1600 cm^{-1}. In the case of gas-phase IR spectroscopy, water vapor can cause many sharp bands between 2000 and 1280 cm^{-1}. Dissolved carbon dioxide displays absorption bands at 2325 cm^{-1}. Depending on the solvent, some areas of the IR spectrum can be interpreted, other areas can be used with caution, and some areas should be avoided altogether. False bands often appear, particularly at the edge of solvent restricted areas. They are an artifact of the background compensation.

IR-Measurement Cells

IR spectrometric analysis can be performed with materials in all three aggregate states. The light permeable measurement windows are made of crystals of NaCl, KBr or LiF. With the exception of liquid cuvettes with filler necks, glass cuvettes with larger layer thickness are also commonly used. To record an IR spectrum, for which, depending on the technique, 1 to 15 mg of substance is required, the following standard techniques are also suitable: measurements of liquid between two NaCl disks, of solids as a suspension in paraffin oil, or in a well homogenized mixture with KBr or KCl in the form of a pellet. When using solvents, band-poor liquids, such as tetrachloride, trichloromethane or carbon disulfide are used, the intrinsic spectra of which should be considered. Nowadays fluorinated carbon materials, such as trichlorotrifluoroethane, are used for the IR spectrometric determination of oil content in water samples or in contaminated soils, in accordance with a DIN regulation.

Golden Gate Attenuated Total Reflection (ATR) Unit

The ATR technique is another specific measurement technique used for studies of viscous substances, paint coatings, fibers and, especially, plastics. ATR (attenuated total reflectionc) means something like diminished or weakened total reflection, which can occur during reflection to absorbent materials. The measuring beam is directed to an optically denser medium on which the test sample can be found, in such a way as to be reflected on the optically thinner sample as a result of total reflection. This process repeats itself, whereby each total reflection causes the measuring beam to be absorbed by the sample at the least infiltration into the sample. The numerous reflections lead to a strong increase in the effective layer thickness. In this way, a spectrum is obtained that can be assessed.

The reflected intensity is measured and gives leeway for conclusions about the absorbing medium. The Golden Gate ATR-unit is a basic optical unit with easily replaceable sample plates. The basic optical unit includes deflection mirrors and a focusing lens system. The standard sample plate has a diamond as an ATR-element. The massive diamond Type IIa (type IIa diamonds contain so little nitrogen that it is not readily detected using IR or UV absorption methods) has an incidence angle of 45° and a size of 2 mm × 2 mm. Due to the large contact surface, very large samples of (up to 8 cm width) can be analyzed, such as plastics. The small aperture, however, only permits very small samples, even micro-samples such as single fibers and particles.

Sample Preparation

One significant advantage of IR spectroscopy is the variability of sample preparation. It makes it possible to record an IR spectrum of virtually all substances in their different physical states. What is important is that the samples are pure (>99%) and the water content is <1%, so that there can be no disturbing absorption bands and that the cell windows are not damaged. It should not be forgotten that for each method a reference measurement (background) must be carried out without a substance.

Solids

Preparation as a KBr Pellet A simple way to measure solids is to use the KBr pellet technique. In this process the solid is pressed for 3 min to a transparent single crystal-like tablet of, for example, 10 mm diameter and 1 mm in thickness in a hydraulic press with 100 times the amount of dry powder KBr. Mix (do not grind) the sample into the KBr powder. The removal of a finished pellet from the mold requires skill and practice, but a broken pellet can still be readily used. KBr is, however, hygroscopic, and when rubbed and pressed traces of moisture must be reckoned with.

Advantages

- Standard method of measurement, so many reference spectra are available in spectral databases
- No peak distortion as real-light spectrum
- High detection limits attainable due to increased layer thickness/high analyte concentration

Disadvantages

- Time-consuming sample preparation
- Hardly suitable for quantitative analysis (control of layer thickness!)
- Not suitable for damp and hygroscopic sample materials

Suitable for

- Dry solids, which can be finely ground
- Inorganics and solid polar organics

Not suitable for

- Polymers and other macromolecular substances
- Liquids
- Hygroscopic and wet samples

Preparation as a Suspension in Nujol Approximately 1 mg of finely powdered solid substance is mixed with a special paraffin oil ("Nujol"), mentioned above, in a small agate mortar. The resulting paste is then placed between two NaCl plates in such a way that a bubble-free film is created.

Advantages

- The method is relatively quick and simple.

Disadvantages

- Disadvantages include interference from mulling agent absorption bands. Nujol is used below 1330 cm^{-1}.

Suitable for

- Dry solids, which can be finely ground
- Very reactive samples; the layer of Nujol can provide a protective coating, preventing sample decomposition during acquisition of the IR spectrum

Not suitable for

- Moist samples

Liquids

Preparation as a Solution The best and by far the most widely used method is inclusion in a solution. This method puts high demands on the solvent: it should be easily purified, and chemically and physically inert. As a matter of course, one must naturally assume optical transparence. In areas of great solvent self-absorption statements can no longer be made about the bands of the solute (restricted areas). Ideal solvents should therefore have absorption-free areas in the IR-field of interest. Frequently used solvents are carbon tetrachloride, carbon disulfide, chloroform, n-hexane and dimethyl formamide having the largest "optical window".

The substance is dissolved in a suitable solvent (l to 5% solution) and injected into the cell with a Pasteur pipette. The openings are sealed with a stopper and the cell is positioned in the measuring beam.

The layer thickness of the IR-cell can be easily up to 1 mm. Weak solvent interactions with the solute bring far less interference to the spectrum than with other methods. Also, apart from the gas-phase technique, the solvent method is the only way to achieve excellent quantitative results.

Preparation of a Liquid Film (Sandwich Technique) A drop of liquid is placed between two flat KBr plates so that a bubble-free liquid film is formed.

Advantages

- Real-light spectrum so, in contrast to the ATR technique, no distortion of bands
- No problems due to solvent bands (restricted areas)

Disadvantages

- Hardly suitable for quantitative analysis with an external standard, since the layer thickness is poorly reproducible
- Volatile solvents and analytes evaporate during the measurement
- Only analytes and solvents may be used which do not attack the saline window.

Suitable for

- Qualitative analysis of medium- and low-volatile liquids, which do not attack the saline window
- Quantitative analysis with internal standard (layer thickness does not matter here)

Not suitable for

- Solids
- Volatile samples
- Quantitative analysis with an external standard

Gases

Due to their low density, gases require large path lengths and a special feeding technique. Usually 10 cm long glass cuvettes with IR-transparent NaCl-windows are used which contain the gas under reduced pressure. In order to obtain reproducible conditions, the gas cells are often filled with nitrogen or an inert gas under equal pressure. Particular attention must be paid during gas analysis with regard to compensation for atmospheric water vapor and carbon dioxide.

Advantages

- Real-light spectrum, so in contrast to the ATR technique, no distortion of bands
- No problems due to solvent bands (restricted areas)
- Compared with other detection methods, the analyzer does not directly interact with the gas. Gas molecules interact only with a beam of light.

Disadvantages

- Requires particular work technique
- Miniaturization difficult to implement

Suitable for

- Qualitative analysis of medium- and low-volatile liquids, which do not attack the saline window
- Quantitative analysis with internal standard (layer thickness does not matter here)

Not suitable for

- Samples with high content of water vapor, moisture

Preparation for use with ATR In contrast to all other methods, ATR is not based on the transmittance, but on the reflection principle. This has the disadvantage that the intensity of the reflected radiation is low. A longer time is required to measure an IR spectrum with a comparable *S/N* ratio. On the other hand, powdery samples, polymers or highly viscous liquids can be measured directly without sample preparation, they are simply applied to the ATR cell and clamped with a diamond anvil. They may have to be crushed so that they fit on the sample stage.

Advantages

- No sample preparation required
- Micro samples possible
- Easy cleaning of the sample holder
- Almost no wrong manipulations possible, rugged device

Disadvantages

- No defined layer thickness (quantitative analysis)
- Bands sometimes distorted (creates problems with the library search)
- Only a sample surface is detected.
- Poor instrumental detection limits due to the small sample size

Suitable for

- Powder
- All liquids except corrosive solutions
- Any form parts with a smooth surface
- Plastics
- Foils and films
- Fibers, textiles

Not suitable for

- Investigation of the inner layers of multilayer foils and films
- Gases
- Hard materials with rough surfaces

8.2.2.3 Material and Methods

Equipment

FTIR-spectrometer (Perkin Elmer with "Golden Gate"-cover), PC for data analysis, KBr disk.

Accessories

Vacuum, vacuum pump, glass wool, Bunsen burner, polystyrene foils (various degrees of thickness, with/without additive), Polyethylene-foils (various degrees of thickness, with/without additive, Saran-household foil, test tube (thick wall), agate mortar/vibratory mill.

Chemicals

Potassium bromide CAS number 7758-02-3 Xi 

Pyrolysis (thermal decomposition)

Crush a few granules with a knife, weigh a few milligrams and fill the chips into a thick-walled test tube.

Put a pad of glass wool into the tube with a pair of tweezers and then plug the whole thing to a vacuum hose. Evacuate to about 15 mbar and carefully heat the sample with a Bunsen burner until it is charred.

Above the glass wool (which is designed to prevent solid particles from entering the upper part of the tube and contaminating the pyrolysis oil) the pyrolysate condenses on the glass walls, especially where the attached vacuum hose cools them further. After (!) cooling, the tube is vented.

Experiment

(A) Determination of the layer thickness of polystyrene/polyethylene foil

Record the spectrum of polystyrene and polyethylene foils and look for regular interference lines in the spectrum. Significant interference maxima are only obtained if the foil/cuvette has very smooth surfaces (no scattering by matt or rough surfaces, respectively). Next both surfaces must be perfectly parallel to the phase boundaries reflecting beam B, as otherwise this beam is deflected and will no longer fall on the detector (Figure 8.39).

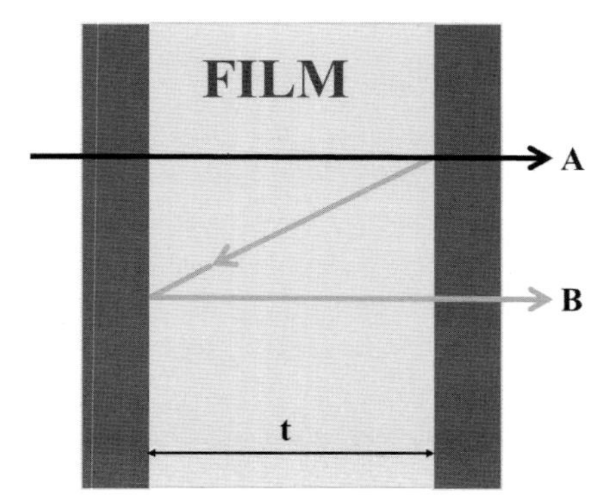

Figure 8.39 Layer thickness determinations by interference. **A** unreflected radiation, **B** reflected radiation, ***t*** film thickness.

Example

From a recorded FTIR reflectance spectrum (Figure 8.40), the number of waves between 3310 and 3920 cm^{-1} is calculated.

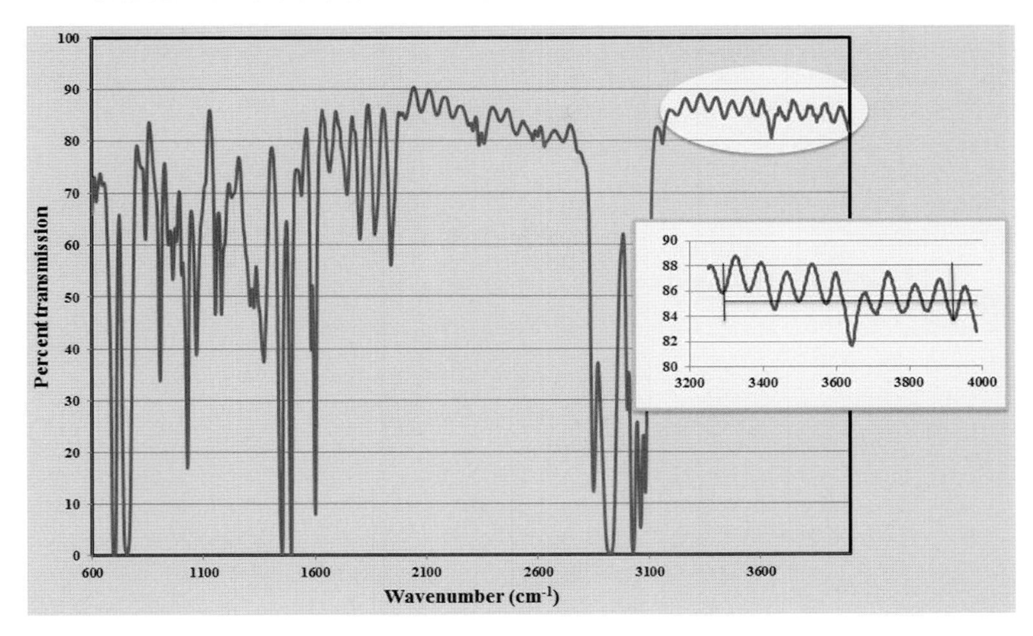

Figure 8.40 Determination of number of maxima/minima from an FTIR spectrum (polystyrene approx. 50 μm thick).

Calculation:

$$d = \frac{k}{2 * n_D(\overline{\nu}_1 - \overline{\nu}_2)\mathrm{cm}^{-1}} \tag{8.43}$$

d Layer thickness
k Number of minima/maxima (count)
$\overline{\nu}_1$ Wavenumber of the first maximum/minimum
$\overline{\nu}_2$ Wavenumber of the last maximum/minimum index of refraction
n_D Refractive index (for air = 1, for polymers see reference works)

Result:

$$d = \frac{9}{2 * 1.59(3920 - 3310)\mathrm{cm}^{-1}} = 0.0046\ \mathrm{cm} = 46\ \mu\mathrm{m}$$

(B) Identification of a plastic foil

1. *Identification of the polymer on which SARAN®-household foil is based.*

   - First, use the basic settings of the device to take a new background spectrum and then the absorbance spectrum of the SARAN foil.

- Reprocess the spectrum (must always be in absorbance for processing):
- Smooth the spectrum and filter out the thin water vapor lines at 3600 and 1600 cm^{-1}
- Cut out the CO_2 peaks at 2340 cm^{-1}
- Correction of the baseline
- Standardization of the most intense bands to <1.0 absorbance by multiplying the spectrum with a matching factor
- Download the processed absorbance spectrum under a new name
- Conversion in transmittance
- Marking of the most important band maxima with the wavenumber
- Print labeled transmittance spectrum

2. *Manual identification of functional groups:*

   Create a peak table of the more intense bands in the range 4000–1400 cm^{-1}. Signals in this area are typical for certain functional groups. These can be identified from the bibliography with the help of tables.

   - Do a literature search on IR spectra of this polymer type and compare the so-called "fingerprint areas" of 1500–400 cm^{-1} which match as closely as possible, in order to further narrow the selection.

3. *Qualitative analysis of the spectra using the spectral database:*

   Download the absorbance spectrum of the foil and run a database comparison with the spectral library available. Repeat the search with different search algorithms.

(C) Identification of a plastic additive

- Create an absorbance spectrum from both films and subtract spectrum 1 (plastic material containing no additives) from spectrum 2 (synthetic material with an additive) with the help of the subtraction subroutine of the spectrometer software.
- Reprocess the difference spectrum (corresponding to the desired additive), convert it to transmittance, label the prominent peaks and print the spectrum.
- Do a library search on the spectrum of the additive.

Which proposals from the hit list seem to you to be the most reasonable?

Use this to identify at least the structural type of the additive.

Volume 3 of the Hummel Atlas of Synthetic Materials [2] contains a collection of IR-spectra of the most common additives.

Search it for examples of the structural type found and compare the fingerprint regions (1500–400 cm^{-1}), to conclusively identify the additive being sought.

(D) Specific sample preparation technique – pyrolysis

Sample preparation

Prepare a "blank" potassium bromide pellet. Then take the glass wool from the glow tube with the tweezers. It is usually soaked with pyrolysis product and on removal also takes up the pyrolysate on the glass wall. Apply the pyrolysate to the KBr pellet using the glass wool.

Record spectrum

Record a background spectrum recorded with an inserted *empty* pellet holder.

Record sample spectrum

The pellet with the pyrolysis oil is placed in the pellet holder and the spectrum is recorded.

When the spectrum is either too insensitive or too noisy, the number of scans must be increased. (However, this is only worthwhile if the number of background scans have been similarly expanded).

- Take a sample from the spectrometer and discard it.
- Identify the polymer with a library search.

8.2.2.4 Questions

(1) What is an IR-inactive vibration?

(2) What distinguishes the bands of liquids from the gaseous bands?

(3) Which basic elements make up an FTIR-photometer?

(4) How does an interferometer work?

(5) How do we eliminate the absorption bands of the solvent from the spectrum of the solute?

(6) Which methods of preparation are used for solid materials?

(7) In the short wavelength range, large layer thicknesses (in the cm-range) must be used. Why?

(8) What are the disadvantages of the ATR-recording techniques? Where does it not work?

(9) How do you evaluate the transferability of the analysis method to other solvent mixtures? What general conditions must these meet, to be able to be analyzed with the same process?

(10) How do we recognize in an IR-spectrum that it was recorded with too large a thickness layer or with too great a concentration of analyte?

(11) What is meant by the "restricted zone" of a solvent? Why is it useless in the case of too large a thickness layer to record the underlying spectrum with a solvent as well?

References

1 Skoog, D.A., and Leary, J.J. (2006) *Principles of Instrumental Analysis*, 6th edn, Brooks Cole, ISBN-10: 0495012017, ISBN-13: 978-0495012016.

2 Hummel, D.O. (1991) *Atlas of Polymer and Plastics Analysis*, Third Completely Revised Edition, VCH Publishers, New York, ISBN-13: 0-89573-973-9.

8.2.3
Near-Infrared (NIR) Spectrometry, Project: "Accelerated Raw Material Intake Control"

8.2.3.1
Analytical Problem

At the pharmaceutical company Krämer Ltd hundreds of different raw materials are delivered on a typical day and they all have to be checked for their identity and quality. Checking the identity of each container is required by law. Usually bulk samples from several containers have to be examined. The incoming inspection of raw materials on this scale can hardly be done using traditional methods ("wet chemistry"). To avoid a jam in incoming goods examination, the management of the company has decided to introduce NIR technology – a very effective and efficient method to ensure quality control. It is expected to achieve reliable identity verification in minutes by using this technology. For this purpose, the administration has commissioned the applications laboratory of the company to create an NIR library and to validate its performance.

Object of Investigation/Investigation System
Raw materials for manufacturing tablets

Samples
Various sugars in powder or crystalline form

Examination Procedure
Near-infrared (NIR) has become increasingly widespread and found new areas of application. This has been supported by the development of new methods and more effective equipment. As opposed to conventional analysis procedures, NIR spectroscopy often offers a clear advantage in speed and the possibility to automate measurement and evaluation. The initial effort and cost of creating the calibration

for several raw materials can be minimized by pre-calibration. Reference analytics continues to be necessary in the transition phase and during the update of the calibrations, but to a clearly lesser degree. Further advantages of this technology are, among others, the contact and destruction free measurement, which with fiber optics can even take place over large distances, and measurement data that are produced within only a few seconds or minutes. In most cases, sample preparation is not necessary. Spectrometric data today are calculated with the help of statistical and mathematical methods that fall under the term chemometrics. Modern chemometric methods are very effective and can be used on a wide basis thanks to the rapid development of computer technology.

NIR spectroscopy is mainly used in the inspection of materials on receipt and in quality control of semi-finished and finished goods, but is also increasingly used in process control. NIR spectroscopy is therefore widely used in the pharmaceutical industry because in less than a minute it is possible to have one qualitative and several quantitative results.

Analytes

Lactose, sucrose, fructose

Keywords

Polarization, diffuse reflection, chemometrics, principal component analysis, PCASpectral libraries

Bibliographical Research (Specific Topics)

Ozaki, Y., McClure, F., and Christy, A.A. (2007) *Near-Infrared Spectroscopy in Food Science and Technology*, Wiley & Sons, ISBN: 0471672017, 9780471672012.

Siesler, H.W. (2006) *Near-Infrared Spectroscopy: Principles, Instruments, Applications*, Wiley & Sons, ISBN: 3527301496, 9783527301492.

Burns, D.A., and Ciurczak, E.W. (2008) *Handbook of Near-Infrared Analysis*, CRC Press, ISBN: 084937393X, 9780849373930.

Schrader, B. (1989) *Raman/Infrared Atlas of Organic Compounds*, 2nd edn VCH, Weinheim.

Lo, S.C., and Brown, C.W. (1992) Near-infrared mixture identification by an automated library searching method: a multivariate approach. *Applied Spectroscopy*, **46** (5), S790–S796.

Project Process/Task

(A) Identification of various types of sugar

- **A.1** Identifying a known sugar
- **A.2** Identifying a sugar with impurities

(B) Quantification of sugar in a mixture

- **B.1** Analysis of different sugar mixtures
- **B.2** Analysis of sugar mixtures outside the calibration range

(C) Project reporting

The results are expected to be summarized in the form of a report (or presentation).

8.2.3.2
Introduction

(See also Section 8.2.2)

Molecular oscillations can be induced by IR light. The oscillation frequency is dependent on the oscillating masses and the type of binding. Through the position of the absorption bands information on single bonds or functional groups can be obtained. A reliable structure determination can only succeed with libraries for spectral comparison.

In the NIR liquids are usually measured with a layer thickness of approximately 1–10 mm in transmission, compared to a maximum of 0.1 mm in the mid-infrared (MIR). The choice of vial material is also less critical. While glass and quartz absorb strongly in the MIR, their absorption in the NIR is negligible. Solids can be measured in diffuse reflectance with virtually no sample preparation. As an example the measurement can be done by immersion of glass fiber probes into the sample.

NIR bands are wider and their contours are usually not as structured as in the MIR. In the case of the middle infrared the spectra can rather be interpreted in terms of partial structures and functional groups. This structural information is basically also included in the NIR. However, by virtue of the numerous possible combinations of the fundamental oscillations of overtone and combination bands, they are generally more difficult to interpret.

Physical Relationships

Near-Infrared Spectroscopy

In the case of NIR spectroscopy, the absorption of electromagnetic radiation is measured in the near-infrared range (about 13 000 to 4000 cm^{-1}). By exciting vibrations in covalently bound molecules (hence vibrational spectroscopy), a reaction occurs in the form of photon energy absorption and thus weakens the IR radiation. The energy of IR photons (approx. 10^{-23} J or 0.1 eV) corresponds to approximately the difference between the different vibrational states of a covalently bound molecule. By absorbing an IR photon, a molecule can pass into an excited state in which one or more atoms vibrate, that is, they oscillate at high frequency around their rest position. The oscillation frequency corresponds to the frequency of the absorbed IR radiation. This frequency can be used to provide information on the bonding structure. In the MIR, fundamental vibrations in the form of deformation and stretching vibrations are essentially excited by molecules in higher energy near-IR combinations and overtones of these vibrations.

Model of the Harmonic Oscillator

Vibrations of chemical bonds can be simply described by the theory of the harmonic oscillator and the spring model. Considering the vibration of a single mass attached to a spring, which is in turn fastened to an immovable object (Figure 8.41), the following was observed:

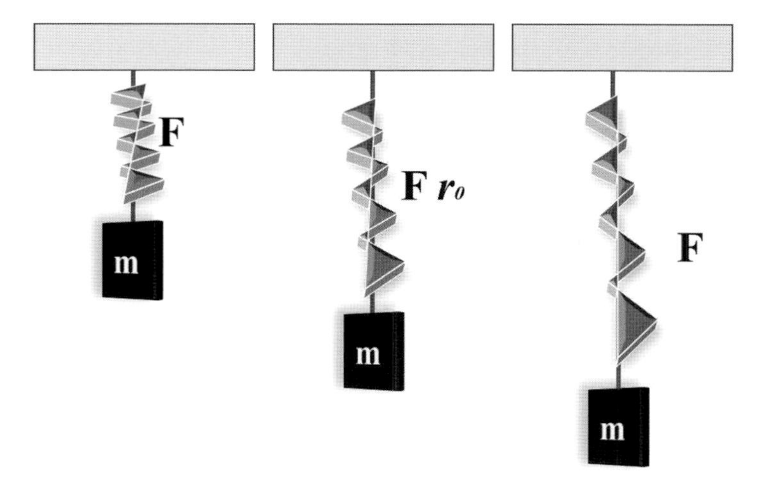

Figure 8.41 Oscillations of a spring (**F** spring, m mass, r_0 equilibrium distance).

If the mass is moved around Δr from its equilibrium position by applying a force along the axis of the spring, then the restoring force F is proportional to the deflection.

$$F = -k\Delta r \tag{8.44}$$

F Force (N)
k Force constant (Nm^{-1})
r Displacement (m)

$$\Delta r = (r - r_0)\,(m)$$

The negative sign indicates that F is a restoring force.

The force constant k is a measure of the "hardness of the spring", or the binding energy.

In a valence or stretching vibration the involved atoms move along the bond axes, that is, there is no change in oscillation angles during oscillation. This behavior is clearly illustrated with a spring (= bond), which connects the two masses (= atoms) connected at their ends. Here the harmonic vibration from classical physics is used (Figure 8.42). During a harmonic oscillation between differently sized masses, these move with different amplitude from their rest position. The larger of the two masses is mentally replaced by a fixed starting point of the spring (m_1 = infinite), and the other mass is, therefore, theoretically reduced to the extent that for the same force constant the same vibration frequency results.

The reduced mass μ (kg or a.m.u) is calculated using:

$$\mu = \frac{m_1 \cdot m_2}{m_1 + m_2} \tag{8.45}$$

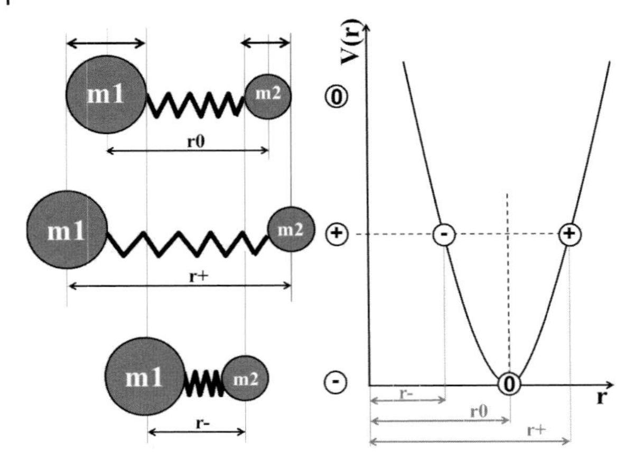

Figure 8.42 Model of the harmonic oscillator (*m* mass, *r-* compressed bond, r_0 bond length at rest, *r+* elongated bond, *V*(*r*) potential energy).

From the model of the harmonic oscillator the following relationship results:

$$k = 4\pi^2 \cdot \mu \cdot \nu^2 \tag{8.46}$$

k Force constant (Nm^{-1})
υ Oscillation frequency (s^{-1})

If Equation (8.46) is solved for the oscillation frequency (υ), one obtains the equation of harmonic oscillations key provision:

$$\nu = \frac{1}{2\pi} \cdot \sqrt{\frac{k}{\mu}} \tag{8.47}$$

From this the following interpretation of vibrational spectra ensues:

- The stronger the bond, the higher the oscillation frequency
- The heavier the oscillating mass, the smaller the frequency

The equations from mechanics do not describe the behavior of a body at the atomic level completely. However, it is possible to use the concept of the linear harmonic oscillator to derive the wave equation of quantum mechanics.

$$E = \left(\nu + \frac{1}{2}\right) \cdot \frac{h}{2\pi} \sqrt{\frac{k}{\mu}} \tag{8.48}$$

E Potential energy
ν Energy level of oscillation
h Planck's constant

Quantum mechanics implies that the energy of molecular resonance can only take certain values:

$$E_v = h\nu \cdot \left(v + \frac{1}{2} \right) \tag{8.49}$$

v Vibrational quantum number ($v = 0, 1, 2 \ldots$)
$h\nu$ Oscillation frequency

An IR photon is absorbed when its frequency coincides with the vibration of a molecule ($\Delta\ v = 1$) or an integral multiple thereof ($\Delta v = 2, 3, \ldots$)

Energy level of the harmonic oscillator	Absorption of one IR-photons
$\Delta E_v = h\nu_{harmonic\ oscillator}$	$\Delta E_v = h\nu_{harmonic\ oscillator}$

The rule for selecting the harmonic means: *Only* $\Delta v = \pm 1$ allowed.

Overtone Oscillation

If the model is transferred from classical to quantum physics, the following energy level scheme results: Above the ground electronic state there are excited states at regular intervals (Figure 8.43a). When a photon is absorbed, the system switches to the first excited state and oscillates at the fundamental frequency. When a photon with twice as much energy collides with a molecule in its ground state, the second excited state can be achieved. The corresponding oscillation frequency is also twice as large and is identified as a first harmonic or (in analogy to acoustic phenomena) called a first overtone.

Anharmonic Oscillator

The molecular vibration is only incompletely described by the harmonic oscillator.

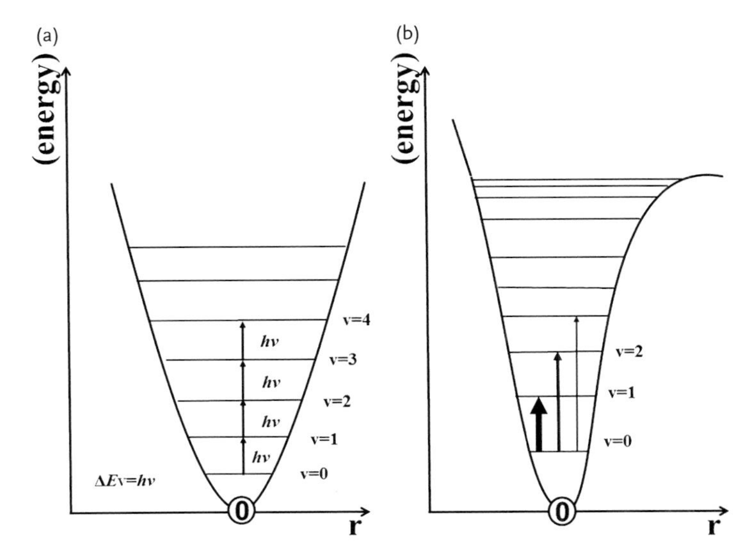

Figure 8.43 Comparison between (a) a harmonic and (b) an anharmonic oscillator.

Approximation to reality and a refinement of the model is achieved with the anharmonic oscillator. Anharmonic oscillation is described as when the restoring force is no longer proportional to the displacement (Figure 8.43b), as opposed to harmonicity where the excited states are equidistant.

With increasing distance, dissociation of the molecule is to be expected, which is reflected in a flatter rise of the potential curve. At distances that are shorter than the rest distance span, different repulsive forces are expected, which are expressed in a steeper profile of the potential curve. This model is called the anharmonic oscillator. The harmonic approximation for excited molecules is only valid to a certain extent from the third vibrational state, since the molecules have such large deflections, that the true power can no longer be approximated by a parabola. The restoring force on the atoms is no longer proportional to their displacement from the equilibrium position that is, the potential energy reaches a limit. This limit is the dissociation energy.

The main difference between the harmonic mean is the selection rule:

Anharmonic oscillator selection rule: $\Delta\nu = \pm\, 1, 2, 3$ allowed, that is, *also overtones* can be excited.

Spectral Libraries

Spectra libraries are databases, in which a large number of spectra of different substances are collected and uniformly archived. These libraries can, for example, be used to identify unknown substances or to confirm the identity of already measured substances. By structural comparison it is possible to make a statement on new molecular structures (mainly in the case of IR).

Usually a "robust library" is required that provides the best possible results. To achieve this it is necessary to "calibrate" the library, whereby differing and sometimes conflicting criteria must be taken into consideration:

- It should, as far as possible, clearly identify a large number of chemically different products. As chemically clearly distinguishable substances also differ spectroscopically, a relatively rough discrimination can be made. This makes the library relatively insensitive to disturbances in the form of spectral artefacts. However, there is the danger of confusing similar substances.
- Similar items should be distinguished as much as possible. These may be substances of a different technical or physical quality or of small chemical difference. Because of the small spectral differences between such similar items, the boundaries for the spectrometric discrimination must be defined more narrowly. This increases the risk of interference by spectral artefacts or variations (e.g., modification of procedure, other supplier, exchange of device components). In these cases, it may happen that the sample is found to be "not identical", although it is still within the specification.

In general, efforts will be made to fulfill both of the above requirements, which, however, cannot necessarily be done in one step. One possibility is to combine similar substances into a class of substances. The assignment of a measured spectrum is carried out first in relation to a particular class within the overall library. In a second step, the measured substance is finally identified by means of sharper discrimination within the class.

There are many factors that can affect the appearance of the spectrum. In addition to external influences such as temperature measurement, parameters such as resolution and characteristics of the device play a role. While some measurement parameters can be specified, the influence of the detector characteristics cannot. The replacement of individual components of the device could, at worst, cause the entire library to have to be re-measured. To avoid this, calibration standards (e.g., in the form of well-defined optical glass filters or polymer foils), are measured at regular intervals. By comparing the spectra of these calibration standards with nominal values it is possible to minimize effects from devices. Through appropriate mathematical correction or standardization processes it is even possible in many cases to transfer the libraries from one device to another. Thus libraries can be maintained in a central laboratory and forwarded to different locations. Such approaches to calibration transfers have up to now been described mainly for quantitative NIR analysis. However, intense efforts are also being made in the field of qualitative analysis and identity control.

The aim of creating a library is to unmistakeably and correctly identify each individual spectrum in a later analysis, which can be understood in the context of the present problem as a test of identity. In so doing, the following definitions are used:

- **True positive:** The spectrum of the sample suspected of being substance X is correctly assigned to the library spectrum of the substance X. This means that the sample corresponds to substance X.
- **True negative:** The spectrum of the sample suspected of being substance X does not match the library spectrum of X. The sample does not correspond to substance X.
- **False positive:** The statement is made, which is described under true positive. It turns out that this statement is not true, and that, therefore, the analysis is incorrect. The sample does not correspond in reality.
- **False negative:** same statement as described under true negative. It turns out that this statement is not true, and that, therefore, the analysis is incorrect. The sample therefore corresponds in reality.

Data Pretreatment

The pre-processing of the measured sensor signals serves to improve the data analysis (among other things, to improve the signal-to-noise ratio (S/N) or to reduce the number of data points) and depends strongly on the evaluation method and the objectives of the data analysis. The choice of evaluation method (calibration method) is a process that takes a lot of experience and which influences the whole course of data analysis. With the help of the calibration data, the evaluation method serves to establish a model for the sensor signals and the various factors. The quality of the model must subsequently be checked using the validation data. In many cases in the subsequent step the measurement and evaluation methods are optimized with the help of the model, until the measurement methodology and the model derived from the evaluations can be used together in routine operation. Also during the routine operation, the measurement methodology and the model must be continuously validated using test samples.

The data pretreatment is a possibility to treat data mathematically in such a way as to improve evaluation. A data pretreatment may also negatively affect the evaluation, however, and must always be questioned. The main methods of data pretreatment are:

- Scaling and centering
- Normalization
- Standardization

Whether and which method of data pre-processing is to be used must be decided on a case to case basis. As a guide, the following rules can be used:

The centering and scaling are used primarily to make the data numerically easier to handle. If an evaluation method does not have basic numerical conditions (e.g., only positive values), the centering and standardizing should not have any positive or negative influences on the evaluation.

For quantitative analysis of multi-sensor systems, the data are often standardized.

For qualitative analysis of multi-sensor systems, the data are often normalized.

Standardization is often used for quantitative analysis using sensor arrays, if the sensors have different transducer principles, or generally show varying degrees of signal changes (transducer). The standardization is based on the fact that sensors with small signal changes (variance) are often just as important as sensors with a large variance. The standardization can be conducted in the following steps:

The standard deviation is calculated for each sensor (column).

Each data element is divided by the corresponding column-wise standard deviation.

Thus, the variances of all sensors are numerically equal.

Mostly the data is not only standardized but additionally also centered. The literature often refers to standardization as auto-scaling.

Normalization is often used in qualitative analysis using sensor arrays. Here it is assumed that the characteristic information of a substance measured is proportionate to the sensor values, while the size of individual sensor values mainly contains quantitative information (in this case disturbing).

Mathematically, the qualitative information thus corresponds to the direction of the input vector (measurement vector, measured values represented as a vector), while the quantitative information lies in the length of the vector. For qualitative classifying analysis, therefore, the sensor values are divided line by line by the standard deviation.

Derivative Spectroscopy

Computing the first or second derivative of a spectrum is called derivative spectroscopy. In IR/NIR spectroscopy, due to the comparatively limited signal-to-noise ratio (compared with other spectral regions such as UV–VIS) higher derivatives than the second are not used.

A band maximum in the original spectrum leads to a zero in the first derivation of characteristics. A shift of the baseline (offset) over the entire spectral area can be eliminated with the help of the first derivative of the spectrum, since its first derivative is equal to zero.

The second derivative corresponds to the curvature of the spectrum. During the derivatization, negative bands of lower half-width result from the original bands, thus increasing the contrast or deconvolution. Furthermore, by using the second derivative a linear slope of the baseline can be eliminated. A disadvantage, however, is the deterioration of the S/N condition because the noise increases with each derivative formation. Numerically, both the derivatization and equalizing with so-called "filters" takes place. For this purpose a data point is combined with a defined number of its left and right-hand neighbors. The value for the original average data point is then replaced on the basis of an arithmetic operation with all data points within the window. Then the window is pushed further point for point and the same process is repeated, starting from the respective raw data. When the window has been pushed through the whole spectrum in this way, the filtering operation is complete. An example of this is equalization (using the floating averaging method). The number of data points included in the window determines the degree of equalization. If the windows are too large, that is, if there is too much equalization, details of the spectrum can get lost. On the one hand closely adjacent bands are no longer resolved properly. On the other hand signal intensity decreases as the number of filter points increases. In the Savitzky–Golay filter the data points are weighted prior to averaging. The weighting factors can be chosen in such a way as to make the results correspond to an approximation of the window region to a nth degree polynomial. The better adjustment to the original data compared to the floating averaging method serves to achieve a more effective equalization, because noise is removed as far as possible, but band contours are better preserved. Derivatives are also calculated with Savitzky–Golay filters, whereby the weighting coefficients are chosen so that an approximation is made to the first or higher derivative of the polynomial.

Normalization

By normalizing spectra, different effects can be achieved. In NIR spectra, parameters such as particle size and bulk density exert influence on the baseline. Shifts can be removed by a suitable normalization. For instance, normalization serves to determine the intensity of the most intense band, whereby the intensity relations between the different bands of the spectrum remain unchanged. There are a number of different approaches to standardizing spectra, where band intensities or areas are taken into consideration. A simple method is, for example, a minimum–maximum normalization, in which the lowest absorption is equal to 0 and the band with the largest absorption is equal to 1. However, fluctuations of a very intense band strongly affect and influence the whole spectrum. Such influences are slighter in the case of area normalization.

Chemometrics

Many branches of chemometrics can be understood as statistical generalizations of one-dimensional data in the statistics of multidimensional data. The algebra is replaced by a matrix algebra. Accordingly, thinking in matrices is of central importance for the understanding of chemometrics. Together with NIR spectroscopy, chemometrics, which is based on mathematical and statistical processes, has also developed. Since NIR spectra can only rarely be evaluated manually (hardly noticeable, overlapping bands), such evaluations are almost always done using chemometrics. Computerized instruments often provide such a large amount of data that without mathematical and statistical methods the results would hardly be analyzable. Instrument computers are increasingly equipped with chemometric programs. Basically, what is noteworthy here is that the software must be trained.

In a first step a calibration model is created with known samples. In a second step this model can then be used to analyze unknown samples of similar composition. In the case of quantitative models this means substances with a similar matrix and within the calibrated concentration range. In the case of qualitative models this means substances with the same identity and comparable quality (e.g., particle size, purity and color).

Data Transformation

Multivariate Data Analysis of an NIR Spectrum

In many cases it may be useful not to consider the data in the original data space, but to transfer them to another domain, that is, to another coordinate system, prior to the analysis.

In NIR spectroscopy, there is frequently a very large amount of data. The multivariate data analysis is used, as there are frequently no suitable undisturbed bands available for assessment, but rather an overlapping of bands. Prior knowledge of the facts is essential to interpret the data obtained. A typical NIR spectrum is hardly selective and is influenced by innumerable physical, chemical and structural variables, which can cause slight spectral differences that are hardly, or not at all, recognizable by the naked eye. With multivariate data analysis the aim is to construct a model which is able to predict the properties of unknown samples.

The objectives of multivariate data analysis can be divided into two areas of application: classification of data and multivariate regression techniques.

Data classification aims at consolidating or even reducing the original amount of data. The relevant information is to be filtered out from the measured values (variables). Measured values with the same information are bundled and put in groups depending on the content. With the aid of the determination of relationships and structures in the data regarding the objects and variables received information is often received about not directly measurable quantities. This information can be used to find quality characteristics or properties in samples or products. Applications are multivariate quality control, process control or also separation of substances. Principal component analysis (PCA) is used for data reduction.

Principal Component Analysis

Data evaluation attempts to draw conclusions on the composition of the sample using a sensor signal and/or the pattern of the sensor signals in sensor arrays. In so doing, PCA is used in most cases.

With PCA from the original measured data (variables) new so-called latent variables are calculated. These new variables are called factors. They represent the principal components (PC). The goal is a data reduction, in order to describe all output variables with a few factors. To identify relationships between the objects, only the principal components have to be examined.

Usually only the most important principal components of the PCA are observed. Thus, the PCA is a projection method, which projects from the high-dimensional sensor space into the low-dimensional principal component space of the most important of the principal components.

The following shows the functioning of PCA without any great mathematical explanations [1].

The individual steps of the example are as follows:

- Various analytes are examined using two sensors.
- The samples are plotted in the sensor space.
- The PCA is illustrated graphically.
- The samples are plotted in the space of the principal components.
- The results of the application are interpreted.

The results obtained in the example are shown in Table 8.7. The (graphical) expansion of the samples in the wavelength space is called a variance. The best way to graphically represent the variance is to draw an ellipsoid around the sample (Figure 8.44).

Table 8.7 Signal response of various analytes.

	Wavelength 1	Wavelength 2
Analyte A		
Signal	Strong	Weak
Analyte B		
Signal	Strong	Strong
Analyte C		
Signal	Weak	Strong
Analyte D		
Signal	Medium	Weak

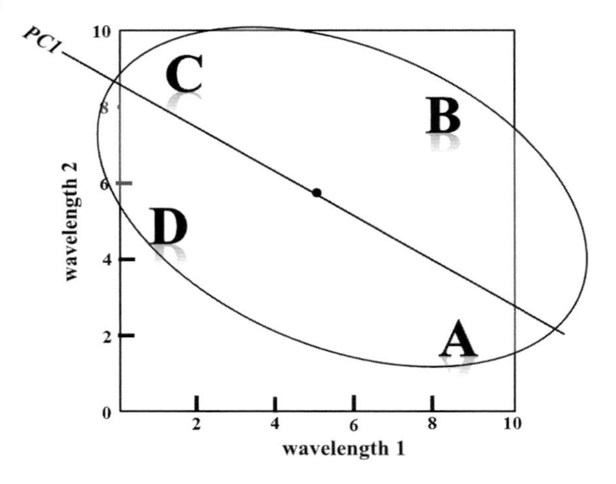

Figure 8.44 Analytes in the wavelength space.

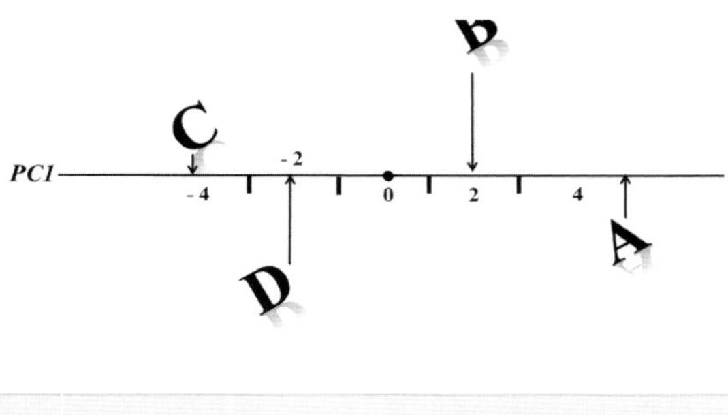

Figure 8.45 Coordinate values of the first principal components.

The first principal component (PC1) is placed in the wavelength space in such a way as to describe the largest proportion of the variance. Graphically, this corresponds to the first principal axis of the ellipsoid. The corresponding coordinate values of the first principal components are obtained by performing for each sample a vertical projection onto the principal component and reading the corresponding coordinate value. By shifting the zero point and rotating the coordinate system, a new axis system arises.

In Figure 8.45, the coordinate values of the analyte are shown graphically for the first principal component. One can see from this figure that all four analytes differ in their principal components. Thus one principal component is sufficient for a qualitative description of the samples in terms of making a distinction. This represents a reduction of the descriptive space from two dimensions to one.

It is always advisable to examine the next principal components, even if one principal component suffices in this case for a qualitative differentiation. The

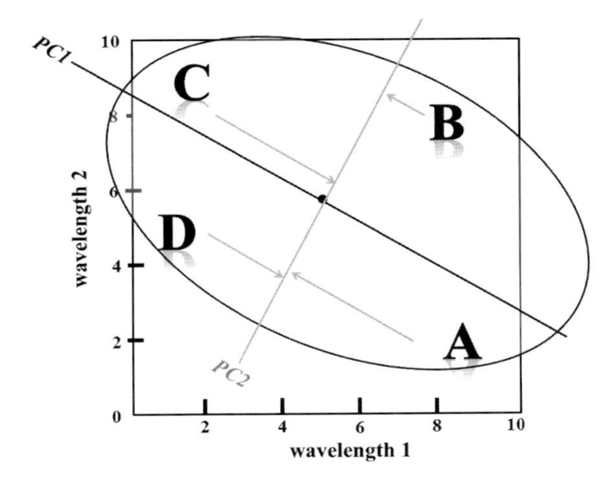

Figure 8.46 The second principal component.

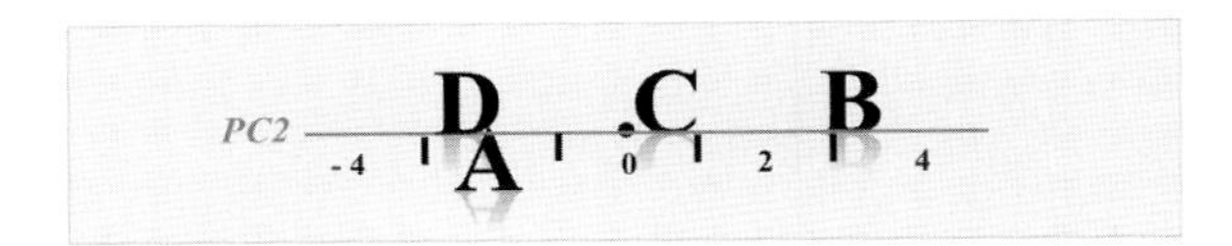

Figure 8.47 Projection onto the second component.

second principal component is orthogonal to the first principal component and extends in the direction in space with the largest proportion of variance, which is not explained by the first principal component. Graphically, this corresponds to the second main axis of the ellipsoid (Figure 8.46).

The corresponding coordinate values of the second principal component are obtained by performing a vertical projection onto the second principal component for each sample and reading the corresponding coordinate value. In Figure 8.47), the coordinate values of the analytes for the second principal component are shown in a diagram. It is obvious that the second principal component cannot distinguish between all analytes (D and A).

The steps carried out above can be carried out in succession for all principal components. Even if, in principle, there are as many principal components as there are sensors, it should be noted that only the first few principal components represent influential information. Often, the higher principal components contain only information about the noise.

In the PCA, the relationships between the sensors and the samples are plotted in different plots

Score Plot

In the so-called score plot, the samples are shown as a diagram in the space of principal components. Thereby, possible patterns and similarities can be

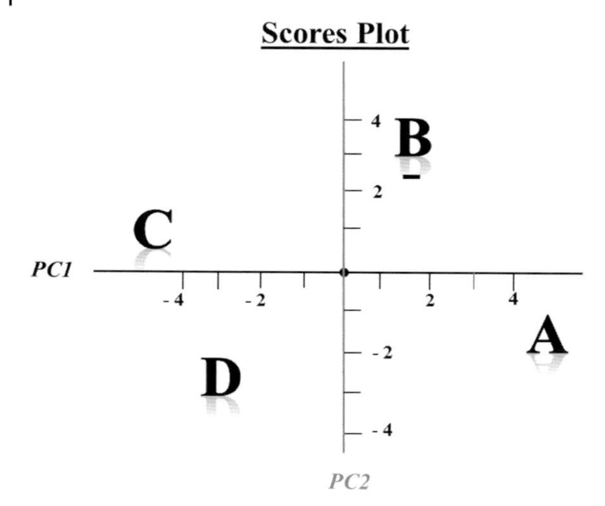

Figure 8.48 Score plot illustration.

identified. The closer data points lie together, the more similar they are, and vice versa. Thereby striking outliers can also be detected.

In Figure 8.48 it can be seen that the first principal component is sufficient to differentiate the four different analytes.

Analysis of NIR-Data Records

First, a spectrum of an analyte is measured over a specific wavelength range. From this spectrum a target characteristic is added, such as, for example, to determine specific bands of an analyte. The spectral features for differentiating certain properties of the sample are small and spread over a wide spectral range and are broad and overlapping, especially in aqueous systems. Characteristics are measured in many samples or, in general, one simply talks about variables. This results in a large data matrix. If this data matrix is evaluated only in a univariate manner, which means looking at only a single variable, several individual results are obtained, which are partly identical and which partly contradict each other, and very quickly one loses track. For this reason, the first goal of multivariate data analysis is *data reduction*. All variables that contain the same information are grouped according to principal components. This results in reduction of the amount of data, since each object is then only described with the few principal components, rather than through several individual variables.

This data reduction leads to simplification. As an example, 100 different variables were used in the original data, these can possibly be reduced to 10 principal components. The samples are then described with only these 10 principal components, which means that only 10 principal component scores must be analyzed per sample instead of 100 individual measurements.

Scores are a measure of the correspondence or distance from the spectrum of the principal component.

From knowledge of the spectral properties depending on their composition a mathematical model is created, which then facilitates the qualitative or quantitative analysis of corresponding samples.

Instrumentation

NIR Spectrometers

- Solids

  Spectrometer with an option for the measurement of solids in diffuse reflection.

  For example, solids can be measured in a Petri dish through the glass bottom. By rotating the dish a relatively large area can be measured. In inspection of incoming goods a fiber optic probe is often used, with which the contents of the container can be measured directly.

- Liquids

  With both of the above-mentioned options and a transflectance adapter even liquids can be measured (in transflection).

  Cuvettes are commonly used for measuring liquids in transmission.

Design of an Interferometer

In the laboratory, and also in process-related environments, spectrometers based on FT technology are mainly used. This is essentially a Michelson or polarization interferometer. For mobile devices or process control equipment mainly those equipped with diode arrays are used.

In the NIR the light source is a halogen lamp. The construction of the rest of the spectrometer is comparable with FTIR spectrometers.

In contrast to Michelson interferometers (see Section 8.2.2) the polarization interferometer consists of two polarizing filters and two birefringent crystals, one of which is moved (Figure 8.49). The light is not split into two beams. Depending on the orientation of the vibrational levels within the light beam a shift of the waves against each other (phase-shift) can be observed in the birefringent crystal, The interference results from the overlapping of the polarization planes.

Generation of the Interferogram

An interferogram is an interference pattern of phase-shifted radiation. As a single-beam polarization interferometer for example, the NIRFlex N-500 produces its interferogram in four steps:

Step 1 – polarization of the light source

The polarizer generates a well-defined polarization of undefined polarized light, which is emitted by the light source. Accordingly, only diagonally polarized light is transmitted.

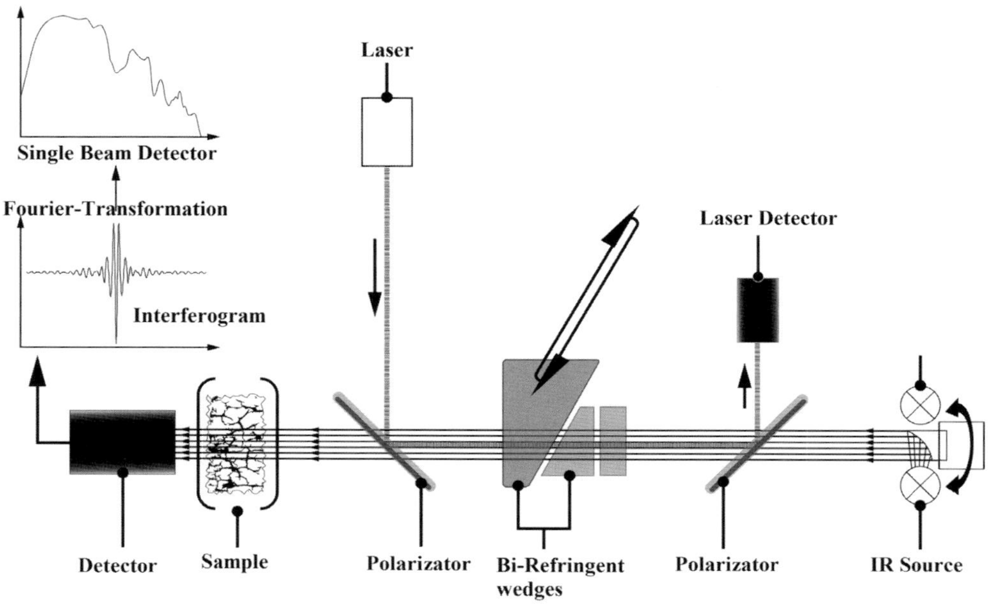

Figure 8.49 FT-NIR spectrometer design.

Step 2–beam splitting and orthogonal polarization

The polarized light enters a birefringent block (compensator). Here the light is broken into two perpendicularly polarized components, which are phase-shifted.

Step 3–generation of the current phase shift

After the compensator two birefringent wedges follow. One wedge is fixed, while the other is constantly moved back and forth by a fast linear actuator. By the motion and the geometrical arrangement, a change in the layer thickness in the beam path results.

This leads to a continuous phase shift between the light beams.

Step 4–beam recombination and generation of the interferogram

A second polarizer converts the phase-shifted beams into a single beam of varying intensity: the interferogram.

Advantages of the FT-NIR Spectrometer in Comparison with the Dispersive NIR Spectrometer

Basically, high resolution spectra can be achieved using the FT technology. What is more important in the routine is shorter measurement times with the same signal/noise ratio (S/N). Through the good wavelength accuracy, calibration can be more easily transferred from one device to another.

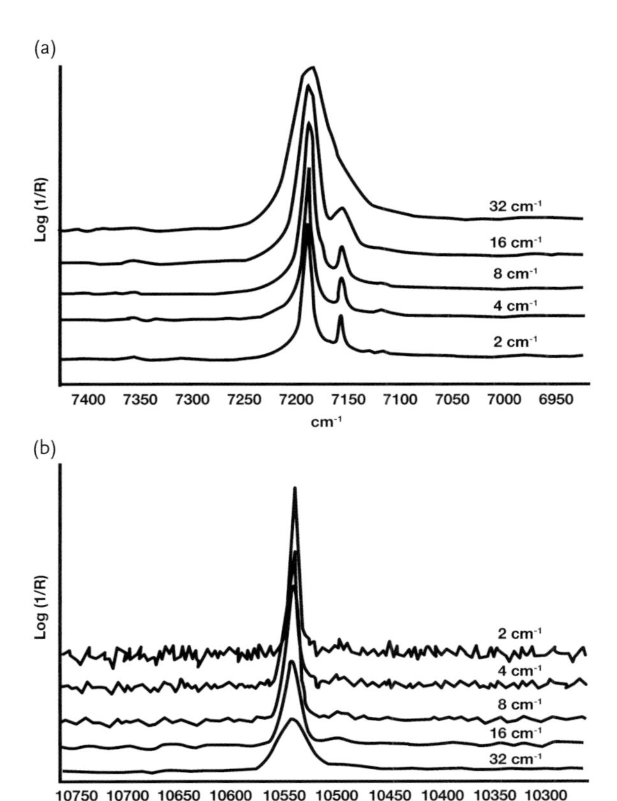

Figure 8.50 (a) Banding resolution and the intensity and (b) noise compared with the resolution.

What is significantly newer than the technology of the Michelson interferometer is that of the polarization interferometer, which provides additional benefits: the light is not split into beamlets. A compact design provides a high light throughput. The linear movement of the crystal can be implemented more easily and is, therefore, less susceptible to disturbances such as vibrations.

The NIR Spectrum

Compared to an IR spectrum, the bands in the NIR are generally less pronounced, that is, wider than their corresponding fundamental vibrations. Relatively narrow peaks as in Figure 8.47 are rather rare, but show the influence of the resolution on the band form. For example, with older devices or diode arrays with low resolution, the bands are broader and less intense ($32\,cm^{-1}$). In Figure 8.50a at $16\,cm^{-1}$ a further band can already be seen, which could not be previously resolved. At $8\,cm^{-1}$ that is well resolved, but one can observe in Figure 8.50b that at the same time the noise increases greatly with the better resolution. The routine, therefore, seldom needs a better resolution than $8\,cm^{-1}$.

Sample Preparation (see Section 8.2.2)

Liquids, Gases

The low intensity of the bands in the NIR is advantageous for liquids, as these with a layer thickness of 0.5 mm up to a few mm can be measured in a cuvette. Gas measurements can involve somewhat more work, depending on the gas density, since the layer thickness must be increased accordingly (e.g., by multiple reflections in mirrors).

Solids

Solids often require no preparation, if they are homogeneous. There is no need for solvents.

Sampling can have a decisive influence on the quality of the results. For example, if a substance separates due to vibration during transportation and if a sample is taken only from one spot, namely from the top, this sample is not representative.

Established sampling procedures should then be used.

The layer should be about 5 mm thick during reflection measurements.

Grain sizes less than around 0.5 mm are no longer noticeable in the spectrum. For larger grains, the intensity of the reflected radiation depends on the orientation of the particles and decreases with their size. Thus it is possible on the one hand, to determine physical parameters such as the grain size (within limits), and, on the other hand, it can be advantageous for the remaining parameters to grind the sample. To be sure, these effects can be minimized by pretreatment data from the spectrum (e.g., by derivation or normalization). However, if possible they should be avoided directly through the sample preparation.

Measurements with a fiber optic probe can be taken directly in the container. In addition to the homogeneity of the sample, the correct pressure is also important. If the sample is compressed, the bulk density increases and with it the intensity of reflection. Multiple measurements should be carried out at various points in the sample. The individual spectra should not differ too much from each other.

8.2.3.3
Material and Methods

Equipment

FT-NIR spectrometer NIRFlex N-500 system (Buchi), with a suitable cell for solids (NIRFlex Solids or NIRFlex Fiber Optic Solids), PC for data analysis,

(mixer/mill), Petri dish attachment

Software

NIRWare Basic Packet, NIRCal

Accessories

Vials for mixing the various sugars (tip: the rotary flask of a rotary evaporator)
Petri dish or the suitable container for the NIR measurement

Chemicals

Lactose	CAS- Number 63-42-3
Saccharine	CAS- Number 57-50-1
Fructose	CAS- Number 57-48-7

Mixtures
Sucrose with 5–10 wt% of another sugar
6 mixtures of sucrose with up to 50% by weight of another sugar

Experimental Procedure
The use of chemometrics software (such as NIRCal) requires practice ("License"), but only limited mathematical understanding.

- Factor analysis is very useful in both qualitative and quantitative analysis (automatic ordering of the variance components).
- Statistics is always the law for large numbers. There must be a sufficient number of samples available.
- The best chemometrics will not help if the logistics are not in order.
- Always do calibration and validation.

Turning on the System
After switching on the NIRFlex N-500 and the PC automatic initialization starts. The main components of the spectrometer are identified and the basic functions reviewed, for example, lamp unit, standard wheel (for the internal device test, SST), laser unit and measuring cell. The initialization time is about 60 s.

Device Test (SST)
After a warm-up period of 10–15 min, the device test (system suitability test) can be started in the operator software. Once triggered, this process is automatic and displays the results.

Examination of the Temperature
The system is equipped with multiple temperature sensors. For each sensor upper and lower temperature limits have been predefined.

Examination of the Detector Linearity
In this test the linearity of the intensity scale of the spectrometer is tested. The absorption of five gray filters in the standard wheel is measured successively and compared with reference values.

Determination of the Signal to Noise Ratio
With this test the relationship between signal strength and noise is determined. Here, two beams of light differing in intensity (maximum intensity 100% and 10% intensity) are analyzed over several wave frequency intervals.

Examination of the Wave Frequency Stability

The stability of the wave frequency of the system is tested with the help of the PMMA reference plate of the reference wheel. An absorption peak of the transmission spectrum is analyzed and compared with its reference value.

Measurement Preparation

In the management console a new application is created. This contains various settings that are most important for this experiment:

- The measurement options (type of measured cell and accessories, scan number, repeated measurements, . . .)
- The display of text fields in the operator (sample name, identity, . . .)
- The assignment of calibration (only to be conducted after the measurement and creating the calibration)

External Reference

At the beginning and approximately every hour, a new, so-called external reference (background) should be measured. This reference spectrum is recorded either without a sample (in the case of transmission) or with a white standard at the sample bench (reflection).

It is referred to as an external reference. The measurement is performed with the empty measuring vessel (or clean probe, and for liquids using a transflectance adapter).

Frequent reference measurements are used to compensate for environmental influences that can affect the spectra. Each cell provides a means for reference measurement, namely the external reference. To measure the transmission or reflectance of a sample the measured intensity spectrum of the sample is divided by a reference spectrum. The operator software automatically indicates any change in measurement parameters or after expiry of the time interval it points out when to measure a new external reference.

Sample Measurement

To measure the sample, the desired application is selected from the operator software. After entering the sample identification, the measurement can be started. A unique sample description is important in order to subsequently and clearly assign the reference values (e.g., identity or quantity). For measurement purposes, the individual samples are filled sequentially into the measuring vessel (at least 5 mm layer thickness) and measured.

Assignment of Reference Values

After measuring all the samples they are retrieved in the management console. With the sample management the various properties such as identity or sugar content can be assigned. The reference values are then entered in the resulting table.

Model Construction

Current chemometrics software is often very powerful. With increasing performance and functionality, complexity also increases. Adjustments to the "user level" are often helpful at this juncture. In the case of NIRCal a "Wizard" helps, in which only a few conditions are queried. The following calculations are made automatically. For example, different wavelength ranges, data preprocessing, and so on are tested. The calibrations with the best results are then displayed.

In the next step, the experienced user can continue to process the calibrations with the "toolbox". Here the important settings and functions are summarized. At the same time the expert has access to the entire functionality of chemometrics on menu and toolbars.

The most important diagrams can be displayed side by side (with "Overview" these are all opened simultaneously) and allow interactions with the mouse. For example, selected spectra are highlighted in all displayed diagrams at the same time.

Experiment

(A) Identification of various types of sugar

Tip: To calculate a model that is as stable as possible, it is necessary to take several influential factors into consideration, which can have an effect on the spectra received. For this reason, conduct multiple measurements on the various pure sugars. If possible, with varying instruments and at varying times (e.g., possibly in the afternoon the room temperature is higher than in the morning). The low variance in the case of a low number of samples can otherwise cause the same substance not to fit into the calibration model at another point in time under changed environmental conditions or to be not correctly identifiable.

Measure the three different sugars and allot each identity to the spectra.

A.1 Identifying a known sugar

If one of the three known types of sugar is measured, it should be true positive identified. A sugar from another manufacturer or finely ground sucrose can possibly be identified, but falls out of the cluster. Then the result is "not ok".

A.2 Identifying a sugar with impurities

Mix a sugar with an "impurity" of 5 to 10 wt% of another sugar. Conduct a threefold measurement and observe the result. The spectra with impurity should be clearly visible outside the cluster of the pure substance.

(B) Quantification of sugar in a mixture

Model construction:

Construct a series of mixtures from two types of sugar (the same sugar as in task A.2). The mixtures should be on the basis of pure sugar and contain up to 50 wt% of the other sugar. The 5–6 mixtures should not be equidistant but well distributed. The exact values of both sugar contents are entered into the chemometrics software.

Conduct multiple measurements for each mixture.

Create a quantitative calibration model.

B.1 Analysis of various sugar mixtures:

Re-measure the sugar mixture from A.2. The result should fit into the quantitative model and provide detailed sugar concentrations.

B.2 Analysis of a sugar mixture outside the calibration range:

Measure both sugars that are not in pure form in the calibration model.

8.2.3.4 Questions

(1) List typical applications of the NIR methodology.
(2) What is a cluster in chemometrics?
(3) How far does the model of the harmonious oscillator contradict reality?
(4) What do the results look like in the case of quantitative and qualitative calibrations, when there are unknown substances (impurities)?
(5) When is a sample preparation necessary?
(6) What is the difference between the bands of liquids and those of gases?

Reference

1 Gauglitz, G., Dieterle, F., and Reichert, M. ChemgaPedia: http://www.chemgapedia.de/vsengine/vlu/vsc/de/ch/13/vlu/daten/multivariate_datenanalyse_sensor/hauptkomponentenanalyse.vlu.html

8.2.4 Atomic Absorption Spectroscopy (AAS), Project: "Recycling of Sewage Sludge in Agriculture"

8.2.4.1 Analytical Problem

Most wastewater treatment processes produce sludge which has to be disposed of. Conventional secondary sewage treatment plants typically generate a primary sludge in the primary sedimentation stage of treatment and a secondary, biological, sludge in the final sedimentation after the biological process. The characteristics of the secondary sludge vary with the type of biological process and it is often mixed with primary sludge before treatment and disposal. Approximately one half of the costs of operating secondary sewage treatment plants in Europe can be associated with sludge treatment and disposal. Land application of raw or treated sewage sludge can significantly reduce the sludge disposal cost component of sewage treatment, as well as providing a large part of the nitrogen and phosphorus requirements of many crops. However, very rarely do urban sewerage systems transport only domestic sewage to treatment plants; industrial effluents and storm-water run-off from roads and other paved areas are frequently discharged into sewers. Thus sewage sludge will contain, in addition to organic waste material, traces of many pollutants used in our modern society. Some of these substances can be phytotoxic and some toxic to humans and/or animals, so it is necessary to control the concentrations of potentially toxic elements in the soil, as well as their rate of application. Depending on the origin of the waste water, it can also contain undesirably high concentrations of heavy metals (such as zinc) and harmful organic compounds, for example, from industrial and commercial discharges of sewage sludge. The application of sewage sludge to land in member countries of the European Economic Commission (EEC) is governed by Council Directive No. 86/278/EEC (Council of the European Communities 1986). This Directive prohibits the sludge from sewage treatment plants from being used in agriculture unless specified requirements are fulfilled, including the testing of the sludge and the soil. The responsibility for enforcing legally appropriate standards for sewage sludge and wastewater treatment plants clearly lies with the locality concerned (municipality) and the beneficiaries (farmers). The community of Winterthur has appointed a contract laboratory to fulfill these statutory requirements for monitoring sewage sludge and soil samples at regular intervals. A suitable AAS method needs to be elaborated, optimized and validated.

Object of Investigation/Investigation System
Sludge waste

Samples
Treated sludge provided as a fertilizer for agriculture

Examination Procedure
Atomic absorption spectroscopy has evolved since the early 1960s to a broad and most commonly used method in trace element analysis. Despite the chemical and physical possibilities of disturbance it is one of the most efficient trace element determination methods. Atomic absorption (AA) occurs when a ground state atom absorbs energy in the form of light of a specific wavelength and is elevated to an excited state. The amount of light energy absorbed at this wavelength will increase as the number of atoms of the selected element in the light path increases. The relationship between the amount of light absorbed and the concentration of analytes present in known standards can be used to determine unknown sample concentrations by measuring the amount of light they absorb. Its strong advantage lies in the possibilities of variation in carrying out the various selectively eligible analytical techniques for a problem, ranging from the simple flame technique to the electrothermic and hybrid as well as cold vapor and electrothermic solid-AAS methods. Many industries require a variety of elemental determinations on

a diverse array of samples. Key markets include: environmental, food, pharmaceutical, petrochemical, chemical/industrial, geochemical/mining.

Analytes
Zinc, calcium

Keywords
Flame atomic absorption spectroscopy, atomizer, radiation sources, absorption and emission, digestion processes

Bibliographical Research (Specific Topics)

Welz, B., Sperling, M., and Resano, M. (March 22, 1999) *Atomic Absorption Spectroscopy*, Third Completely Revised Edition, Wiley-VCH Verlag GmbH, ISBN-13: 978-3527285716.

Ebdon, L., Evans, E.H., Fisher, A.S., and Hill, S.J. (1998) *An Introduction to Analytical Atomic Spectroscopy*, 2nd edn, John Wiley & Sons, ISBN-13: 978-0471974185.

Varma, A. (March 17, 1990) *Handbook Furnace Atomic Absorption Spectroscopy*, 1st edn, CRC-Press, ISBN-13: 978-0849332432.

Project Process/Task

(A) **Processing sewage sludge samples**
Sludge samples should be dried and subjected to a microwave digestion as a "dry matter".

(B) **External calibration**
A calibration graph should be determined with zinc standard solutions.

(C) **Zn and Ca contents in digested sludge samples are to be determined**

(D) **Method validation**
Validate the method (detection limit, determination limit, linear range, reproducibility).

(E) **Project reporting**
The results are to be submitted in the form of a report (or presentation). Comparisons with bibliographical references are to be made.

8.2.4.2
Introduction

Historical Facts
One could describe *Johannes Marcus Marci of Cronland* (1595–1667), Professor of Medicine at the University of Prague, as the first spectroscopic scientist who explained the origin of a rainbow on the basis of deflection and diffusion of light in water drops, in a book published in 1648. The phenomenon of light absorption was first mainly tested on crystals and liquids. Thereby it was observed that the original radiation intensity falls into three parts, reflected, through-flow and absorbed intensity. The history of absorption spectroscopy is closely related to the

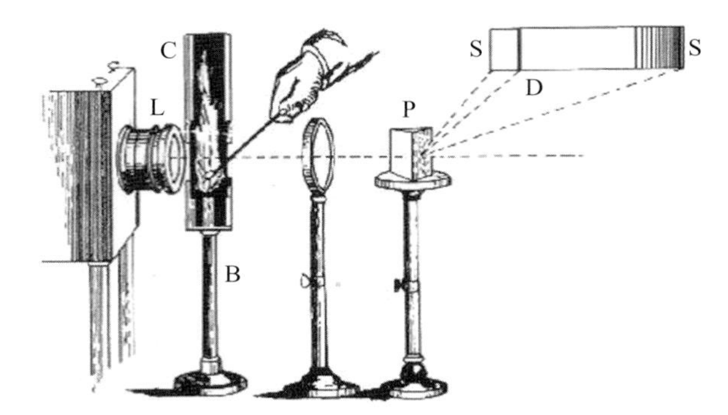

Figure 8.51 Experimental set-up of Kirchhoff and Bunsen. The radiation bundled by the lens L of a lamp radiates the flame of a Bunsen burner B, in which sodium chloride is introduced with the help of a spoon. The radiation bundle spectrally broken by prism P is observed on the screen S. The sodium D-lines thereby appear as black interruptions in the otherwise continuous spectrum. (Source: B. Welz, *Atomic Absorption Spectroscopy*, Verlag Chemie 1972).

observation of sunlight. As early as 1802 *Wollastone* discovered the black lines in the sun's spectrum, which were later examined in detail by *Frauenhofer* [1].

The fundamental physical principle of atomic absorption spectrometry (AAS) is based on the general law formulated by *Kirchhoff*, according to which each substance can absorb radiation of the wavelength at which it also emits radiation. *Kichhoff* and *Bunsen* worked out basic relationships during their systematic examination of line reversal in alkali metal and alkaline earth metal spectra [2]. Figure 8.51 shows the historical experimental design.

The law established by *Planck* (1900) pertaining to the quantum-like absorption and emission of radiation [3] provides the basis for his specific and quantitative use of the connection found by *Kirchhoff*.

Atomic Spectra

Atoms can be "excited" by the addition of energy quanta. The energy absorbed is emitted in the form of radiation when atoms return to their original state.

Types of energy absorption/emission

1. Thermal excitation

   Energy emission in the form of an emission spectrum

2. Electronic excitation

   Energy emission in the form of an emission spectrum

3. Excitation of a gas with light

   Free atoms (in the gas phase) can be conveyed to higher states of energy by the input of energy. Hereby electrons of the outer shell are raised to a higher

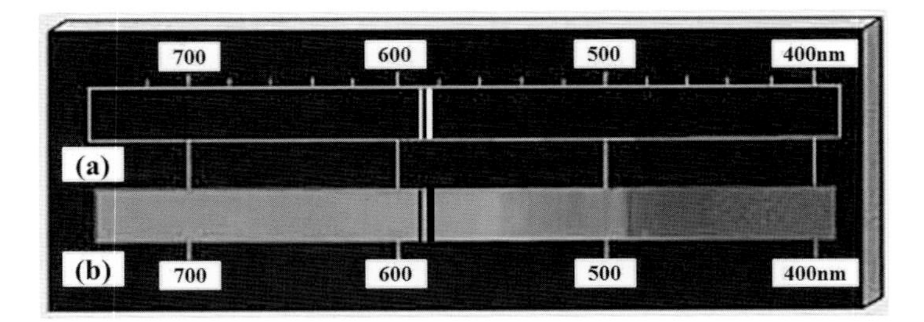

Figure 8.52 Absorption and emission spectra of sodium. (a) Characteristic emission lines of sodium: two bright lines in the center appear in the yellow part of the spectrum; (b) absorption spectrum of sodium: two dark lines appear at exactly the same wavelengths as the bright lines in the sodium emission spectrum.

energy level and the atom finds itself in the so-called "excited" state. The input of energy can occur either by collision processes with other particles or by electromagnetic radiation. As only certain energy amounts can be absorbed by the atom (in its basic state), only radiation of certain wavelengths can be absorbed. Absorption and emission spectra of an element differ by the number of lines observed. As in AAS lower atomization temperatures are used, most atoms are in their original state and almost only transitions based on the original states (so-called resonance transitions) are observed. In atom emission spectrometry temperatures of up to 8000 K are applied, so that the varying excitation conditions including the formation of excited ions can be achieved. For this reason, atomic emission spectra clearly have more lines than atomic absorption spectra.

As an example, consider the element sodium, whose emission spectrum appears in Figure 8.52. When heated to high temperatures, a sample of sodium vapor emits visible light strongly at just two wavelengths, 589.9 and 589.6 nm, lying in the yellow part of the spectrum. When a continuous spectrum is passed through some relatively cool sodium vapor, two sharp, dark absorption lines appear at precisely the same wavelengths. The emission and absorption spectra of sodium are compared in Figure 8.52, clearly showing the relation between emission and absorption features.

One can image the whole as a bar code. Each package that is scanned in a store by the cashier has a bar code (Figure 8.53), that contains all the information required to identify the product. Each identical product has the same bar code. In the same way, the atoms or ions of each element have a unique spectrum and the pattern of spectral lines offers us a characteristic "finger print".

Atomic Absorption Spectrometry

AAS is an analytical technique with which element composition can be determined. Thereby, the fact that the atoms of each element from its electronic ground state can absorb a characteristic energy and reach an electronically excited state is

Figure 8.53 Pattern of a bar code.

exploited. To analyze a sample of a certain element, light which only contains the desired wavelengths is thus required. This is supplied most easily by the excited atoms of the same element. In order to analyze copper, for example, a lamp containing copper is needed. The light of this lamp is partially absorbed by the process described above, and this weakens the light intensity at the detector. For example, the flame chamber or the atomization unit (atomizer), acts as an absorption cell. The weakened intensity of the primary light is measured and results in the signal. AAS distinguishes the following techniques: flame, graphite tube furnace, hybrid and cold-vapor and, recently, the FIAS-technique.

The most common techniques are:

Flame Atomic Absorption Spectroscopy

Atomic absorption (AA) occurs when a ground state atom absorbs energy in the form of light of a specific wavelength and is elevated to an excited state. The amount of light energy absorbed at this wavelength will increase as the number of atoms of the selected element in the light path increases. The relationship between the amount of light absorbed and the concentration of analytes present in known standards can be used to determine unknown sample concentrations by measuring the amount of light they absorb.

Graphite Furnace Atomic Absorption Spectroscopy

With graphite furnace atomic absorption (GFAA), the sample is introduced directly into a graphite tube, which is then heated in a programmed series of steps to remove the solvent and major matrix components and to atomize the remaining sample. All of the analyte is atomized, and the atoms are retained within the tube (and the light path, which passes through the tube) for an extended period of time. As a result, sensitivity and detection limits are significantly improved over flame AA. Graphite furnace analysis times are longer than those for flame sampling, and fewer elements can be determined using GFAA. However, the enhanced sensitivity of GFAA, and its ability to analyze very small samples, significantly expands the capabilities of atomic absorption. GFAA allows the determination of over 40 elements in microliter sample volumes with detection limits typically 100 to 1000 times better than those of flame AA systems.

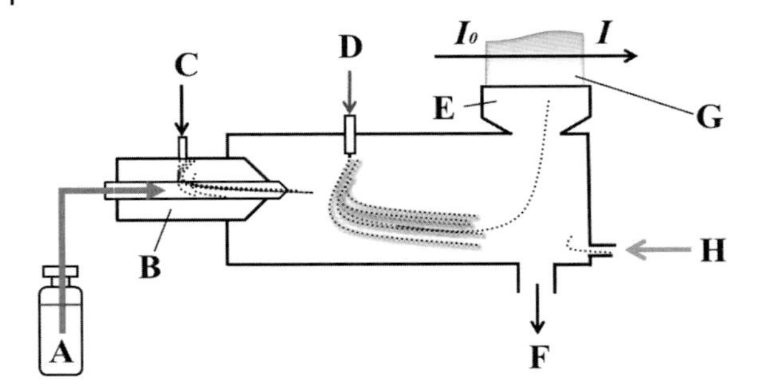

Figure 8.54 Burner with atomizer: **A** sample, **B** nebulizer, **C** nebulizer gas, **D** auxiliary gas, **E** burner, **F** to waste, **G** flame, **H** fuel gas.

Atomic Absorption Spectrometer

The atomic absorption spectrometer has four basic parts: interchangeable lamps that emit light with element-specific wavelengths, a sample aspirator, a flame or furnace apparatus for volatilizing the sample, and a photon detector. In order to analyze for any given element, a lamp is chosen that produces a wavelength of light that is absorbed by that element. Sample solutions are aspirated into the flame. If any ions of the given element are present in the flame, they will absorb light produced by the lamp before it reaches the detector, as shown in Figure 8.54. Finally, the detector records over the absorption signal as a weakening of the intensity of the primary light after corresponding strengthening and conversion. In flame AAS a light path that is as long as possible through the absorbing flame zone is required, among other things, to achieve the highest possible sensitivity. This requirement is achieved through the use of slit burners, which are 10 cm or more in length. The light path crosses the longitudinal axis of the flame. The sample solution is pneumatically atomized in a mixing chamber. The resulting aerosol is passed from there, together with the fuel gas mixture of fuel gas and oxidizing gas, into the stretched flame of the burner slot after mixing. What is particularly advantageous is an acetylene–air mixture with a combustion gas velocity around $1.6\,m\,s^{-1}$ and temperatures around 2300 °C.

Higher temperatures are achieved with acetylene–oxygen or nitrous oxide–acetylene flames. In the mixing chamber larger liquid droplets are deposited on deflector surfaces. Only the fine aerosol reaches the flame.

Sources of Radiation

The high selectivity of AAS can be traced back to the fact that for each element a special hollow cathode lamp (HCL) is used. It consists of a glass cylinder filled with inert gas under a low pressure of a few mbar with a metal cathode and anode. The cathode in the form of a half cylinder, surrounded by a protective glass flash, consists of the metal to be determined or is filled with this metal; the anode is a strong wire made of tungsten metal or nickel. At some 100 V voltage, a glow discharge occurs, which leads to the origination of the emission spectrum of the respective element. Thereby a flow of positive gas-ions impacts the cathode, metal atoms released from the cathode and finally excited to radiate. Another commonly used primary radiation source is the electrodeless discharge lamp (EDL) where some milligrams of an element are melted in a quartz bulb. Here the excitation occurs through the coil of a high frequency generator. More intense (by a factor of 100) radiation emissions are achieved with these lamps but they are only suitable for more volatile elements. With both radiation sources only the spectra of inert gas and of the element from which the cathode is manufactured or with which the lamp is filled are generated. Multi-element lamps contain several powdery metals, whereby they should, as far as possible, meet the same requirements regarding radiation intensity and selectivity for each element. Thus the use of multi-element lamps is limited.

Atomization

The goal of atomization is the generation of free atoms in the gas phase. Atomization equipment used in AAS is, apart from graphite tube furnaces, especially flames, for example, the air–acetylene flame, with which temperatures of some 2550 K are achieved. In contrast to the atom emission spectrometry the flame here only has the task of vaporizing the sample and decomposing it thermally (Figure 8.55). Particular attention must be paid to the chemical and physical processes in the flame in analytical applications. The solution of the analyte is sprayed into special atomizers with mixing chambers and the gas flow leads the resulting cloud droplets into the combustion chamber. Larger droplets are deposited as condensate. In the combustion chamber the fuel gas and an additional oxidant are mixed. While in the flame the solvent (water) evaporates and solid particles in the aerosol (gas stream) are converted to free atoms. The formation of oxides can occur as an unwanted reaction. A reduction in sensitivity and spectral disturbances occurs through the formation of molecules, of excited molecules, ions, excited atoms and radicals. Optimization of the flame aims at minimizing these side reactions. Hence, atomization has the task of bringing the highest possible amount of the element to be determined into the atomic state. A reduction in sensitivity and spectral disturbances occurs through the formation of molecules, of excited

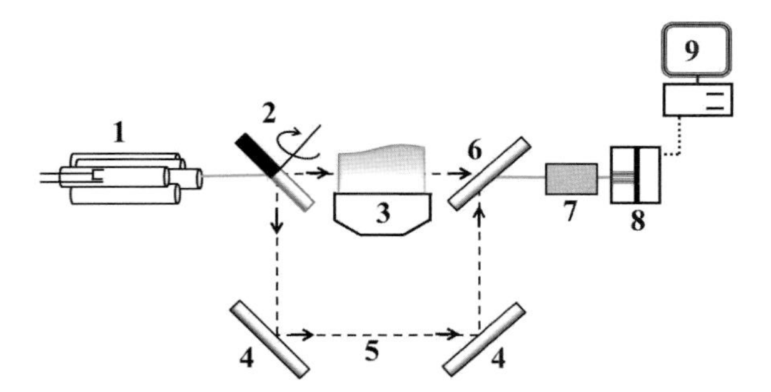

Figure 8.55 Diagram of an atomic absorption spectrometer: **1** hollow cathode lamp, **2** chopper, **3** burner, **4** mirror, **5** reference beam, **6** half-silvered mirror, **7** monochromator, **8** detector, **9** data acquisition and analysis system.

Table 8.8 Gas mixture in AAS.

Gas mixture	Temperature (°C)
Air–methane	1850–1900
Air–gas	1700–1900
Air–hydrogen	2000–2050
Air–acetylene	2125–2400
Nitrous oxide–acetylene	2600–2800

molecules, ions, excited atoms and radicals. Different gas mixtures are used to obtain the optimum flame temperature required each time (Table 8.8).

The disadvantage of this method is the need for a relatively large amount of analyte, since the efficiency of the atomizer is only about 1%. The detection limits for most elements are in the lower ppm range. Instead of flames, a graphite tube furnace can also be used for electrochemical atomization of a sample. The argon protective gas atmosphere prevents burning of the graphite tube. The sample solution is fed into a tube from above using a microliter syringe. It is first dried by a pre-selected temperature program and then further heated to remove, at this stage, disturbing substances by thermal means prior to the actual atomization. After atomization heating of the tube takes place before the next analysis.

For the hybrid and the cold vapor technique a quartz cell heated to 850–1000 °C is used as an atomizer. The elements arsenic, antimony, bismuth, selenium, tellurium and tin can evaporate after reduction, for example, with sodium borohydride and are thus freed from interfering substances. In the heated quartz cell atomization then takes place as a result of thermal decomposition. Today, this technique is practiced mainly in connection with flow-injection as the FIAS technique.

Interferences

AAS distinguishes between spectral and non-spectral interferences. Spectral interferences result in a non-specific background. They arise due to the absorption of radiation by molecules of gaseous substances or through radiation scattering of particles. Furthermore, even overlapping of atomic lines of different elements is to be expected. Such overlaps can often be avoided by choosing a different atomic line.

Physical Interferences

Transport interferences are disturbances in the transfer of the sample into the atomization device and they occur during application of the flame technique. With the use of atomizers it is important to note that the rate of suction, the effectiveness of the atomizer and the droplet size distribution, and thus the measurement signal, depend on the density, viscosity and surface tension of the solvent. For solutions with up to 1% salt content it is often sufficient to use the same solvent (mixture) for the sample and the standard.

Another way to compensate for disturbances of transport is to use the standard addition method.

Chemical Interferences

The best known example of a chemical interference is in the determination of calcium in the presence of phosphate.

The relatively stable calcium phosphate is formed, which is not dissociated in the flame and therefore causes no absorption at the Ca-line. There are several ways of solving the problem: An element can be added, which together with phosphate forms a salt more difficult to dissolve than calcium. Lanthanum serves this purpose very well. Alternatively, the flame temperature can be increased because the temperature is obviously not sufficient to atomize the calcium entirely. Nowadays, this is done by using a nitrous oxide–acetylene flame, which makes the troublesome use of lanthanum superfluous.

Ionization Interferences

Easily ionizable species do not often exist in their original state, especially in hot flames, but rather as cations. The ions have different electronic properties and therefore absorb light at wavelengths other than the neutral atoms and, therefore, cannot be measured by AAS. The interference can be controlled by the addition of still more easily ionizable elements. An excess of a very easily ionizable element generates a lot of free electrons, which prevent the analyte from being ionized. For example, barium can be measured by adding a surplus of potassium.

Background Correction with a Continuum Source

Spectral interferences always occur when absorption of radiation from the hollow cathode lamp is possible by the overlapping of sharp atomic lines and broad molecular lines of the accompanying substances (matrix) or by an indirect influence of the atomizing mechanisms. This so-called background absorption is primarily caused by molecules of gaseous substances and by radiation scattering on particles. However, even in cases of line interference, it can be simply overcome by choosing to perform the analysis using another line that has no interference with other lines.

The method with a continuum source is the most common method for background corrections. In this procedure, emission from a deuterium lamp and an HCL lamp alternately pass through the graphite tube analyzer. It is essential to keep the slit width of the monochromator sufficiently wide in order to pass a wide bandwidth of the deuterium lamp radiation. In this case, the absorbance by analyte atoms is negligible and absorbance can be accredited to molecular species in the matrix. The absorbance of the beam from the deuterium lamp is then subtracted from the analyte beam (HCL) and thus a background correction is obtained.

Process

An important task of the AAS technique is now to eliminate this basic disturbance with the possibilities of background compensation described. Also an optimized temperature program can help ensure that the background signal is largely separated from the atomic signal. Trace concentrations of lead in blood samples are preferably carried out with the flameless AAS technique because of the low detection limits. Another work technique in the AAS-method is solid AAS. In the case of the "boat technique" sample volumes between 1 and 10 mg are directly vaporized within a few seconds and quantitatively atomized in the graphite tube. Solid samples in powder form, granulates and fibers can be analyzed directly in the graphite furnace without any need for sample preparation. The direct solid-AAS is mainly used for easily vaporizable elements such as Zn, Cd, Ag, Cu, Hg, Tl, Pb, Se, Te. In contrast to the boat technique, only about 10% of the sample is put into the flame and thus into the absorption space using a nebulizer-burner system. Selenium trace analysis can be implemented with both the graphite furnace and with hybrid technology. For serum samples, of which usually only a limited sample volume is available, it is advisable to use the raw graphite furnace technique. In water samples, for which mostly larger volumes can be used, the hybrid technology can be used advantageously.

Standard Addition

There are, however, influences on the signal which cannot be compensated so easily. If, for example, the solution is more viscous than the standard needed for calibration, the nebulizer works less efficiently and the measured signal is too low because less analyte is transported into the flame. These effects are generally referred to as matrix effects. However, there is a good way to get around the problem of the matrix: the standard addition method. An unknown concentration of an analyte is measured by adding to the sample to be determined known amounts of analyte several times. The intersection of the resulting straight line with the y-axis indicates the desired concentration (see Section 4.3.3.3).

Characteristic Concentration

The characteristic concentration (char. conc.) indicates how high the concentration must be to generate 1% of absorbance (= 0.0044 absorbance). This value indicates sensitivity but makes no statement about the limit of detection, because the calculation does not take the signal noise into account.

$$\text{char.conc.}\left(\text{mg L}^{-1}\right)=\frac{0.0044\ \text{conc.}_{\text{standard}}\left(\text{mg L}^{-1}\right)}{\text{measured absorbance}} \qquad (8.50)$$

8.2.4.3
Material and Methods

Equipment

AAS, AAnalyst 100(Perkin-Elmer), microwave oven, PC for data processing

Chemicals

- Nitric acid 68% puriss CAS- Number 7697-37-2 C
- Hydrogen peroxide 35% CAS- Number 7722-84-1 O, C
- Muriatic acid 1% puriss CAS- Number 7647-01-0 C

Reagents

Samples for Microwave Digestion

Sample for analysis A	approx. 100 mg dry matter (note the exact weight)
Sample for analysis B	approx. 100 mg dry matter
Sample for analysis C	approx. 100 mg dry matter
Spiked sample D:	weigh approx. 100 mg dry matter + 0.03 mg Zn (=0.3 ml of a 100 ppm-solution) + 2 mg Ca (2 ml 1000 ppm-solution)
Blind solution E1	digestion vessel empty
Blind solution E2	digestion vessel empty

Samples

To prepare the test samples the digestion solutions with 1% HCl are quantitatively transferred through a folded filter into a volumetric flask of 100 ml: First, the contents of the Teflon digestion vessel (caution: concentrated acids) are poured through the filter into the flask, and then the digestion vessel is rinsed 3 times with a few mL of 1% HCl, and also the rinsing solution is transferred through the filter into the 100 ml measuring flask. At the end the flask is filled up to the mark with 1% HCl.

As the Ca content in the sewage sludge is higher than that of zinc, the test samples for the Ca measurement must be diluted 5 times, so that the signals are in the linear work range of the spectrometer:

Measurement sample F	mix 20.0 ml of the sample A with 1% HCl and fill to 100.0 ml.
Measurement sample G	mix 20.0 ml of the sample B with 1% HCl and fill to 100.0 ml.
Measurement sample H	mix 20.0 ml of the sample C with 1% HCl and fill to 100.0 ml.
Spiked sample I	mix 20.0 ml of the spiked sample D with 1% HCI and fill to 100.0 ml.
Blind solution E3	mix 20.0 ml of the sample E1 with 1% HCI and fill to 100.0 ml.
Blind solution E4	mix 20.0 ml of the sample E2 with 1% HCI and fill to 100.0 ml.

As blank solution 1% HCl is suctioned up each time. The blind solutions E1, E2, E3 and E4 are measured and assessed like samples.

Standard Solutions

Ready-to-use Titrisol zinc-standard 1.000 ± 0.002 g Zn

Ready-to-use calcium-standard solution CertiPUR®, Article 119778 (Merck Millipore)

Zinc solutions

0.1, 0.2, 0.3, 0.5, 0.7,1.0, 1.3, 1.7, 2.0 and 5.0 mg $_{Zn}$ l^{-1}

Calcium solutions

1, 2, 3, 4, 6, 8, 10, 13, 17 and 20 mg $_{Ca}$ l^{-1}

Experiment

(A) Processing sewage sludge samples

The samples should be digested in a microwave oven. Thereby the vessels E1 and E2 (blanks) are treated exactly the same as the other four (reagent quantities, etc.).

(B) External calibration

B.1 First prepare 10 standard solutions by diluting the zinc stock solution in a 100 ml volumetric flask. Analogously, 10 calcium standards are to be prepared.

B.2 Determine the absorption of these standards and display the calibration curve.

B.3 Now record the measurement functions for both analytes using the analysis software of the device or Excel.

(C) Zn and Ca contents in digested sludge samples are to be determined

C.1 Measurements: Analyze the samples A to D, E1 and E2 for the Zn-determination, and/or F to I, E3 and E4 for the determination with the method of external calibration.

C.2 Blank correction: if significant zinc content is found in the blind solutions E1 and E2, and significant calcium content is found in the blind solutions E3 and E4, the results (C.1) should be corrected accordingly.

C.3 Determination of recovery rates: for A, B and D now convert the measured concentration (mg l^{-1}) to the zinc content of the sludge (mg (kg dry matter)$^{-1}$) and/or for F to I to the calcium content. How much more zinc or calcium has been absolutely found in the spiked sample D and/or I (in mg) than on average in the samples A and B, or F to H, respectively?

C.4 Determining the final result: finally, correct the content found in A, B, F, G and H by the recovery rate and determine the mean of the Zn-, and/or Ca-analysis with its standard deviation.

(D) Method validation

Validate the method (detection limit, determination limit, linear range, reproducibility).

8.2.4.4
Questions

(1) What parts does an AAS-Spectrometer consist of?
(2) Describe possibilities of atomization.
(3) What are the important parameters in AAS?
(4) Which individual steps of the analytical method are, in your opinion, mainly responsible for the uncertainty of the final results in the AAS methodology?
(5) Do you see anywhere in the method used sources of systematic deviations? How would they affect the final results and how could they be eliminated?
(6) How could one further lower the detection limit of the process?
(7) What strongly limits the use of AAS and lowers its sensitivity? In your opinion, what would be the main areas of application of AAS in pharmaceutics?
(8) What other method do you know of that could substitute AAS?

References

1 Tyndall, J. (1898) *Six Lectures on Light*, D. Appleton & Co, New York.
2 Kirchhoff, G., and Bunsen, R. (1860) *Philosophical Magazine*, **20** (4), 89 ff.
3 Planck, M. (1900) *Annalen der Physik*, (1), 9 ff.
4 Skoog, D.A., and Leary, J.J. (2006) *Principles of Instrumental Analysis*, 6th edn, Brooks Cole, ISBN-10: 0495012017, ISBN-13: 978-0495012016.

8.3
Electrophoretic Separation Methods

8.3.1
Capillary Electrophoresis, Project: "Preservatives in Cosmetics: Friend or Foe"

8.3.1.1
Analytical Problem

Every consumer expects long-term stability from a product. It should not change its color or smell, or have formed any toxic substances even a few weeks after purchase. As in food, bacteria and fungi can also spread in cosmetics. They not only shorten shelf life, but can also have serious health consequences, because a drug that is effective against microorganisms may also have effects on human skin.

For almost ten years, all ingredients used in cosmetic products sold in Europe have had to be specified on the packaging in descending order of their concentration. The naming of the ingredients is consistent with the international INCI (International Nomenclature of Cosmetic Ingredients) nomenclature. With the help of this nomenclature, it is possible to identify the substances in a product used in all EU Member States. Quantities of substances that are listed in the Cosmetics Ordinance as subject to declaration are checked as part of official controls. The detection and determination of preservatives is a highly demanding analytical task. Since not all substances are officially registered, specific methods are required for analysis, depending on the problem at hand. The Solari Testing Institute provides consultation to manufacturers of cosmetic products on analytics-related issues. The foundation operates a powerful analytical laboratory. Currently, a new method for the identification and quantification of the most common preservative in cosmetic products needs to be developed, optimized and validated.

Object of Investigation/Investigation System

Personal care products

Samples

Creams, lotions

Examination Procedure

Capillary electrophoresis (CE) is an instrumental analysis method that combines the advantages of electrophoretic separation methods with the ability to automate and more easily and directly quantify. It is used successfully in the analysis of samples over a wide molecular weight range, that is, analytes of almost any size can be separated using this technology: from the simple ion to macromolecules, viruses, cell organelles and even whole cells. The focus in the field of polyelectrolytes, such as nucleic acids and proteins is, based on the new technology of capillary electrophoresis, increasingly shifting to low molecular weight substances to include separations of inorganic and organic ions. Capillary electrophoresis is especially useful when water-soluble, charged substances are to be separated. By the addition of surfactants and micelle formation of acetonitrile to the buffer solution one can get around this restriction, and thus nonpolar, uncharged components can also be separated using micellar electro-kinetic chromatography. The main applications of electrophoretic methods lie in biochemical, clinical chemistry and, to a lesser extent, in food analysis. The total area of electro-kinetic methods of separation plays a central role in biochemical analytics and analytics in clinical chemistry [1].

Analytes

Benzyl alcohol, phenoxyethanol, dehydroacetic acid, sorbic acid, benzoic acid

Keywords

Electrophoretic separation principles, basic equations of electrophoresis, micellar electrokinetic chromatography (MEKC), electro-osmosis, electrophoretic mobility, isoelectric focusing, carrier (free) electrophoresis, capillary electrophoresis, disc electrophoresis, properties of surfactants, micelle formation

Bibliographical Research (Specific Topics)

Rathore, A.S., and Guttman, A. (2004) *Electrokinetic Phenomena: Principles and Applications in Analytical Chemistry and Microchip Technology*, Marcel Dekker, Inc., New York, Basel, ISBN: 0824743067.

Westermeier, R. (1998) *Electrophoresis in Practice: A Guide to Theory and Practice*, Oxford University Press, USA, ISBN-13: 978-01-9963640-2.

Schmitt-Kopplin, P. (2008) *Capillary Electrophoresis; Methods and Protocols (Methods in Molecular Biology)*, Humana Press, ISBN-13: 978-15-8829539-2.

(2005) *Analysis and Detection by Capillary Electrophoresis, Volume 45 (Comprehensive Analytical Chemistry)*, Elsevier Science, ISBN-13: 978-04-4451718-0.

Weinberger, R. (2000) *Practical Electrophoresis*, 2nd edn, Academic Press, ISBN-13: 978-01-2742356-2.

Project Process/Task

(A) Separation of five preservatives

Five most commonly used preservatives are to be separated by modifying the buffer solution baselines.

(B) Importance of the buffer

By modifying the pH-value, the concentration of buffer and surfactant (to a lesser extent, also by the proportion of organic solvent) the separation property is to be investigated.

(C) Identification and quantification of the preservatives

With the appropriate buffer solution the preservatives are to be identified quantified and validated by comparison with standard compounds.

(D) Assay of a preservative in various cosmetics

Determine the content of the preservative in a hand cream, a hair conditioner and a body lotion.

(E) Project reporting

The results are to be submitted in the form of a report (or presentation). Comparisons with bibliographical references are to be made.

8.3.1.2 Introduction

For the quantitative determination of preservatives in cosmetics HPLC is usually applied. Because of the complex matrix of the samples purification is only possible to a very limited extent. This is why all the eluent-soluble components go on the analytical column during investigation and make long rinsing and reconditioning necessary. Quaternary alkyl ammonium compounds, which are often found in shampoos and similar products, unite with the stationary phase of the column and give these new, but rarely desired, separation properties. As a pre-column offers no protection in this case, the separation column must be replaced after a relatively short time. Many ingredients in cosmetic products display UV absorption, making demanding gradients necessary for a clean separation of the preservative-peaks from the matrix. Precisely in this case capillary electrophoresis demonstrates an advantage over HPLC.

A side effect of electrophoretic separation is the formation of heat during the passage of current through the solution. This heat increases with the square of the intensity of the electric current. The heat causes convection currents which disturb the interface, and not so much the separations themselves. This heat rising from the current flow requires that the capillary be cooled. Despite the good heat dissipation of a capillary, an additional thermostating of the capillary should be done, in the light of increasing reproducibility. Contrary to other electrophoretic separations capillary electrophoresis (CE) allows a more effective cooling, and thus a separation at higher field strengths, by the favorable ratio of surface area to volume when using capillaries.

Historical Overview

Initial experiments on the migration of ions in an electric field were carried out as early as the 19th century based on the theoretical foundations of *Kohlrausch* 1897 (theory of ion movement) [2]. In 1923 J. Kendall succeeded in separating rare earths and some acids as the first example of an isotachophoretic separation [3].

The authorship of electrophoresis is assigned to the Swedish researcher Arne Tiselius (1902 to 1971). He was the first person to use the technique in analytical chemistry. In 1948 he was awarded the Nobel Prize for his work: *"for his research on electrophoresis and adsorption analysis, especially for his discoveries concerning the complex nature of the serum proteins"*. The method he developed was known as the "moving boundary" method. In the course of time, different variations of electrophoresis have been developed. Disc electrophoresis (discontinuity of the buffers used, discontinuity in the gel- and buffer system) was described for the first time in 1959 and the polyacrylamide gel made its debut in the same year. The introduction of quartz capillaries by Jorgenson/Luke (1983) [4] was a further milestone. In their breadth of variety these electrophoretic methods became the most important tools in biochemical analysis, because they are very suitable for separating peptides, proteins, polynucleotides, and other biologically interesting macromolecules.

Electrophoresis

Separation methods based on different rates of migration and movements of electrically charged particles in an electrolyte solution under the influence of an electric field are grouped under the methodological term electrophoresis. If the electrophoresis takes place in free solution, this oldest version is known as carrier-free electrophoresis. A carrier-electrophoresis takes place when the analyte solution is applied on carrier material impregnated with buffer solution and the electrophoretic separation takes place within and under the influence of this carrier material. Ions move at a constant speed in an electric field. Although electrophoresis is a separation technique, a moving (mobile) phase, such as for

Table 8.9 Electrophoretic separations utilize the following terms compared to their "counterpart" in HPLC.

Electrophoresis terms	Chromatography terms
Migration time	Retention time
Buffer solution/BGE	Mobile phase
Column coating	Stationary phase
Electropherogram	Chromatogram
Theoretical plates	Theoretical plates
EOF, electro-osmotic flow	Flow rate
Precision	Precision
Efficiency	Efficiency
Resolution	Resolution
Band broadening	Band broadening
Current/voltage	Back pressure

example, a buffer solution and an inert (stationary) phase, for example, in the form of the capillary wall in the CE, is often encountered. The separation is not, however, based on the distribution between two phases and thus also not on the distribution equilibrium which arises between two phases. It is a separation technique, but not chromatography (Table 8.9), providing different information. Rather, the analytes remain in a phase (e.g., dissolved in a buffer solution), migrate in an applied electrical field, according to how charged they are, to the anode ("+"- pole) or cathode ("−"- pole) and are thus separated. One variation of CE, micellar electrokinetic capillary chromatography (MEKC) is an exception. This is a chromatographic technique, since in this case the separation is based on the analyte distribution between two phases (micelles and buffer solution).

Electrvic force (F_E)

Transport of ions in a solution under the influence of an electric field, hereby the acting electric force is defined as F_E:

$$F_E = q \cdot E = z \cdot e \cdot E \tag{8.51}$$

z Effective charge number of the analyte "(e.g., −2, −1, +1, +2)"
e Elementary charge (1602.10^{-19} C)
E Electrical field strength (V/cm)
q Charge (C)

Counterforce: Friction force (F_R)

$$F_R = 6 \cdot \pi \cdot R \cdot \eta \cdot v = \eta \cdot v = k\eta\upsilon \tag{8.52}$$

R Hydrodynamic radius (hydrated ion as idealized sphere) (cm)
η Viscosity of the medium (Pa·s)
v Electrophoretic velocity of the component [cm/s]
k constant (6πR idealized sphere)

Electrophoretic Mobility

The electrophoretic mobility μ_{ep} (also known as electrophoretic motion in the electrophoretic literature) is defined as the ratio of the electrophoretic velocity v and the electric field strength E. Since in equilibrium $F_E = -F_R$ applies, the following results:

$$\mu_{ep} = \frac{v}{E} = \frac{e_0}{6\pi} \cdot \frac{1}{\eta} \cdot \frac{z}{R} \tag{8.53}$$

Factor 1 constant ($\frac{e_0}{6\pi}$)

Factor 2 buffer ($\frac{1}{\eta}$)

Factor 3 ion (analyte) ($\frac{z}{R}$)

- v Electrophoretic velocity of the component (cm/s)
- E Field strength (V/cm)
- R Radius (hydrates ion as an idealized sphere) (cm)
- η Viscosity of the medium (Pa·s)
- z Effective charge number of the analyte
- e_0 Elementary charge (1602.10^{-19} C)

The first factor is a constant, the second only depends on the buffer properties, and the third are properties of the analyte. It becomes clear that electrophoresis separates molecules according to their charge and radius (i.e., their size).

Electrophoretic Velocity

The electrophoretic velocity of an ion in an electric field is defined as the product of the ion mobility and the applied field strength:

$$v = \mu^* E = \frac{qE}{6\pi R\eta} = \frac{ze_0 E}{6\pi R\eta} \tag{8.54}$$

- v Electrophoretic velocity of the component (cm s^{-1})
- E Electric field strength (V cm^{-1})
- z Effective charge number z of the analyte
- e_o Elementary charge (1602×10^{-19} C)
- R Hydrodynamic radius (hydrated ion as an idealized sphere)
- η Viscosity of the medium (Pa s)
- q Charge

Increasing the voltage, and thus increasing the field strength, always leads to an increase in the electrophoretic velocity of the ions and thus quicker analysis.

Buffer System

The quartz capillaries used in CE have ionizable silanol groups on their inner surface. If the capillary is filled with a buffer, the silanol groups dissociate, depending on the pH-value of the buffer used and thus the capillary surface is negatively

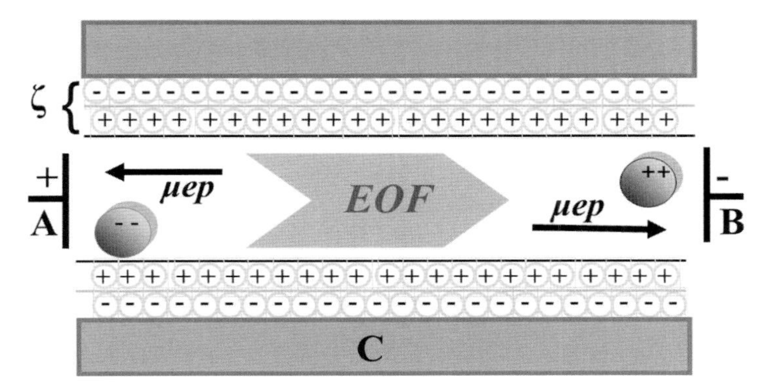

Figure 8.56 Electro-osmotic flow (EOF). **A** anode, **B** cathode, **C** capillary wall, ζ zeta potential (which describes the decrease in the charge density of the capillary to the interior of the capillary), μ_{ep} electrophoretic mobility.

charged. Due to electrostatic interactions, cations move to the capillary wall where they form an electrochemical double layer. This electrochemical double layer is described by the zeta potential ζ. If a voltage is now applied across the capillary, cations migrate from the diffuse part of the double layer to the cathode. In so doing, these cations transport water molecules to the cathode in their hydration shell. The result is a net flow of the buffer in the capillary to the cathode, which is called electro-osmotic flow (EOF) (Figure 8.56).

EOF depends significantly on the type of buffer used, on its dielectric constant and viscosity. The zeta potential is further determined by the degree of dissociation of the silanol groups and thus by the pH-value of the buffer. It also depends on the concentration of the separation buffer.

Electro-osmosis

In CE, the electro-osmotic flow (EOF) plays an important role. Its strength depends on the pH-value of the electrolyte and on the charge near the capillary surface. The EOF occurs as a result of the interfacial phenomenon between the internal capillary wall and the electrolyte solution when an electric field is applied.

The positively charged liquid column moves in the electric field, similarly to a mechanical pump, towards the cathode.

The following equation is used to describe the velocity of the electro-osmotic flow v_{eof}:

$$v_{eof} = \mu_{eof} E \tag{8.55}$$

v_{eof} Velocity of the electroosmotic flow
μ_{eof} Electroosmotic mobility
E Field strength ($V\,cm^{-1}$)

The electro-osmotic flow (EOF) can be described in a simplified manner by the Helmholtz equation:

$$\mu_{eof} = \frac{\xi\varepsilon}{4\pi\eta} \tag{8.56}$$

ξ Zeta-potential (mV)
ε Dielectric constant
η Viscosity of the medium (Pa s)

It is inversely proportional to the viscosity η of the electrolyte, proportional to its dielectric constant ε, the applied field strength E and the number of charges on the capillary wall, that is, to the thus constructed ζ-potential. While electrophoresis causes the separation of particles of different mobility, electro-osmosis causes a flow (EOF) of buffer solution in an electric field.

Resolution and Number of Theoretical Plates

Analogous to chromatography, in CE too the separation efficiency (performance) of the method can be described by the resolution (R_S) of two components.

The resolution of two peaks is defined as the migration time (t_M) interval between the peak maxima divided by the average peak width (W):

$$R_S = \frac{(t_{M1} - t_{M2})}{W} \tag{8.57}$$

Equation (8.57) also generally applies to chromatography and shows that N is related to the retention time (here migration time) and peak width, and equation (8.58) shows how one can influence that.

$$N = 16\left(\frac{t_R}{W}\right)^2 = 5.54\left(\frac{t_R}{W_{1/2}}\right)^2 \tag{8.58}$$

N Number of theoretical plates
t Migration time of the components (s)
W Peak base width W (s)
W ½ Peak width at half height (s)

The electrophoretic separation is based on the migration of charged particles in the electric field. By applying a voltage across the isolation gap a flow of current occurs in the buffer. Due to the electrical resistance of the heat buffer, this causes heat to develop in the separating segment, known as Joule heat. The dispersion or peak broadening has an essentially negative effect on the separation efficiency of capillary electrophoresis. Dispersion can have various causes. Under the simplified assumption that only the molecular diffusion exerts a significant influence on the peak broadening, one obtains for the number of theoretical plates N, the expression:

$$N = \frac{U(\mu_{ep} + \mu_{eof})}{2D} \tag{8.59}$$

$(\mu_{ep} + \mu_{eof})$ Total mobility
U Applied voltage (V)
D Diffusion coefficient ($cm^2\,s^{-1}$)

$A \quad \mu_{tot} = \mu_{ep} + \mu_{eof}$

$B \quad \mu_{tot} = \mu_{eof}$

$C \quad \mu_{tot} = \mu_{ep} - \mu_{eof}$

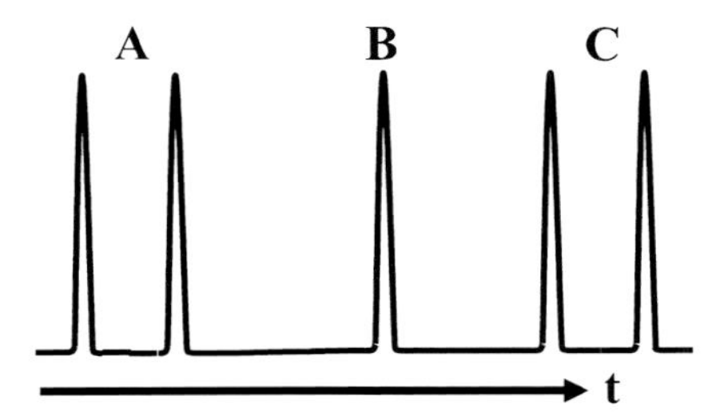

Figure 8.57 Migration time. **A** cations, **B** neutral species, **C** anions.

This shows that for high separation voltage and analyte molecules with small diffusion coefficients (macromolecules) high separation efficiencies are achieved.

To assess the separation efficiency of CE, not only the efficiency of the method is of importance, but also the selectivity for two different sample components (Figure 8.57).

Selectivity

The selectivity in CE is determined by the mobilities of the analytes. If there is a large difference in the mobilities of two analytes, a good resolution is achieved during separation. To change the selectivity of a separation additional substances are dissolved in the separation buffer. With a skillful selection of the chemicals used additional interactions between analyte and buffer additive can be exploited to achieve a better separation of the analytes. The types of interactions that have been applied in practice are ion pairing, complex formation, distribution between solvent and micelle and the formation of host–guest complexes.

Thereby, the following classes of compounds are applied as a buffer:

- Surfactants (anionic, cationic, neutral)
- Zwitterions
- Linear hydrophilic polymers
- Organic solvents
- Chiral selectors
- Complexing agents
- Micelle surfactants (e.g., SDS)

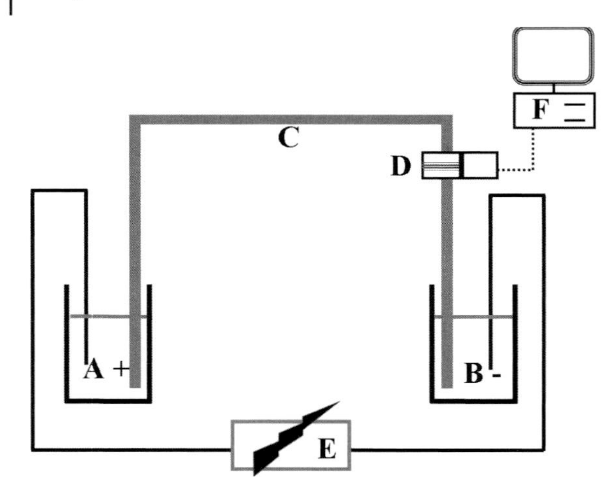

Figure 8.58 Diagram of a CE apparatus. **A** inlet electrolyte (anode), **B** outlet electrolyte (cathode), **C** capillary, **D** detector, **E** high voltage (1–30 kV), **F** data acquisition.

Capillary Electrophoresis

Description of a CE System

The main components of a CE apparatus are the capillaries filled with buffer, the high voltage supply, the injection system and the detector (Figure 8.58). The assessment of electrophoretic separations is carried out in the case of the carrier electrophoresis through electropherograms (similar to thin-layer chromatograms). The HV power supply provides DC voltages of 0–30 kV. In addition, timer functions for controlling the injection and analysis can be present. The most commonly used buffer systems are phosphate (pH around the neutral point) and borate buffer (pH around 9). Excessive buffer concentrations should be avoided since this results in an increased current flow which in turn leads to increased heat development. On the other hand, the buffer concentration should not be less than the analyte concentration, since this leads to an effect that corresponds to the overloading of an HPLC-column (depleted buffer capacity). Furthermore, the buffer solution contains all the additives required for the relevant electrophoresis modus such as methanol (or another organic modifier), SDS, cyclodextrins (or other complexing agents) and additional salts (electrolytes) that affect the ionic strength.

Capillaries

The separation capillaries used are largely made of polyimide-coated quartz (fused silica), which due to its mechanical robustness and better handling is given preference over Teflon or glass capillaries. Customary dimensions are internal diameters of between 25 and100 μm, external diameters of between 130 and 500 μm and capillary lengths of 10 to 100 cm (compare GC: 5–50 m, ID 220–500 μm). The resulting theoretical plate number N varies between 500 000 and 1 000 000. A capillary of 50 cm in length and with a diameter of 50 μm has a volume of about 1 μl. Apart from untreated fused silica capillaries, it is possible to use capillaries with the most varied "coating". The coating serves to reduce the electro-osmotic flow and to avoid adsorption effects on the capillary wall.

High Voltage Supply

Typical separation voltages in CE are between 1 and 30 kV. The currents that thereby arise in the capillaries can reach up to 350 μA. It has proved advantageous if voltage or current strength can be kept constant independently of each other.

Injection

An injection as in the GC or HPLC is not possible in the CE, since the injection volumes are in the range of approximately 1/100 of the capillary volume. This corresponds to about 1–10 nl in standard capillaries. The injection techniques in CE therefore work on the principle of submitting a liquid quantity (at least 1 μl), and transporting aliquots thereof into the capillary. The three most common methods are the differential pressure (hydrodynamic) techniques, electro-kinetic injection and the gravity method.

Differential pressure (hydrodynamic) injection: by applying pressure at the inlet or a vacuum at the outlet of the capillary a sample can be added to the column. The injected sample volume is determined by the amount of pressure or vacuum and by the time. This method of injection can be technically tedious and time-consuming, but it offers very good values for reproducibility and discrimination.

Electrokinetic injection: with this injection method the sample (5–50 μl) is migrated into the capillary for a certain time by applying a voltage (impulse approx. 5 kV). The reproducibility of this method is relatively high, but a discrimination of the sample occurs. If there is no electro-osmotic flow present, only charged sample components with the correct polarity are moved into the column. Even in the presence of an electro-osmotic flow, based on their different electrophoresis motilities, the different sample components are applied to the capillary in varying degrees (molecules with high mobility better than molecules with less mobility).

Gravity injection: sample can also be applied to the capillary by raising or lowering the buffer tank about 50–20 cm, creating hydrostatic flow by the difference in levels. There is no discrimination here. However, reproducibility is less than in the first case.

Detection

Detection presents a problem in CE, since the detectors should be adjusted to the tiny column dimensions and the very sharp bands. Due to the very small capillary diameter of 10 to 100 μm and the resulting low detection volumes in the pL to nL range, CE places special, very high demands on usable detection systems. In CE shorter response times and higher sensitivities are required than for chromatographic methods. A typical band has a width of 0.2 mm in the possibly 600 000 theoretical plates in a 100 μm capillary, which corresponds to an elution volume of 1.5 nl. Mainly optical systems (UV and fluorescence) are used, as these are commercially available and the adjustment to the capillary system has been technically well solved. However, the use of electrochemical and mass spectrometric detectors is also on the increase. Detection is conducted "on-column", as a rule. For this purpose, a window is burnt, etched or scratched into the polyimide film, through which detection can be carried out. An alternative to direct detection is the indirect methods by which the buffer solution provides a basic signal (e.g., a fluorescent component in the buffer), which is then destroyed by the analyte. In general, this method is, of course, less sensitive, but very universal, for example, for compounds that show no fluorescence. Other detection options are electrochemical detectors (especially conductivity detectors), radioactivity detectors, thermo-optical detectors, refractive index, and conductivity detectors. What is also particularly interesting is the coupling of CE and MS, as the low flow rates of CE particularly favor this coupling, as opposed to the conventional LC-MS-coupling (see Section 8.1.4).

Electrophoresis Techniques

Gel Electrophoresis (GE)

Gel electrophoresis separates macromolecules due to differences in charge, molecular weight and folding. The charge of the molecules causes migration in the electric field, and the gel impairs large molecules more during transport than small molecules. What is important in assessing the molecular size is the cross-sectional area of the molecule that lies vertically to the flow direction (called the end face or frontal area), which determines flow resistance. In other words, a molecule with a specific molecular weight in the folded state (small volume, small frontal area) offers a lower flow resistance than in the denatured, unfolded state. Thus it is not the molecular weight alone that affects the separation, but rather it is the volume or the cross-sectional area and the hydrodynamic diameter associated with it. In addition, the charge of the analyte influences its path in the gel. For this reason, there is no simple relationship between the path and the molecular weight in gel electrophoresis. It frequently helps to compare pure substances that can be run as a standard for the separation. A frequently used gel consists of polyacrylamide. A known application of sodium dodecyl sulfate polyacrylamide (PAGE) is the separation of DNA fragments. Associated with this is the polymerase chain reaction (PCR), which can replicate sought fragments of a DNA in a targeted quantity. The amplified DNA-fragments are separated onto a gel and then stained with ethidium bromide. This fluorescent dye stores itself in all DNA strands, independent of the sequence, and makes them detectable with UV-excitation at $\lambda = 254\,\text{nm}$. Such chemical staining methods are by far the most commonly used detection methods in gel electrophoresis.

As standards, known DNA fragments are added to the sample and, by comparing the distance migrated, it can be detected whether a certain gene segment is present in the sample. This contains applications of the so-called genetic fingerprint to include even proof of genetically modified food.

In conventional electrophoresis column-shaped gels or flat-bed gels are used to minimize convection, while with a capillary the high ratio of surface to volume appears to be anti-convective. Apart from preventing convection, gels such as polyacrylamide or agarose also have unique screening properties, which are essential in determining size and/or molecular weight of proteins or peptides in SDS-PAGE (sodium dodecyl sulfate polyacrylamide gel electrophoresis).

During gel electrophoresis, separation takes place mostly in a freshly prepared gel. The precursor substances used in the gel (monomers or dissolved oligomers and polymers) usually dissolve when heated in water. The solution is poured into a vessel, where it cools and forms a solid gel by cross-linking polymer strands. Water or a buffer solution, can move in the gel, whereby the liquid-filled gel pores represent a defined resistance for the analytes. As a result, substances are both separated according to their charge (determines the migration towards the cathode or anode) as well as according to their size.

During gel production (the "casting"of the gel) a so-called gel comb can be used to slit pockets at one end of the gel into which the samples are pipetted. As in DC or TLC, multiple samples can be run parallelly. Here a track is usually reserved for a standard mixture, that is, a sample with a mixture of molecules of known molecular weight. After casting, the samples are pipetted into the pockets and a defined voltage is applied between both ends of the gel. The analytes migrate according to their charge towards an electrode and separate due to different charge and size.

Capillary Zone Electrophoresis (CZE)

Capillary zone electrophoresis is the most common electrophoresis technique. It should be noted that in many cases, CZE is meant when one speaks of CE. In capillary zone electrophoresis, the separation occurs in accordance with the different mobility of ions in the electric field. This causes the ions to move towards the cathode at different speeds. Anions also move to the cathode because the electro-osmotic flow is usually greater than their rate of migration towards the anode. The sample application here is, however, a band as narrow as possible. After applying the electric field each component moves because of its mobility, so that ideally pure zones with only one component are formed. The aim is thus to maximize the differences in migration rates, and to simultaneously minimize zone broadening due to dispersion. The capillary zone electrophoresis is a method that takes advantage of electro-osmotic flow. By this effect, it is possible to study both cations, anions, and neutral particles in an analytical run. The total mobility is obtained for the individual species as the sum of the electrophoretic mobility and the electro-osmotic mobility. As in the case of the cations, the two mobilities are added, they migrate fastest. Within the cations separation results due to the different electrophoretic mobilities. After the cations the neutral particles reach the detector. They move with the speed of the electro-osmotic flow. A separation into individual components usually does not take place. The third group eluting are the anions, because in their case, the electrophoretic and the electro-osmotic movements are opposed. Within the anions the components with the lowest electrophoretic mobility are the first to elute.

Micellar Electrokinetic Capillary Chromatography (MEKC)

MEKC (Figure 8.59) is a type of chromatography, which can be carried out in a CE apparatus (with a conventional CE capillary). In the case of MEKC, separation takes place at the varying migration velocities of micelles surrounding the analytes, but also by the distribution equilibrium of the analytes between the micelle and the buffer solution [5]. With this form of separation in the capillary charged surfactants are dissolved in concentrations above their critical micelle concentration (CMC) in the buffer. The analytes can be distributed between the interior of the micelle and the buffer. Depending on the location of this distribution equilibrium, the migration of the analytes is affected by the migration of the micelle. If surfactants are dissolved in concentrations in the buffer below their CMC, interactions of the surfactant monomers with the sample components can also occur if structural and/or electronic requirements are a given. The buffer solution, which moves through the capillary due to the EOF, represents the mobile phase. The second phase is given by the micelles.

The most common detergent so far is sodium dodecyl sulfate (SDS). If this is added to the solution, micelles form above a certain concentration, the CMC, whereby the hydrophobic residues of the detergent are addressed to the interior of the micelle. The hydrophilic heads are oriented outwards. Separation is based on the fact that the analytes scatter between the mobile and pseudo-stationary phase according to their polarity. Nonpolar molecules mainly stay in the hydrophobic interior of the micelles and are, therefore, retained, whereas polar molecules remain primarily in the buffer solution and elute faster. This corresponds to the basic principle of reversed-phase LC, which makes it clear that the MEKC is rightly called chromatography.

In this case, therefore, the movement of the analytes through the capillary is not (exclusively) based on their own charge. Rather the buffer solution (mobile phase) migrates towards the cathode due to the EOF. The slower movement of the micelles (pseudo-stationary phase) is due to their negative surface charge (migration towards the cathode) and their size (less speed than EOF). If, besides uncharged analytes, ions are also present in the sample, these of course move toward either the cathode or anode according to their charge.

The advantage of MEKC over HPLC is often in clearly narrower peaks and shorter analysis times. The drawbacks include the previously mentioned difficult reproducibility of migration or retention times in CE in general [6].

Capillary Iso-Electrical Focusing (CIEF)

A carrier ampholyte is injected into the capillary together with the sample. This can be, for example, a mixture of different polyaminocarboxylic acids, which may all be present as zwitterions and all have a different isoelectric point (IP). After applying the voltage, positively charged ampholytes (higher IP) migrate to the cathode, negatively charged ampholytes (lower IP) migrate to the anode and thus build a pH-gradient. Ampholytes with a low pK must be on the anode side at the beginning and those with a high pK must be on the cathode side. The sample is injected somewhere in the mixture and the field is applied. The amphoteric sample components now start to migrate. Components with a negative net charge migrate to the

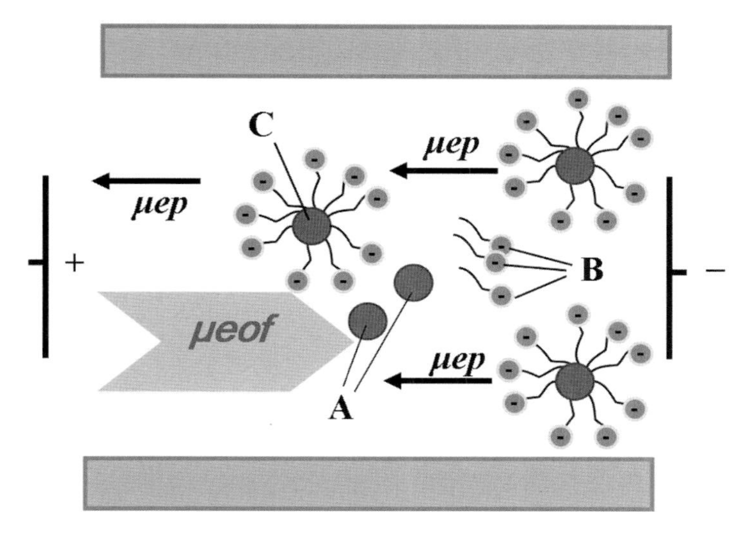

Figure 8.59 Micellar electrokinetic capillary chromatography (MEKC). **A** analyte, **B** surfactant monomers, **C** micelles.

anode, thereby enter zones with a smaller pH-value and are increasingly protonated. Eventually a point is reached where the net charge is zero (the IP) and the component stops migrating. In the same way each component migrates to its IP, then migration ends. The end of the separation is indicated by the decreasing flow of current. The field is then switched off and the capillary content is transported out of the capillary under pressure, whereby the components pass the detector sorted according to their pH-values. The method is particularly suitable for proteins, since the charge is independent of the pH-value. One drawback, however, is that proteins are least soluble at the IP and there is a risk that they will precipitate out.

8.3.1.3 Material and Methods

Equipment

For the method described here each device can be used, provided that it has the required UV detection sensitivity. A photodiode array or a fast-scanning detector is certainly an advantage for identifying the peaks, but is not essential.

Suggestion:
Capillary: fused silica, uncoated
Effective length: 35 cm, inner diameter: 75 µm

Parameters

Detection	Fast-scanning of 200 nm > 350 nm
Voltage	17 kV, resultant current: 60 µA
Injection	Hydrodynamic injection (duration: device-specific)
Temperature:	25 °

Chemicals

Benzyl alcohol	CAS- Number 100-51-6	Xn
Phenoxyethanol	CAS- Number 122-99-6	Xn
Dehydroacetic acid	CAS- Number 520-45-6	Xn
Sorbic acid	CAS- Number 110-44-1	Xi
Benzoic acid	CAS- Number 65-85-0	Xn
Acetonitrile	CAS- Number 75-05-8	F, Xn
Sodiumdodecyl sulfate	CAS- Number 151-21-3	F, Xn
Tromethamine (TRIS)	CAS- Number 77-86-1	Xi

Solutions

Buffer Solution

60 mM of sodium dodecyl sulfate and 30 mM of Tris are dissolved in a mixture of 7 parts of deionized water and 3 parts of acetonitrile (v/v) solution, with acetic acid at pH 9.

Standard Solutions

1, 2, 3, 5, 7, 10, 13, 20, and 50 $mg_{\text{preservatives}}\,l^{-1}$

Experiment

Technical Requirements

Since with the method described here high concentrations of acetonitrile are used in the buffer, you should check before that all parts of the electrophoresis equipment coming into contact with the buffer solution are resistant to this solvent.

Please note:

- Degas buffer since dissolved oxygen is bad for the stability of the current.
- Observe warm-up period of the equipment.
- For optimal operations, all solutions including sample solutions should be filtered through a 0.45 μm membrane.

Cleaning the Capillary Column

Rinsing:

- Water
- NaOH (0.5 M)
- HCl (1 M)
- NaOH (0.5 M)

(A) Separation of five preservatives

(The composition of the buffer solution is a key consideration in developing a method for capillary electrophoresis, because, unlike HPLC, the buffer solution assumes the function of the stationary *and* mobile phases.)

Modify the proposed buffer solution until the baselines of the five preservatives are separated.

(B) Importance of the buffer

By changing the pH-value, the concentration of buffer and surfactant (to a lesser extent also by the proportion of organic solvent) the separation characteristic is to be illustrated.

(C) Identification and quantification of the preservatives

C.1 With the appropriate buffer solution the preservatives are to be identified and quantified.

Measure each standard solution three times and determine the analytical sensitivity and the intercept of the five test functions with their standard deviations using a linear regression.

Calculate the detection limits of the method.

C.2 The method is to be validated.

(D) Assay of a preservative in various cosmetics

Determine the content of the preservative in a hand cream (oil/water emulsion), a hair treatment (water/oil emulsion) and a body lotion (oil/water emulsion).

Processing Samples

- Hand cream

0.5 g of the sample is placed in a centrifugal glass with 3.0 ml of saturated solution of common salt and 3.0 ml of acetonitrile and vigorously shaken. It is centrifuged for phase separation. 1.0 ml of the supernatant solution is mixed with 2.0 ml of water and used for analysis.

- Hair treatment

1.00 g of the sample is placed in a centrifuge tube with 4.0 ml of acetonitrile and vigorously shaken. Then it is centrifuged for 5 min. 1.0 ml of the supernatant solution is mixed with 2.0 ml of water and used for analysis.

- Body lotion

  Processing: 2.00 g of the sample is processed as described under hand cream.

8.3.1.4
Questions

(1) Describe the components of a capillary electrophoresis system.
(2) Describe the workflow of a CE-analysis
(3) What is electrophoretic mobility based on?
(4) What are the most frequently used capillaries in CE?
(5) What is isoelectric focusing?
(6) Give the advantages and disadvantages of CE.
(7) Give the advantages of CE as opposed to HPLC.
(8) Name the possibilities of CE-application.
(9) What is understood by electro-osmosis?
(10) What is the difference between a hydrodynamic and an electro-kinetic injection?

References

1 Karger, B.L., Cohen, A.S., and Guttman, A. (1989) High-performance capillary electrophoresis in the biological sciences. *Journal of Chromatography*, **492**, 585–614.

2 Kohlrausch, F. (1897) *Annals of Physics*, **62**, 209.

3 Kendall, J., and Crittenden, E.D. (1923) *Proceedings of the National Academy of Sciences of the United States of America*, **9**, 75.

4 Jorgenson, J.W., and Lukas, K.D. (1983) Capillary zone electrophoresis. *Science*, **1883**, 222.

5 Terabe, S., Otsuka, K., Ichikawa, K., Tsuchiya, A., and Ando, T. (1984) Electrokinetic separations with micellar solutions and open-tubular capillaries. *Analytical Chemistry*, **56**, 111–113.

6 Lauer, H.H., and Ooms, J.B. (1991) Advances in capillary electrophoresis: the challenges to liquid chromatography and conventional electrophoresis. *Analytical Chimica Acta*, **250**, 45–60.

8.4
Automation

8.4.1
Flow Injection Analysis (FIA), Project: "Phenol-like Flavor in Beer: a Quality Parameter to be Mastered"

8.4.1.1
Analytical Problem

The quality-control team of the "Old Brewery" strictly controls each step of production, from the arrival of the raw materials to the analysis of the beer samples produced. These are taken randomly from the bottles in the holding room. Control procedures include operations such as examination of the raw materials upon their arrival at the brewery and before the commencement of brewing. Water analysis prior to brewing, the control of the various brewing steps and the analysis of the finished product are also vital analytical steps. During the past few months the beer tasting panel has noted the presence of abnormal flavor characteristics which were best described as "medicinal" or "phenol-like". On recent occasions this off-taste has been of such intensity as to cause customer complaints and withdrawal of beer from the market. The quality control laboratory team of the "Old Brewery" is therefore facing an increased demand for phenol analysis. The performance of batch analysis was not satisfactory, and the management decided to implement a better method. The phenol samples of the brewery are analyzed manually according to the U.S. Environmental Protection Agency (*EPA*) method 9065 "Phenolics (Spectrophotometric, manual 4-AAP)" [1]. Performing batch chemistry measurements manually by mixing samples and reagents and waiting for color development using manual sampling is slow, tedious, and requires a great deal of skill on the part of the analyst if precise and accurate results are to be achieved. Eventually, the demand for a large number of analytical determinations per laboratory became apparent. The management has now decided to implement a more efficient way of tackling the rising number of phenolic analyses. Continuous flow analysis (CFA) was thought to be a sound basis for an improved method. An alternative strategy for quantitative analysis could be flow injection analysis (FIA), which has several advantageous features, including versatility, simplicity, and low cost. These methods were investigated, compared and validated.

Object of Investigation/Investigation System

Beverage manufacturing operation

Samples

All sources of potential (phenol)-impurities in beer production are under examination: water, the beer fermentation process during production, and beer as a final product.

Examination Procedure

A significant number of chemical assays worldwide are still carried out manually, by means of volumetric glassware that has remained virtually unchanged in the last 200 years. Automation of reagent-based chemical analysis has been implemented whenever many samples have to be processed, or when continuous monitoring is desired. Downscaling of automated assays is the current trend, as it decreases sample and reagent consumption and waste generation. In the water and environmental analytics, in clinical-chemical analytics, and in industrial analytics, where large numbers of samples must be analyzed, automation measures generally have high priority. The main applications, in addition to environmental analysis, are the analysis of food and process control.

Analytes
Phenol, phenolic components

Keywords
Automated analysis, batch and continuous flow analysis (CFA), flow injection analysis (FIA), phenol index, dispersion (according to Růžička), photometry

Bibliographical Research (Specific Topics)

Růžička, J., and Hansen, E.H. (1975) *Analitica Chimica Acta*, **78**, 145–157.
Ramos, L.A., *et al.* (2005) *Journal of Chemical Education*, **82** (5), 1815–1819.
Economou, A., Papargyris, D., and Stratis, J. (2004) *Journal of Chemical Education*, **81** (3), 406–410.
Schoonen, J.W., and Sales, G.M. (2002) *Analytical and Bioanalytical Chemistry*, **372**, 822–828.

Project Process/Task

(A) Continuous flow analysis (CFA)

A.1 Determine the sample throughput of the manual determination of phenol (EPA 9065). The calculated sample throughput serves as a basis of comparison (analysis/work day).

A.2 Now run the phenol determination by continuous flow analysis according to the (EPA 9066) method [2].

A.3 Using 10 standard solutions record a calibration graph.

A.4 Validate the CFA method.

A.5 Determine the concentration of three unknown samples.

(B) Flow injection analysis (FIA)

The following measurements are to be carried out:

B.1 Determination of dispersion by means of patent blue V [3]

B.2 Effects on dispersion are to be determined.

B.3 Using 10 standard solutions, record a calibration curve

B.4 Optimization of the FIA method

B.5 The optimized FIA method is to be validated.

B.6 The sample throughput and sensitivity of the three methods is to be compared. The influence on the sensitivity of detection is to be determined.

(C) Project reporting

The results are to be submitted in the form of a report (or presentation).

8.4.1.2 Introduction

Automation in the Chemical Laboratory

In a traditional wet-chemical method, a known amount of an analyte and a reagent are brought together manually and detected after a certain reaction time. Then the measurement data are finally recorded and evaluated. The ever increasing number of samples to be analyzed and the need for an automated method of wet chemical analysis, coupled with well-reproducible measurement results with high sample throughput and low reagent consumption, have led to the development of flow analysis. These automated methods are usually based on manual methods, but, compared to manual procedures, they have a significantly higher sample throughput.

Flow Analysis Method

ISO standards define flow analysis methods as: "Automation of wet chemical determination methods with high frequency analysis". A distinction is made between the continuous flow analysis (CFA) and the flow injection analysis (FIA).

- **Continuous flow analysis (CFA)**

  The basic principle of CFA is to eliminate the hand mixing of reagents, substituting a continuously flowing stream of liquid reagents circulating through a closed system of tubing fed with a continuous flow of analyte (Figure 8.60). Flow operations are easy to control in space and time, since using closed tubing avoids evaporation of liquids, provides exactly repeatable path(s) through which solutions move and provides a high reproducibility in mixing and product formation. The flow then passes through a flow-cell in a detection system and the emerging stream runs to waste. Neither physical (flow homogenization) nor chemical equilibrium is attained by the time the signal is recorded (Figure 8.61).

- **Flow Injection Analysis (FIA)**

  Flow injection analysis was first introduced in 1975 by Růžička and Hansen [3] for automating wet chemical methods. Savings in sample volume and time

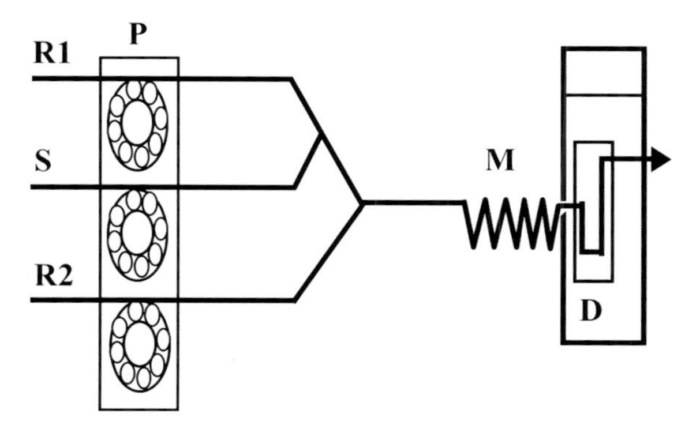

Figure 8.60 Schematic diagram of a CFA-system. **R** reagent, **S** sample, **P** pump, **M** mixing, **D** detector.

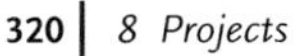

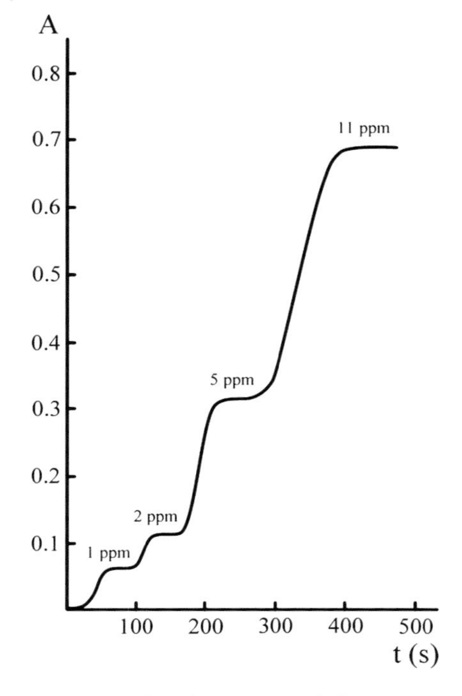

Figure 8.61 Signal response in CFA.

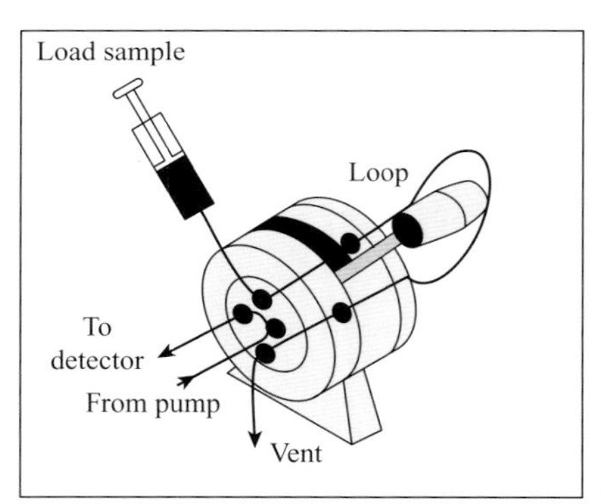

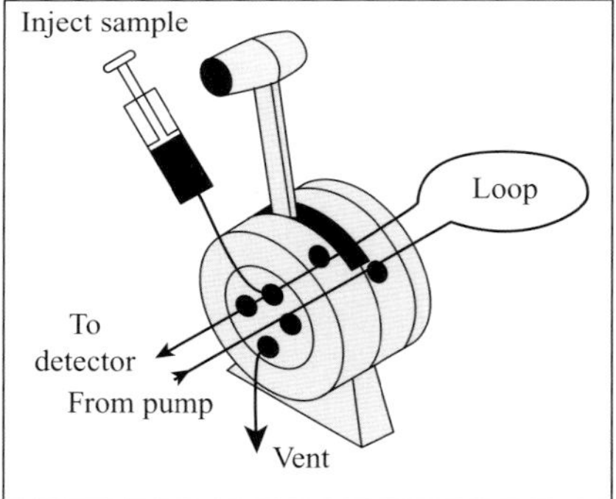

Figure 8.62 Example of injection in FIA.

led to the methodological innovation represented by FIA. In FIA, a small, fixed volume of liquid sample is injected as a discrete zone, using an injection loop, into a liquid carrier, which flows through a narrow bore tube or conduit (Figure 8.62). The sample zone is progressively dispersed into the carrier, initially by convection, and later by axial and radial diffusion, as it is transported along the conduit. Reagents may be added at various confluence points and these mix with the sample zone under the influence of radial dispersion, to produce

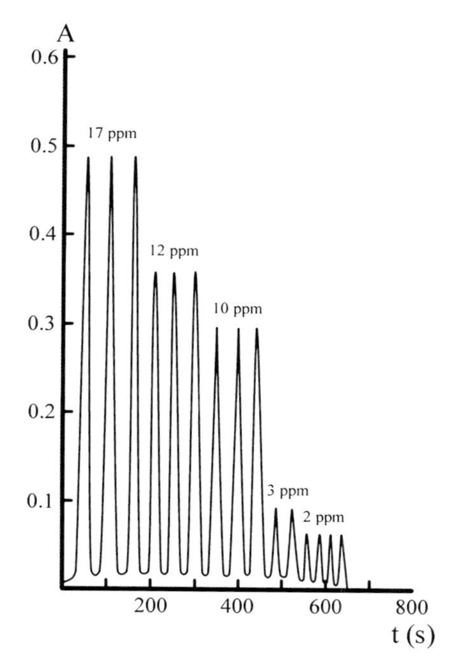

Figure 8.63 Signal response in FIA.

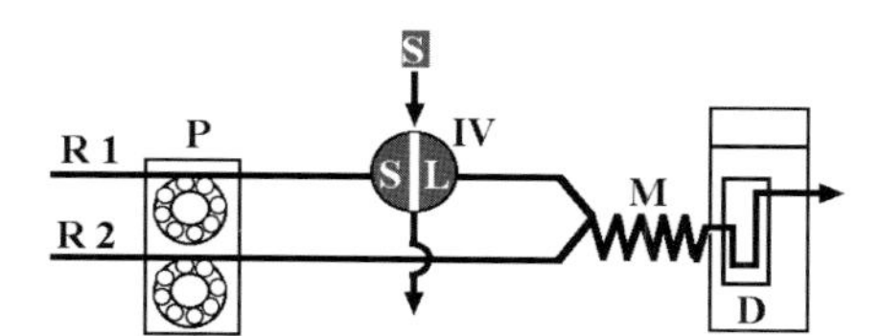

Figure 8.64 Diagram of an FIA system. **R** reagent, **S** sample, **P** pump, **IV** injection valve, **SL** sample loop, **M** mixing, **D** detector.

reactive or detectable species which can be sensed by any one of a variety of flow-through detection devices. The height of the peak-shaped signal thus obtained can be used to quantify the analyte by comparison with the peaks obtained for solutions containing known concentrations of the analyte (Figure 8.63).

Components of a Simple FIA System

A FIA analyzer consists of four main components: Propulsion, injection, mixing/ reaction and detection [4] (Figure 8.64). The liquid delivery unit is a critical component. There are several technical solutions for propelling the liquid with a constant flow. Three types of delivery devices utilized are: peristaltic pump, gas pressure, and piston pump. A necessary requirement for FIA is to inject a defined and reproducible volume of the sample into the carrier stream. This can be accomplished with injection valves actuated manually or electrically or by hydrodynamic injection.

The physical mixing occurs most frequently in plastic tubes, which can be coiled, knitted, or knotted. Since the dispersion processes do not reach a stationary state (equilibrium), it is necessary to keep the measurement conditions (e.g., transport distance, flow rate, sample volume) stable, in order to ensure reproducible measurements. Depending on the chemistry involved the function of the detector is the measurement of the signal produced by the analyte/reagent product (e.g., UV and visible spectrophotometry, luminescent detection, electrochemical detection). The detector is connected to a data acquisition device or to a recorder.

Toxicity of Phenol

Poisoning by ingestion often occurs if absorbed through skin and mucous membranes. The maximum allowable workplace concentration (MAC) amounts to 5 ppm phenol (EU, 2000).

The limits set for the individual phenolic compounds are present at $10\,\mu g\,l^{-1}$ for water and 0.3 to $1\,mg\,kg^{-1}$ for soil. The limit for the maximum allowable concentration of the sum of all phenolic compounds in drinking water is (2000) $0.5\,mg\,l^{-1}$ in the EU.

Phenolics in Water

Phenols, sometimes called phenolics, are a class of chemical compounds consisting of a hydroxy group (–OH) attached to an aromatic hydrocarbon group. The simplest of the class is phenol (C_6H_5OH). The brewing industry will often take water supplied by the local water authority. This water will meet the requirements of water intended for human consumption, but may be variable (within specification limits) regarding suspended solids, trace organic compounds (such as phenol, chlorophenol, cresol, and guaiacol) and possible bacterial contamination from the distribution system. Less than one part per billion of some phenols is detectable by odor and taste in water. Since water is the largest amount of raw material in beer manufacture it is logical to expect that phenolic water would have an effect on beer character.

Phenolics in Beer

There are a number of phenolic-type compounds, such as phenol and chlorophenol, that normally occur in wort and beer and whose presence in beer in above minimal concentrations causes abnormal flavors. A flavor component may be completely normal when present in a limited concentration but cause an abnormal flavor when present in higher concentration. Some of these phenolic-type compounds have very low taste thresholds. Another source of phenolic flavors in beer comes from yeast. Some yeast strains produce phenolic flavors as a normal part of their metabolism.

Phenol Analysis

Phenol offers numerous possibilities for derivatization, so that an analysis of individual components in complex media such as sewage is seldom desirable and

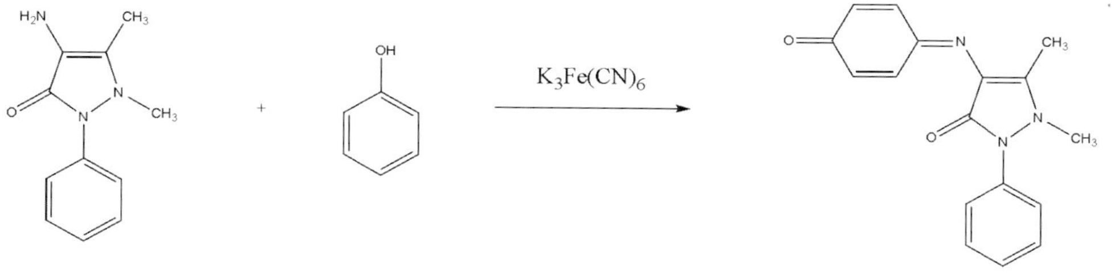

$K_3Fe(CN)_6$

4-Aminoantipyrine Phenol Colored phenol complex

Figure 8.65 Phenol color reaction.

very expensive. For this reason, the analysis of phenol and its derivatives is usually recorded as a sum parameter.

A standard colorimetric procedure based on the reaction of phenols with 4-aminoantipyrine (4-AAP) is frequently used for determination of phenols in aqueous samples. This procedure can be used in a manual or automated way.

There are two methods that are capable of measuring phenolic materials in the aqueous phase (without solvent extraction) using phenol as a standard. These are the manual EPA method 9065 and the U.S. Environmental Protection Agency (EPA) method 9066 "Phenolics (Spectrophotometric, automated 4-AAP)". Methods for the analysis of phenols employ the reagent 4-aminoantipyrine (4-AAP), which reacts in the presence of potassium ferricyanide at a pH of 10 with phenolic compounds to produce a dark red product, an antipyrine dye (Figure 8.65). The amount of color produced is a function of the concentration of phenolic material.

The concentration of the phenolic compounds is determined by measuring the absorbance of the sample at a wavelength of 460 to 520 nm, depending on the method. The methods are calibrated using a series of standards containing the single compound phenol (see Section 8.2.1).

Dispersion in a Flow System

Dispersion is the name given to the ratio of the concentration before and after transport through the FIA system. This classification was introduced in 1977 by Růžička [3].

Dispersion is also an approximate measure of the dilution of the sample in an FIA system.

Mathematical calculation of the dispersion (D) according to Růžička:

$$D = \frac{C_0}{C_{max}} = \frac{\text{Signal}_0}{\text{Signal}_{max}} \tag{8.60}$$

C_0 Concentration of the undiluted sample

C_{max} Concentration of the sample after passing through the system

Targeted Dispersion (D)

FIA systems with limited dispersion ($D \leq 3$) are used when measuring a parameter of the analyte directly in the absence of processes other than transport (e.g. an atomic absorption photometer or a flow-cell with pH electrodes). Medium dispersion systems ($D = 3$–10) are designed for studies involving processes additional to the transport, usually a chemical reaction. FIA systems with large dispersion ($D \geq 10$–15) are characterized by the high degree of mixing between the carrier reagent and the sample, resulting in a well-defined concentration gradient. The residence time is rather long, so in some cases chemical equilibrium is attained.

8.4.1.3 Material and Methods

Equipment

The CFA/FIA system consists mainly of the following elements: reagent and analyte vessels (ordinary beakers), a static mixer, a peristaltic pump, an injection valve and a spectrophotometer. The output signal of the photometer can be recorded by a computer with data acquisition or by a recorder.

Proposal (other variants are possible)

- peristaltic pump Ismatec MS-4 Reglo, with 4 wheels
- silicone tubes of different lengths (mixing/reaction length)
- 6 port Rheodyne Teflon valve Type 50 (injection valve)
- UV–VIS spectrometer
- 3 Hellma flow-through cuvettes ($d = 10$ mm, volume 30, 80 and 390 μl)
- PC with Windows XP
- interface IOtech (Personal Daq/3000) Multifunction Data Acquisition Module
- data acquisition program DASYLab (Data Acquisition System Laboratory)

Chemicals

Phenol	CAS- Number 108-95-2	T, C
Ammonium chloride NH_4Cl	CAS- Number 12125-02-9	Xn
Ammonium hydroxide NH_4OH	CAS- Number 1336-21-6	C, N
Potassium(III) hexacyanoferrate	CAS- Number 13746-66-2	–
4-Aminoantipyrine	CAS- Number 83-07-8	Xn
Potassium chloride	CAS- Number 7447-40-7	–

Sodium hydroxide CAS- Number 1310-73-2 C 

Patent Blue V CAS- Number 20262-76-4

Reagents

Environmental Protection Agency (EPA) 9065 Method (Manual Analysis)
- Buffer solution
  Buffer solution was prepared by dissolving 16.9 g NH_4Cl in 143 ml concentrated NH_4OH and diluting to 250 ml with water.
- Potassium ferricyanide solution
  8.0 g potassium ferricyanide are dissolved in 100 ml of water.
- 4-aminoantipyrine solution
  2.0 g 4-aminoantipyrine are dissolved in 100 ml of water.

Environmental Protection Agency (EPA) 9066 Method (CFA/FIA)
- Buffered potassium solution

  2.0 g of potassium hexacyanoferrate, 3.1 g boric acid and 3.75 g of potassium chloride are dissolved in 800 ml of water. The pH was adjusted to 10.3 with 1 N sodium hydroxide and the flask was filled to 1000 ml.

- 4-aminoantipyrine solution

  0.65 g 4-aminoantipyrine are dissolved in 800 ml of water and filled to the 1000 ml mark.

Phenol Calibration Solutions for Both Methods
- Phenol stock solution
  1.0 g phenol was dissolved in 500 ml water and made up to 1000 ml. 0.5 ml of conc. H_2SO_4 was added as a preservative (1.0 ml = 1.0 mg of phenol).
- Working solutions
  From the stock solution 10 calibration standards are made, depending on the intended measuring range.

Color Solution for Dispersion Seasurements (Patent Blue V)
2 mg of Patent Blue V are dissolved in 1000 ml of water.

Experiment [5, 6]
(A) Continuous Flow Analysis (CFA)

A.1 Determination of the sample throughput of the manual determination of phenol (EPA 9065) as a basis for comparison (analysis/hour).

Examine the influence of the complex formation time on the color development by changing the mixing time. 0.2 ml of buffer solution are added to 10 ml of sample, and mixed. The pH of the sample should be

10±0.2. After mixing 0.2 ml 4-aminoantipyrine solution are added, and after further mixing 0.2 ml potassium ferricyanide solution are added and mixed. The absorbance is read at $\lambda = 510$ nm. Standard curves measured after four different mixing times (5, 10, 15, and 30 min) are prepared by plotting the absorbance values versus the corresponding phenol concentration using the modified direct photometric EPA 9065 method.

The rate of sample throughput per hour was determined from the batch analysis rate at room temperature (22–24 °C, mixing time 15 min). These values represent the actual sample throughput per hour of the "Old Brewery".

A.2 After validation of the manual method, CFA measurements are carried out according to EPA method 9066. The equipment configuration is shown in Figure 8.60.

Evaluate the effect of changes in flow rate (1.0, 3.0, and 5.0 ml min^{-1}) and reaction length (20, 100, and 200 cm).

Set flow of the solutions at 1 ml min^{-1} (approximately)

Record absorption at $\lambda = 505$ or 520 nm.

A.3 Record a calibration curve after optimization of the method (sensitivity, throughput), using 10 standard solutions. Standard solutions covering the desired range are successively injected three times each and the calibration function is created by plotting the peak signal against concentration.

A.4 Validate the CFA method.

A.5 Determine three unknown samples.

The samples, which must be diluted beforehand, as required, are also injected several times.

(B) Flow injection analysis (FIA)

Change the set-up of the apparatus from the CFA to FIA according to Figure 8.64.

B.1 and B.2 Determination of dispersion by means of patent blue V.

Diffusion is to be optimized between 3 and 10 through the parameters of the flow velocity and the reaction length. The detection of the solution occurs at $\lambda = 610$. First, the absorption maximum (c_{max}) is measured. The carrier solution is water and the patent blue solution is injected via the injection loop.

Fill injection pump with a syringe (loading position) with the already diluted patent blue solution of 2 mg l^{-1}. Start the measurement by tilting the switch from load to inject. Make a note of the absorption maximum. After the measurement, move the switch back to the loading position. Next, c_0 is determined. Only pump the patent blue solution through the detector. The measurement signal corresponds to c_0.

Now calculate the dispersion according to Equation (8.60)

Influence of different flow-through cells (chamber volumes 30, 80, and 390 μl) on the dispersion is to be determined. Work with a flow rate of solutions of $1\,ml\,min^{-1}$.

B.3 Record a calibration curve using 10 phenol standard solutions and determine the working range.

B.4 Try to optimize the sensitivity of detection by varying the parameters.

B.5 Validate the optimized FIA method.

B.6 The concentration of three unknown samples is to be determined.

B.7 Proposals for improvement of the existing equipment are to be considered.

B.8 The sample throughput and sensitivity of the three methods (manual, CFA, and FIA) are to be compared.

8.4.1.4 Questions

(1) What norms do you know?

(2) What is the significance of norms, for example, in the food industry?

(3) What is the essential difference between the CFA and FIA methods?

(4) How is the dispersion factor defined?

(5) How does the step response of the CFA signal change when using flow cuvettes with the same light path (10 mm) but with different chamber volumes?

(6) What factors limit detection in the system used?

(7) How does flow rate influence dispersion?

(8) How does the length of the pump and/or reaction time influence dispersion?

(9) What indicates the blank value?

(10) How is sensitivity defined?

(11) Precision, accuracy: Give examples.

(12) How can sensitivity in the system under investigation be increased?

(13) When is a measurement process described as selective?

(14) How is the selectivity of a method defined?

(15) What is the goal of validation?

References

1 U.S. Environmental Protection Agency (EPA) (September 1985) Method 9065 Phenolics (*Spectrophotometric, manual 4-AAP with distillation*).

2 U.S. Environmental Protection Agency (EPA) (September 1986) Method 9066, Phenolics (*Colorimetric, automated 4-AAP with distillation*).

3 Růžička, J., and Hansen, E.H. (1975) *Analitica Chimica Acta*, **78**, 145–157.

4 Valcarcel, M., and Luque de Castro, M.D. (1987) *Flow Injection Analysis, Principles and Applications*, Halsted Press, Ellis Horwood.

5 Petrozzi, S. (2009) An open-ended design experiment: development from batch to automated flow-injection analysis (FIA) for phenolics determination. *Journal of Chemical Education*, **86** (11), 2009.

6 Petrozzi, S. (2009) An open-ended design experiment: development from batch to automated flow-injection analysis (FIA) for phenolics determination – lab documentation. *Journal of Chemical Education – Online)*, http://www.jce.divched.org/Journal/Issues/2009/Nov/abs1311.htmL

8.5
Mass Analytical Determination Methods

8.5.1
Karl Fischer Water Determination, Project: “Water Content as a Quality Parameter”

8.5.1.1
Analytical Problem

Biodiesel is produced from vegetable oils (such as rape, soya and jatropha oil) and obtained from animal fats by esterification with methanol. It consists of a mixture of fatty acid methyl esters (FAME). In 2005 in Germany the consumption of biodiesel was 1.8 million tons. As a replacement for conventional diesel, it offers some advantages such as better lubricating properties, but also some disadvantages such as increased water solubility.

The presence of water in biodiesel reduces the calorific value and enhances corrosion. Moreover, water promotes the growth of microorganisms and increases the probability that oxidation products will be formed during long-term storage. These oxidation products can cause disturbances in the injection system and in the engine itself. In view of this, the EN 14214 standard specifies a maximum water content of 500 ppm for biodiesel. Many biodiesel fuels contain additives that can interfere during direct coulometric Karl Fischer titration. The analytical department of Luptakova Ltd. has been commissioned to investigate this process interference.

Object of Investigation/Investigation System
Fuels

Sample
Biodiesel

Examination Procedure
The determination of water content is of fundamental importance in the analysis of many raw materials and finished products. Thus, for example, the presence of water in products such as kerosene, transformer insulating oils or automotive brake fluid has serious consequences.

Moreover, the amounts of water contained in ingredients of pharmaceutical products must be known with a great deal of precision, as the durability, stability and efficacy of a drug depend strongly on its water content.

Even in the production of food that complies with the quality requirements, the raw materials and end product must be systematically controlled.

For this reason, the determination of water content is crucial in many areas, and the method of Karl Fischer has proven over time to be the easiest and quickest practicable method, which also provides the most accurate and reproducible results.

Analyte
Water

Keywords
Coulometric Karl Fischer titration, volumetric Karl Fischer titration, trace moisture determination

Bibliographical Research (Specific Topics)

Scholz, E. (1984) *Karl Fischer Titration. Determination of Water*, Springer, Berlin, ISBN: 3-540-13734-3.

Mitchell, J., and Smith, D.M. (1980) *Aquametry Part III. A Treatise on Methods for the Determination of Water (The Karl Fischer Reagent)*, John Wiley & Sons Inc., New York, ISBN: 0-471-02266-7.

Bruttel, P., and Schlink, R. (2006) Water Determination by Karl Fischer Titration, Metrohm Ltd., Metrohm Monograph (Article number: 8.026.5013), CH-Herisau (Source of the theoretical part of this project).

Project Process/Task

(A) Determining the water content of a biodiesel
Determine the water content of a biodiesel. Check the measurement device with liquid water standard.

(B) Determining the recovery rate
Calculate the recovery rate by means of adding a liquid water standard ($1\,\mathrm{mg\,g^{-1}}$).

(C) Influence of additives on the direct titration
A possible falsification of the determination is to be examined by adding various diesel additives.

(D) Validation of the method without additives
Validate the method with the main characteristic equipment precision.

(E) Project reporting
The results are to be submitted in the form of a report (or presentation). Comparisons with bibliographical references are to be made.

8.5.1.2
Introduction

Besides pH measurement, weighing and acid–base titration, the determination of the water content is one of the most frequently made measurements in laboratories around the world. In addition to methods requiring complex apparatus, such as infrared reflection spectrometry, gas chromatography or microwave spectroscopy, two methods in particular have been able to establish themselves:

- *Drying methods* (drying ovens, infrared lamps and infrared balances)

  This method is commonly found in various standards, but suffers from the following disadvantages:

  In principle the loss on drying is determined, and not necessarily the water content. Apart from water, other volatile components of the sample and/or decomposition products are also determined. In order to obtain comparable results, work must be carried out under strictly controlled conditions (temperature and duration of the drying process).

  It takes a long time to obtain the analytical results (several hours in a drying oven). This can lead to bottlenecks, for example, in the monitoring of manufacturing processes.

- *Titration methods*

  In contrast to drying, this is a specific and selective method. If no side reactions occur, only water will be determined. The method is rapid (determination normally takes only a few minutes), it can be validated and, therefore, fully documented. Since its introduction more than 60 years ago [1], it has become indispensable in many laboratories. The name Karl Fischer (1901–1958), KF for short, is therefore well known in most laboratories.

KF titration allows the determination of both free and bound water, for example, surface water on crystals and the water contained inside them. The method works over a wide concentration range from ppm up to 100% and supplies reproducible and correct results.

Reagents

When developing his new analytical method, Karl Fischer took into account the well-known Bunsen reaction, which is used for the determination of sulfur dioxide in aqueous solutions:

$$SO_2 + I_2 + 2\,H_2O \rightarrow H_2SO_4 + 2\,HI$$

This reaction can also be used for the determination of water if sulfur dioxide is present in excess and the acids so produced are neutralized by a base. The selection of pyridine, the base used by Karl Fischer, was actually due to chance: "pyridine was just standing on the shelf". This led to the formulation of the classical KF reagent (KFR), a solution of iodine and sulfur dioxide in a mixture of pyridine and methanol.

Original Formulation according to Karl Fischer

Dissolve 254 g iodine in 5 l of anhydrous methanol. Add 790 g pyridine, mix thoroughly and then add 192 g liquid SO_2.

1 ml KFR corresponds to approx. 3 mg H_2O.

The titer of this combined reagent was not very stable – titer determinations had to be carried out virtually every day. Help was provided by the use of two-component reagents. The titration solution consisted of 30 g iodine in 1 l methanol. For the reaction solution 60 g SO_2 was introduced into a mixture of 300 ml each of methanol and pyridine. The titer of the titration solution prepared in this way (approx. 1.7 mg $H_2O\,ml^{-1}$) only decreased by approximately 1.3% over a two-month period and could be regarded as being relatively stable.

However, for most applications it is advisable to work with a one-component reagent with a relatively stable titer. To obtain such a reagent, the methanol was replaced by ethylene glycol monomethyl ether (Cellosolve).

Further Developments

The noxious odor and toxicity of pyridine was a source of annoyance for many users of the KF method. Work had to be carried out in the fume cupboard, which was helpful to the user but not to the environment. Research was carried out into the use of other bases in order to obtain pyridine-free reagents. During this work it was also found that the KF reaction only proceeded rapidly and stoichiometrically at a pH between 5 and 7.

Among the bases tested with varying success were diethanolamine, sodium acetate and sodium salicylate. The ideal substance was finally found: imidazole ($C_3H_4N_2$), which is similar to pyridine but odorless and has a higher basicity [2].

The latest trend is to replace the toxic methanol by ethanol. Today there is general differentiation between one-component and two-component reagents; the two compounds are called titrant and solvent in volumetry, while in coulometry one has anode and cathode solutions.

A wide range of KF reagents and auxiliary solutions are offered by various manufacturers. They are intended for particular applications, that is, custom-made:

- One-component and two-component reagents for volumetry and coulometry.
- Pyridine-containing and pyridine-free reagents.
- Special reagents for water determination in aldehydes and ketones.
- Stable volumetric KF reagents with a titer of 1 mg $H_2O\,ml^{-1}$ or 2 mg $H_2O\,ml^{-1}$ for the determination of low water contents.
- Solvent additives for fats and oils, with halogenated hydrocarbons (usually undesirable) or without these (e.g., with decanol).
- Solvent additives for polar substances (e.g., salts, sugars).
- Buffer substances or buffer solutions for strongly alkaline or strongly acidic samples.
- Certified water standards from 0.1–10 mg $(ml\ H_2O)^{-1}$ with a density of $1\,g\,ml^{-1}$.

This wide range of possibilities allows the user to determine the water content in practically any matrix [3]. It is certainly no longer worth preparing the reagents yourself!

Chemical Reactions

Karl Fischer presented the following equation for water determination with his reagent:

$$2\,H_2O + SO_2 \times (C_5H_5N)_2 + I_2 + 2\,C_5H_5N \rightarrow (C_5H_5N)_2 \times H_2SO_4 + 2\,C_5H_5N \times HI$$

This would result in molar ratios of

$$H_2O : I_2 : SO_2 : \text{pyridine} = 2:1:1:4$$

under the assumption that methanol only functions as a solvent and pyridine forms additive compounds with the acids.

However, this molar ratio was incorrect, as established by the studies of Smith, Bryant and Mitchell [4]. These authors recognized the participation of methanol and determined the following molar ratios:

$$H_2O : I_2 : SO_2 : \text{pyridine} : CH_3OH = 1:1:1:3:1$$

In a first partial reaction water forms the (hypothetical) pyridine sulfur trioxide, which then reacts further with methanol:

$$H_2O + I_2 + SO_2 + 3\,C_5H_5N \rightarrow 2\,C_5H_5N \times HI + C_5H_5N \times SO_3$$

$$C_5H_5N \times SO_3 + CH_3OH \rightarrow C_5H_5N \times HSO_4CH_3$$

The authors subsequently revised this equation by formulating iodine and SO_2 as pyridine adducts.

Based on the pH dependence of the reaction constants Verhoef and Barendrecht [5] later realized that it is not the SO_2 in the KF reagent that forms the reactive component, but rather the monomethyl sulfite ion that is formed from SO_2 and methanol:

$$2\,CH_3OH + SO_2 \rightleftarrows CH_3OH_2^+ + SO_3CH_3^-$$

They also established that pyridine does not take part in the reaction but only acts as a buffer substance (such that at the same pH the same reaction rate is achieved with sodium salicylate). Pyridine can, therefore, be replaced by another suitable base (RN). The added base results in a considerable displacement of the equilibrium to the right:

$$CH_3OH + SO_2 + RN \Rightarrow [RNH]SO_3CH_3$$

This means that the KF reaction in methanolic solution can now be formulated as follows:

$$H_2O + I_2 + [RNH]^+SO_3CH_3^- + 2\,RN \rightarrow [RNH]^+SO_4CH_3^- + 2[RNH]^+I^-$$

Influence of the Water Content

The water content of the working medium influences the stoichiometry of the KF reaction. If it exceeds $1\,mol\,l^{-1}$ ($18\,g\,l^{-1}$) then the reaction behavior changes in favor of the Bunsen reaction for aqueous solutions. This means that 2 H_2O are consumed for one I_2 or one SO_2. It is uneconomic to titrate such a high water content in a large volume of sample. In this case it is important to dilute the sample and/or to use a small amount of sample. If two-component KF reagents are used then the water capacity of the solvent must be taken into consideration, particularly for subsequent titrations. The water capacity of these solvents is normally about $5–7\,mg\,ml^{-1}$, for a solvent volume of 25 ml this means a maximum 125–175 mgH_2O.

Influence of the Organic Solvent

Theoretically non-alcoholic KF reagents can also be used, but in this case the stoichiometry also changes. In protic solvents, for example, methanol or ethanol, the ratio is $H_2O : I_2 = 1:1$; in aprotic solvents, for example, DMF, the ratio is $H_2O : I_2 = 2:1$.

Intermediate values (uneven ratios) have been observed, for example, with propanol, butanol and other long-chain alcohols. This has to do with the reactivity of the alcohol used. In methanol about 100%, in propanol approximately 80% and in butanol only approximately 50% is present as sulfite ester. With the KF reagents normally used an alteration in the stoichiometry is not to be expected, because the titer determination and water determination are carried out in the same titration medium and the alcohol fraction in the sample solution continuously increases during the course of the titration.

Volumetry

It is only since the introduction of piston burets by Metrohm in the 1950s that it has been possible to carry out volumetric Karl Fischer titrations without requiring a cumbersome set-up. The introduction of the Exchange Units in the 1960s brought further improvements.

Higher (absolute) water contents are preferably determined by volumetric titration. Volumetry also has the advantages that solid or pasty samples can be introduced directly into the titration vessel and work can be carried out with a variety of suitable organic solvents specially adapted to a particular sample.

Of course it is also possible to determine small amounts of water volumetrically, for example, in solvents. If an appropriate amount of sample and diluted KF reagents are used then the lower determination limit is approximately 50–100 ppm H_2O.

Despite great progress in the manufacture of KF reagents, these do not have a stable titer. One of the minor disadvantages of volumetry is, therefore, that the titer has to be determined at regular intervals.

Coulometry

Instead of a buret electric current is used to generate a reagent – an "electronic buret", so to speak. The current releases the stoichiometrically corresponding amount of iodine from the iodide-containing KF reagent by electrolysis.

Faraday's law applies:

$$m = \frac{MQ}{zF} \tag{8.61}$$

m Mass of converted substance (g)
M Molar mass (g mol^{-1})
Q Measured amount of charge (A s^{-1})
z Number of exchanged electrons (equivalence number, charge number)
F Electrochemical equivalent ($1\text{ F} = 96\,485\text{ C mol}^{-1}$, $1\text{ C} = 1\text{ A s}$)

Example for iodine: $2\text{ I}^- - 2\text{ e}^- \rightarrow I_2$

$$z = 2, M = 253.8\ (I_2)$$

126.9 g iodine is released by 96 485 A in 1 s, or 1.315 mg iodine is generated by 100 mA in 10 s.

Requirements for coulometric titration are:

- The process must take place with 100% current efficiency, that is, there must be no side reactions.
- Oxidation or reduction must lead to a defined oxidation stage.

Current KF coulometers meet these requirements with modern coulometric reagents. They work according to the galvanostatic coulometry principle, that is, with a constant current.

The same chemical processes take place as in a volumetric KF titration, that is, 1 H_2O consumes 1 I_2. As the iodine is generated electrolytically from the iodide in the KF solution this means that the coulometric determination of water is an absolute method – a titer does not need to be determined.

In order to generate iodine at the anode (Metrohm KF) coulometers work with variable current strengths and pulse lengths (100, 200 and 400 mA – with diaphragm-less cells with a constant current strength of 400 mA). Higher current strengths have not been able to establish themselves – side reactions occur and heat is produced. Working with variable pulse lengths allows precise "iodine addition", even in the region of the titration endpoint.

The instrument measures the time and current flow that is required to reach the titration endpoint. The instrument measures the time and current flow that is required to reach the titration endpoint. The product "time × current" is directly proportional to the amount of iodine generated and, therefore, to the amount of water determined.

Coulometric water determination is primarily used for the determination of small amounts of water. Larger amounts of water require a lot of time and/or may exceed the water capacity of the KF reagent, which could lead to incorrect results. Modern KF coulometers typically have a determination range of 10 µg–200 mg H_2O with a resolution of 0.1 µg H_2O.

Detection Methods

To determine the end of the KF reaction – that is the titration endpoint at which all the H_2O has been converted – a detection or indication method is required. Titrimetric volumetric analysis stands and falls with correct endpoint recognition. This means that the indication method is crucial as it determines to a great extent the reproducibility and, above all, the correctness of the results.

- Visual or Photometric Indication

  Karl Fischer himself only had the possibility of determining the titration endpoint visually. This method requires a high degree of experience and can normally only be used for uncolored sample solutions. When excess iodine is present the solution turns increasingly yellow to brown (the latter with a large excess of iodine). This already indicates the difficulty of deciding the coloration point to which the titration is to be continued. In addition, the coloration in polar solvents (e.g., methanol) differs from that in nonpolar solvents (e.g., chloroform, DMF). Further disadvantages of these methods are that they can hardly be automated or validated. They fell out of use rapidly as electrometric indication methods became available toward the end of the 1940s.

- Electrometric Indication Methods

  In the meantime indication methods using two polarized electrodes have established themselves. Double Pt-wire or double Pt-ring electrodes are commonly used.

Biamperometric Indication (U_{pol} or Dead Stop)

In this case a constant voltage of 500 mV maximum is applied to the electrodes and the resulting current is measured. This method was first used mainly for aqueous iodometric titrations. The following reactions occur at the electrodes:

Cathode: $I_2 + 2\,e^- \rightarrow 2\,I^-$ (reduction)

Anode: $2\,I^- - 2\,e^- \rightarrow I_2$ (oxidation)

The relationship between the voltage, current and concentration of the depolarizer (iodine or triiodide) is similar to that found in polarography, that is, initially only low voltages are required to produce a flow of current. Then the voltage increases steeply – as a function of the iodine concentration – and, when the limiting diffusion current has been achieved, the current increases only slightly. If the voltage is then further increased decomposition of either the solvent or conductive salt could occur.

This means that a current flows for as long as iodine is present. When the iodine has been consumed (e.g., by titration with thiosulfate) the current drops abruptly to 0 (which is where the name "dead stop" comes from), the resistance between the electrodes increases strongly and the electrodes are again polarized. In KF titration the reverse occurs. If excess H_2O is present only a minimal current flows; this increases to a few µA when excess iodine is present. The "water loss" resulting from electrochemical conversion can be ignored. At 50 µA for 30 s it is theoretically approximately 1 µg H_2O.

Bivoltametric Indication (I_{pol})

In this case a small direct or alternating current is applied between the electrodes and the resulting voltage is measured. AC is much more sensitive than DC, which is why only AC is usually used for coulometric titrations. The shape of the titration curve is similar to that obtained by biamperometry; however, the jumps are usually more marked and larger. If the sample solution and titrant are electrochemically active then peak-shaped or V-shaped titration curves are obtained. If only one partner is electrochemically active (the KFR with excess iodine) then L-shaped curves are obtained.

The resulting voltages for excess iodine or water depend, of course, on the applied current (I_{pol}). Three examples (double Pt-wire electrode):

$I_{pol} = 50\,\mu A$, U approx. 100 mV for excess iodine, approx. 680 mV for excess H_2O
$I_{pol} = 10\,\mu A$, U approx. 5 mV for excess iodine, approx. 580 mV for excess H_2O
$I_{pol} = 1\,\mu A$, U approx. 0 mV for excess iodine, approx. 350 mV for excess H_2O

The titration endpoint is determined by entering a switch-off voltage; reagent addition ceases when this limit is reached. This "ideal endpoint" is best found experimentally. If a too low voltage is selected, then the amount of excess iodine required to achieve the switch-off point will be too large. If it is too high then the titration will not start because no free iodine is required for this voltage to be achieved (decomposition of the solvent or electrolyte).

Once all the water has reacted, the indicator electrode detects the very slight excess of iodine and the Karl Fischer titration stops. In comparison to the visual methods, the electrometric methods have the advantage that the titration is always carried out to the same slight excess of iodine and in this way a better reproducibility and accuracy can be achieved.

Instruments

Volumetric Karl Fischer Titration

Different manufactures offer instruments for volumetric KF titration, ranging from simple instruments for routine determinations up to high-end equipment with numerous advantages. What all instruments have in common is that they are easy to operate and produce rapid, accurate and reproducible results.

The Karl Fischer titrators are automated titration systems that carry out the titration automatically, that is, the titration rate and volume increments are controlled according to the signal measured by the indication system. In addition, the sample weight can be transmitted directly to the titration system via a connected balance. Results, curves, parameters and the configuration data can either be printed out or transferred to a PC database.

The titrators have both bivoltametric (I_{pol}) and biamperometric (U_{pol}) endpoint detection. For both types of detection the control parameters can be freely selected. This has the advantage that the control parameters can be optimized for difficult samples and special reagents, for example, for the addition of solubility promoters. Good results are normally obtained with the default settings.

Slowly reacting reagents (e.g., those containing pyridine) require an additional stop delay when the endpoint is reached, that is, the endpoint voltage or the endpoint current must be maintained for a few seconds when the endpoint has been reached. In the course of the development of new reagents and improvement of the instruments, this stop delay has been replaced by the drift-stop criterion. In addition to a predefined voltage (or current), a certain drift value must also be achieved. The drift is the amount of consumed KF reagent per unit time to keep the titration cell dry. In general, the drift value for titration cells titrated to dryness is of the order of a few $\mu l\,min^{-1}$. The aim of the absolute drift-stop correction is to achieve the drift value measured before the titration. In practice it has been shown that the achievement of an identical drift can considerably increase the titration time, which is why the use of a relative drift-stop criterion is recommended. If a relative drift stop is used then the titration will be stopped at a slightly higher drift than the initial drift.

If very small amounts of water are to be determined, or if a long titration time is necessary, then work should be carried out using automatic drift correction. In this case the drift value is multiplied by the titration time and the value obtained is subtracted from the added volume.

In addition to the indication system, the resolution of the buret makes a contribution to the accuracy and reproducibility of the results. In principle, all commonly available KF reagents can be used with the current KF titrators. Since these reagents do not have a stable titer, it must be determined regularly. The interval between the titer determinations depends on the choice of reagent as well as on how impervious the reagent bottle is.

Leading reagent manufacturers offer certified water standards for determining the titer. It is also possible to determine the titer with ultrapure water. In this case work has to be carried out with very small volumes and handling takes some getting used to. This is why the use of certified standards is recommended; it is possible to back-weigh the added amount and volume dosing is no longer necessary. Solid substances such as disodium tartrate dihydrate can also be used, but care must always be taken that the substance is completely dissolved. Standard solutions are available with water contents of $10.0 \pm 0.1\,\text{mg}\,\text{g}^{-1}$ (for volumetric water determination), $1.0 \pm 0.03\,\text{mg}\,\text{g}^{-1}$ or $0.1 \pm 0.005\,\text{mg}\,\text{g}^{-1}$ (for coulometric determinations) and are offered in ampoules.

When handling the standards proceed as follows:

Immediately after the ampoule is opened a syringe is approximately half-filled with the standard solution. The syringe is rinsed by inverting it and ejecting the complete contents (air and standard solution). This is to ensure that the inner walls of the syringe are completely wetted with the solution. The syringe is then filled with the whole of the remaining ampoule contents without entraining any air. After the needle has been wiped off the syringe is tared on a balance and the KF titrator is started. Part of the standard solution is then injected at one go. There are two ways of doing this:

- The standard solution is injected directly beneath the surface of the KF solution.
- It is injected without immersing the needle. The last drop is drawn back into the syringe.

The syringe is then pulled out and back-weighed. The sample weight can be transmitted directly from the balance to the KF titrator or entered manually.

Particular attention should also be given to the titration cell. Water is present everywhere: At a relative humidity of 60%, 0.5 ml air at 25 °C, contain approximately 7 μg H_2O. This is why it is essential that the titration cell is impervious; it should be checked at regular intervals. Special attention should be paid to the sealing rings of the titration cell and the molecular sieve – check the drift! The titrator automatically conditions the titration cell before and between determinations. It is advisable to swirl the cell about from time to time in order to remove any water adhering to the walls. It is important that a constant drift value that is as low as possible is achieved.

Coulometry

In contrast to volumetric Karl Fischer titration, in KF coulometry no titrant is added but the required iodine is generated directly in the iodide-containing electrolyte by electrochemical means ("electronic buret"). There is a strictly quantitative relationship between the amount of electric charge and the amount of iodine generated. This relationship allows the extremely precise addition of the iodine.

KF coulometers work with variable current strengths and pulse lengths (100, 200, and 400 mA) for generating iodine at the anode. Higher current strengths have not proven themselves in practice – side reactions can occur and excess heat is produced. Working with variable pulse lengths allows precise "iodine dosification", even in the region of the titration endpoint.

The instrument measures the time and current flow that is required to reach the titration endpoint. The product of time and current is directly proportional to the amount of iodine generated and, therefore, the amount of water determined.

Coulometric water determination is primarily used for the determination of small amounts of water. Larger amounts of water require a lot of time and/or may exceed the water capacity of the KF reagent, which could lead to incorrect results. Metrohm KF coulometers, for example, have a determination range of 10 μg–200 mg H_2O with a resolution of 0.1 μg H_2O.

The iodine is produced by a generator electrode. There are two different types of generator electrodes:

Generator electrode without diaphragm [6–8]

The generator electrode without diaphragm is easy to handle and easy to clean. Only a single reagent is required. However, care must be taken that only those reagents that are intended for generator electrodes without diaphragm are used. As no humidity can adhere to the diaphragm there is a further advantage – the electrode is rapidly ready for use.

Generator electrode with diaphragm

A generator electrode with diaphragm should be used whenever your samples contain ketones and aldehydes, as the special reagents used for them are only available for generator electrodes with diaphragms. An electrode with diaphragm should also be used for reagents with a low conductivity (the addition of, e.g., chloroform, reduces the conductivity) and when measurements are to be made in the lower trace range. Reagents for coulometric water determination consist of an anolyte that is filled into the titration vessel and a catholyte that is filled into the cathode compartment. Both liquids should have about the same level in order to prevent pressure differences from forcing the anolyte into the cathode compartment and vice versa. The endpoint is determined by a double platinum electrode and is largely similar to that in volumetric Karl Fischer titration.

As the coulometric KF method is an absolute method this means that no titer has to be determined. It must only be ensured that the reaction that generates the iodine takes place to 100%. Although this is the case with the reagents that are available, the whole system should be checked from time to time with the aid of certified water standards with a water content of 1.0 ± 0.03 or $0.1 \pm 0.005\,mg\,g^{-1}$. In coulometry, cells made completely of glass are used as titration cells, as cells with a plastic cover have a certain permeability to water vapor. This does not have much influence on volumetric KF titrations, as these are normally used to determine higher water contents than in KF coulometry. Care must also be taken that the cell is as impervious as possible.

Karl Fischer Oven

Many substances only release their water slowly or at elevated temperatures. This means that they are not suitable for direct Karl Fischer titration. A further problem is the low solubility of certain samples in alcohols. In such cases traditional methods recommend complicated sample preparation procedures or the use of solubility promoters, which represent a health hazard. Other substances react with the KF reagents to release water or consume iodine; this falsifies the results.

These problems can be avoided by using a heating method. The substance under investigation is heated in an oven located upstream from the cell and the released water is transferred by a flow of dry carrier gas to the titration cell where it is determined by Karl Fischer titration. As only the water enters the titration cell and the sample itself does not come into contact with the KF reagent, this means that side reactions and matrix effects are ruled out. Work is carried out either with a conventional tube oven or with an oven using the vial technique.

Automation in Karl Fischer Titration

The use of a sample changer is well worth while for large numbers of samples. The advantages are not only the amount of time saved by the laboratory personnel: automated systems control the operating procedures and improve both reproducibility and accuracy. In automated KF titration there is an additional difficulty; normal KF sample changers do not work in closed systems and penetration by atmospheric humidity cannot be prevented. For a water content >1% this interference can be corrected by the subtraction of a previously determined blank value. In principle, the use of such sample changers can only be recommended for volumetric determinations.

Side Reactions

KF water determination is only specific if no side reactions with the KF reagents take place. This means that no water should be released in side reactions, nor should the sample consume or release iodine. Most side reactions can be suppressed by suitable measures. The most important interfering substances are:

- Carbonates, hydroxides and oxides
- Aldehydes and ketones
- Silanols and siloxanes
- Metal peroxides
- Reducing agents
- Mercaptans simulate higher water content and are oxidized to disulfide
- Oxidizing agents

Sample Preparation Techniques

Each sample preparation is preceded by sampling. Particularly with inhomogeneous samples – and there are a lot of them, even if it is not always obvious – this initial step is important for the correctness and accuracy of the results. Correct sampling is, therefore, an essential requirement for the subsequent determination! Many standardized methods provide detailed instructions concerning this subject. This may also apply for in-house instructions. If no instructions are available then the following points must be observed:

The water content of the sample taken must be the same as the average water content of the original sample, that is, the sample must be representative.

Atmospheric humidity is an ever-present source of contamination, and not only in tropical countries.

Hygroscopic samples have a large amount of water adhering to their surfaces.

Samples that easily release water or become weathered have only a small amount of water on their surfaces.

Liquids can release water if the water is not present in dissolved form but as a dispersion (precipitation or flotation, adherence to vessel walls–e.g., ointments, creams, and lotions). The same applies for samples in which the solubility limit for water is reduced on cooling.

The more heterogeneous the water distribution in the sample, the larger the sample to be taken. Sampling must take place as quickly as possible and the water content of the sample must not change during transport and storage.

General

Many samples can be weighed out and/or injected directly into the titration vessel or titration cell, provided that they are soluble in the solvent used and do not undergo side reactions with the KFR.

In contrast, other samples require one or more preparation steps. The most important sample preparation techniques are described below. Further information can be found under the individual classes of substances.

It is very important that the sample preparation technique used does not introduce any additional water into the sample and that no water losses occur by heating the sample too strongly.

Dilution

For reasons of accuracy (sample size) samples with a high water content (>40%) are diluted with an inert dry solvent before determination. Methanol can normally be used for this purpose. If the sample contains emulsified fats, oils, or both then a dry mixture of 60% methanol and 40% decanol (or, if permitted, chloroform) should be used. The procedure is as follows:

Approximately 1 g sample is weighed out exactly into a dry Erlenmeyer flask fitted with a septum stopper. This is treated with approximately 20–25 g dry solvent (mixture), the flask is then sealed and the contents mixed. Part of the mixture is drawn off in a dry syringe and the syringe and its contents are tared. Some of the mixture is injected into the titration vessel and titrated. The amount of sample injected is determined by back-weighing the syringe. If a dry solvent is used then it is not absolutely necessary to take its blank value into account.

Here is an example:

Sample weight	1.575 g
Methanol weight	22.725 g
Injected mixture	1.500 g
KFR titer	4.855 mg $H_2O\,ml^{-1}$
KFR consumption	6.255 ml

Calculation

Diluted sample	22.725 g + 1.575 g = 24.300 g (100%)
Sample fraction	1.575 g : 24.3 g = 0.0648
Injected sample	0.0648 × 1500 mg = 97.22 mg
H_2O found	6.255 ml × 4.855 mg ml^{-1} = 30.37 mg
H_2O in sample	30.37 × 100/97.22 = 31.24%

The following samples can be treated in this way: ointments, cosmetic lotions and emulsions, alcohol-containing samples.

Solvent Extraction

This can take place in two different ways – external or during the titration in the titration vessel. It is usually carried out either warm or hot, less frequently at room temperature. If an external extraction is carried out then please remember that all the water can never be extracted, an average water content between the sample and the solvent is obtained. Example:

10 g sample with a water content of 10% is extracted with 50 ml methanol (density approx. 0.8 g cm^{-3}) with a water content of 100 ppm. The H_2O content of the mixture is 1004 mg/50 g, corresponding to 20.08 mg H_2O g^{-1} or 2.008% H_2O.

This also means that the larger the ratio between the sample and dry solvent, the less water will be extracted from the sample. The blank value of the solvent must, of course, be taken into consideration when calculating the water content.

Better values are obtained if the extraction is carried out directly in the titration vessel and the water is titrated immediately. A titration vessel equipped with a heating jacket and a reflux condenser is used and the titration is carried out, for example, at 50 °C. Work can be carried out with or without an extraction time (set at the titrator) as required. If an extraction time is used then please note that a sufficiently long stop delay must be observed for the endpoint.

Homogenization

In liquids water can float, adhere to the inner walls of the sample bottle or be deposited on the bottom. This occurs particularly when the sample is taken at a high temperature and allowed to cool down to room temperature before the analysis. The water can again be distributed homogeneously by immersing the bottle in an ultrasonic bath.

The separated water can also be dissolved again by the addition of an exactly weighed-out amount of a solubility promoter (e.g., isopropanol), whose water content has been accurately determined (blank value). This method is suitable for pastes and viscous liquids.

Solids are usually inhomogeneous and must be thoroughly comminuted and homogenized. The procedure depends on the constitution of the sample:

- *Coarse particles, hard and tough*

  Ball mill or analytical mill (closed) with water cooling. This method is especially suitable for pills

- *Coarse, fatty*

  Grating and high-frequency comminution.

- *Coarse, soft, inhomogeneous*

  Comminute with a meat grinder or knife, then high-frequency comminution.

- *Pasty, inhomogeneous*

  Homogenization with a blender, then possibly high-frequency comminution.

- *High-frequency comminution*

  High-frequency comminution can be carried out directly in the sealed titration vessel. Any alteration to the water content caused by atmospheric humidity is eliminated.

The Karl Fischer Oven Method
A dry stream of gas is passed through the heated sample; in each case the released water is transferred to the titration vessel and immediately titrated. The heating temperature should be set as high as possible – just high enough so that the sample does not decompose and no interfering secondary constituents distill over. The heating temperature and duration must be optimized in preliminary tests. The appropriate blank values must be subtracted.

This method is used for samples that cannot be extracted (or cannot be completely extracted) with a suitable solvent or that undergo side reactions with the KFR.

Norms

American Society for Testing and Materials

- ASTM D 3401-97 Standard Test Method for Water in Halogenated Organic Solvents and Their Admixtures

- ASTM D 4928-89 Standard Test Method for Water in Crude Oils by Coulometric Karl Fischer Titration

- ASTM D 5460-98 Standard Test Method for Rubber Compounding Materials – Water in Rubber Additives

- ASTM D 6304-04a Standard Test Method for Determination of Water in Petroleum Products, Lubricating Oils, and Additives by Coulometric Karl Fischer Titration

- ASTM D6869-03 Standard Test Method for Coulometric and Volumetric Determination of Humidity in Plastics Using the Karl Fischer Reaction (the Reaction of Iodine with Water)

- ASTM E 1064a-04 Standard Test Method for Water in Organic Liquids by Coulometric Karl Fischer Titration

British Standard

- BS 6829:1.5:1990 Analysis of surface active agents (raw materials). Part 1. General Methods. Section 1.5 Methods for determination of water content

International Organization for Standardization

- ISO TC 158/SC 2 Methods of analysis of natural gas and natural gas substitutes.
  Direct determination of water by Karl Fischer method, coulometric method
- ISO 10101-1: 1993 Natural gas–Determination of water by Karl Fischer method.
  Part 1: Introduction
- ISO 10101-3: 1993 Natural gas–Determination of water by Karl Fischer method. Part 3: Coulometric procedure
- ISO 10337: 1997 Crude petroleum–Determination of water–Coulometric Karl Fischer titration method

8.5.1.3
Material and Methods

Work Media

Two types of reagents are available on the market (Figure 8.66).

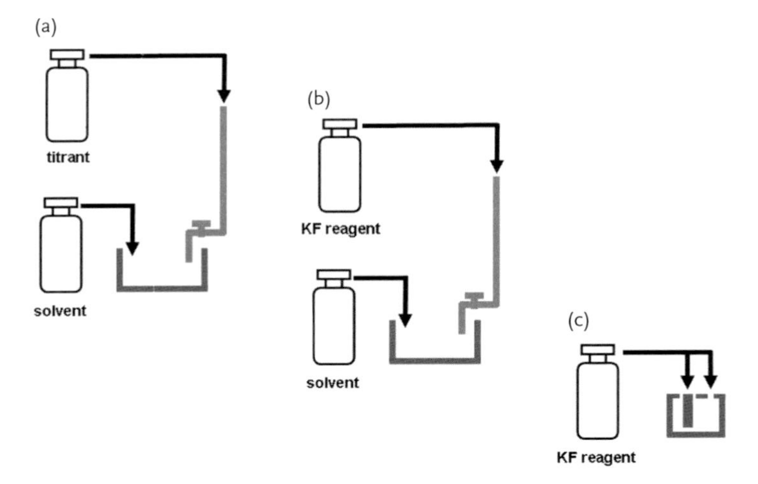

Figure 8.66 (a) volumetric determination: separate reagent for use as solvent and titrant, (b) volumetric determination: combined reagent, (c) coulometric determination: combined reagent (anode and cathode solutions).

One-Component Reagents

In such reagents the basic chemicals iodine, sulfur dioxide and the base are all available in only one solvent. The person using the solvent is free to choose the work medium (solvent). In doing so they should, however, not forget that the stoichiometric 1 : 1-implementation of the Karl Fischer reaction only applies when more than 25% methanol is present in the reagent mixture. The use of a methanol-free medium is, however, possible when calibrating the reagent and when conducting titrations in the same medium.

Two-Component Reagents

Solvent Solvents available on the market have a high buffer capacity and solvating power. They contain sulfur dioxide, a base and methanol.

Their essential advantages are:

- Faster titration thanks to the increased reaction kinetics, which is particularly welcome in analyses with high aqueous content.
- Improved repeatability thanks to a stable reaction medium, a constant pH-value and a stable sulfur dioxide concentration. When titrating samples, that cause side reactions (aldehydes, ketones and silanols), it is imperative to use a special solvent. Most manufacturers of reagents add the letter K in the trade name of such products.

Titrant It contains dissolved iodine in methanol. The three titres of 1, 2 and 5 mg water (ml titrant)$^{-1}$ are common. Although a larger volume than the flask size of the buret can be titrated, the titrant and/or the water quantity to be titrated (sample quantity) should, as far as possible, be so chosen that a titrant quantity less than corresponds to this volume would also suffice, as this serves to reduce the duration of analysis and achieve a better repeatability. For the titration of samples that cause side reactions, (aldehydes, ketones and silanols) a special titrant must be used.

We recommend the given maximum speeds listed below for the individual solvents and titrants used. However, the conditions can be changed depending on solvent additives (such as chloroform), or samples.

Equipment

831 KF Coulometer (Metrohm) cell without diaphragm, 728 Magnetic stirrer

Parameters

Settings	EP at U	50 mV
	Control range	70 mV
	Max. rate	max. μg min^{-1}
	Min. rate	15 μg min^{-1}
	Stop criterion	rel. Drift
	Rel. drift	5 μg min^{-1}
Titration parameter	Start drift	10 μg min^{-1}
	I_{pol}	10 μA

Experiment

Generator Electrode without Diaphragm
Fill about 100 ml of the appropriate reagent into the cell.

Condition and Drift
Before the actual determination can be started, the coulometer must be dried. A constant drift in the range of $\leq 4\,\mu g\,min^{-1}$ is good. Lower values are certainly possible. If higher, stable values occur, the results are usually still good, since the drift can be compensated.

Sample Addition
Pastes, greases can be inserted into the cell with a syringe without a needle. For this, the joint opening can be used. If you also want to suck up even more, you can use the opening with the septum.

Size of Sample
The sample size should be small, so that as many samples as possible can be titrated in the same electrolyte solution and for the titration time to be short. Make sure, however, that the sample contains at least 50 μg H2O.

Chemicals

HYDRANAL®- Coulomat AG-H – T, F, N
(Sigma Aldrich)

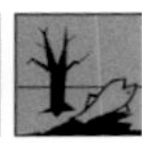

Grade reagent for coulometric Karl Fischer titration for cells without diaphragm

Composition
- pentane-1-ol
- diethanolamine
- imidazole-hydrobromide
- 1*H*-imidazole, monohydriodide
- methanol
- sulfur dioxide
- iodine

Standard Solution
HYDRANAL®-Water Standard 0.1 (Sigma Aldrich)

Standard for coulometric and volumetric Karl Fischer titration (1 g contains 1 mg = 0.1% H_2O)

Optimization of a KF Titration

Analysis Fill about 100 ml Hydranal Coulomat AG-H into the cell and condition until the drift is steady and below $10\,\mu\gamma\,min^{-1}$. Prior to each injection rinse the syringe with the sample solution. Inject 2–5 g sample into the cell (the exact sample mass is determined by difference weighing) and start the determination.

(A) Determining the water content of a biodiesel

Determine the water content of a biodiesel. Check the measurement device with liquid water standard.

(B) Determining the recovery rate

Calculate the recovery rate by adding a liquid water standard ($1\,mg\,g^{-1}$).

(C) Influence of additives on the direct titration

Search the bibliography for possible additives that are added to diesel. Possible adulteration of the KF determination is to be examined by "spiking" with various additives.

(D) Validation of the method without additives

Validate the method with the main characteristic equipment precision.

8.5.1.4 Questions

(1) Why is it important to know the water content of food?
(2) Identify a great difficulty in determining the water content in gaseous samples.
(3) Why is the Karl Fischer determination nowadays mostly conducted electrochemically?
(4) Explain the broad application spectrum of the Karl Fischer determination.
(5) Name possible interferences that can lead to false positive determinations.

References

1 Fischer, K. (1935) *Angewandte Chemie*, **48**, 394–396.
2 Scholz, E. (1984) *Karl Fischer Titration. Determination of Water*, Springer, Berlin, ISBN: 3-540-13734-3.
3 Jones, A.G. (1951) A review of some developments in the use of the Karl Fischer reagent. *The Analyst*, **76**, 5–12.
4 Smith, D.M., Bryant, W.M.D., and Mitchell, J., Jr. (1939) Analytical procedures employing Karl Fischer reagent, I. Nature of the reagent. *Journal of the American Chemical Society*, **61**, 2407–2412.
5 Verhoef, J.C., and Barendrecht, E. (1977) Mechanism and reaction rate of the Karl Fischer titration reaction. *Journal of Electroanalytical Chemistry*, **75**, 705–717.
6 Larsson, W., and Cedergren, A. (2005) Coulometric Karl Fischer titration of trace water in diaphragma-free cells. *Talanta*, **65**, 1349–1354.
7 Cedergen, A., and Jonsson, S. (2001) Progress in Karl Fischer coulometry using diaphragm-free cells. *Analytical Chemistry*, **73**, 5611–5615.
8 Nordmark, U., and Cedergren, A. (2000) Conditions for accurate Karl Fischer coulometry using diaphragm-free cells. *Analytical Chemistry*, **72**, 172–179.

8.6
General Analytical Methods

8.6.1
Nitrogen and Protein Determination according to Kjeldahl, Project: "Official Control at the Swiss Alps Dairy Ltd"

8.6.1.1
Analytical Problem

Milk proteins have, like proteins from meat, fish and eggs, very high nutritional value, since they contain all the amino acids in a satisfactory relationship. These substances are crucial for an organism. In addition to their essential role in the formation of tissue, recent studies have shown that certain milk proteins act biologically on various systems of an organism: the immune system, the nervous system, the circulatory system (through lowering blood pressure), digestion (through promoting absorption of calcium), or on the cell system (antibacterial effect). Within official inspections the total protein content declared on the respective milk packaging (according to Swiss food law) should be checked. This periodic procedure serves as an integrated part of consumer protection and will be carried out by the responsible authorities.

Swiss Food Ordinance (LGV Switzerland) of 23 November 2005 (updated 1 May 2009) – Section 6 Food labelling type. 26. pre-packaged foods – 1. Anyone supplying pre-packed food must indicate to the consumers:

a. the physical description;
b. the composition (ingredients);
c. durability;
d. the origin;
e. the application of genetic engineering or special technological processes for production (such as irradiation);
f. instructions for proper use.

Object of Investigation/Investigation System
Dairy Products

Samples
Whole milk, partially skimmed milk, skimmed milk

Examination Procedure
The Kjeldahl method for nitrogen determination is a widespread method developed in 1883. The method was named after the Danish chemist Johan Gustav Kjeldahl (1849–1900). At the Carlsberg brewery in Copenhagen Kjeldahl was assigned to scientifically observe the processes involved in beer making. While studying proteins during malt production, he developed a method of determining

nitrogen content that was faster and more accurate than any method available at the time. Kjeldahl's visionary idea of providing a simple method for nitrogen and protein determinations, which also can be carried out by non-academic laboratory personnel remained basically unchanged over the past 130 years. Nitrogen can be determined with this method in many different nitrogenous compounds. This is reflected in the broad field of application of the method: the food industry, environmental analysis, the pharmaceutical and chemical industries. This method of determination is an officially recognized method (AOAC–Association of Official Analytical Chemists- 991.20, ISO 8968-2:2001) and is widely used.

Analyte

Nitrogen

Analyte 1: nitrogen contained in milk proteins

Analyte 2: nitrogen, not originating from proteins, called non-protein nitrogen.

Keywords

The Dumas combustion, catalysis, digestion process, steam distillation, protein content in food, protein determination

Bibliographical Research (Specific Topics)

AOAC 991.20

ISO 8968-2: **2001**

Bradstreet, R.B. (1965) *Kjeldahl Method for Organic Nitrogen*, Academic Press.

Egli, H. (2008) *Kjeldahl Guide, Büchi Labortechnik AG, CH-9230 Flawil*, ISBN 978-3-033-01688-0.

Project Process/Task

(A) Determination of protein content in milk

The determination of different milk samples, calculation of the nitrogen and protein content and comparison with information on the label is to be performed.

(B) The influence of various parameters on protein determination is to be investigated

B.1 Digestion

Influences: Sample size, amount of catalyst, volume 98% sulfuric acid, different temperature ramps, digestion temperature.

B.2 Distillation

Influences: Amounts of H_2O and 32% NaOH added and distillation time.

B.3 Titration

Decide on a suitable concentration of titrant based on expected titrant consumption.

(C) The method validation using standard solutions

Validation of distillation: Distillations of samples of ammonium sulfate or ammonium dihydrogenphosphate and report recovery rates.

Validation of digestion: Digestion and distillation of glycine samples.

(D) Project reporting

The results are to be submitted in the form of a report (or presentation). Comparisons with bibliographical references are to be made.

Discussion of the true protein content in the light of non-protein nitrogen (NPN) and discussion of the problem of adulterated milk samples (e.g., with melamine) is to be carried out.

8.6.1.2
Introduction

Method

The Kjeldahl nitrogen determination consists of the following three steps [1] (Figure 8.67):

1. Kjeldahl digestion

   In the digestion step the organically bound nitrogen is converted to ammonium sulfate in a sulfuric acid environment. This oxidative digestion of the sample containing C, H, N and O leads also to the conversion of water and organic carbon into water and carbon dioxide.

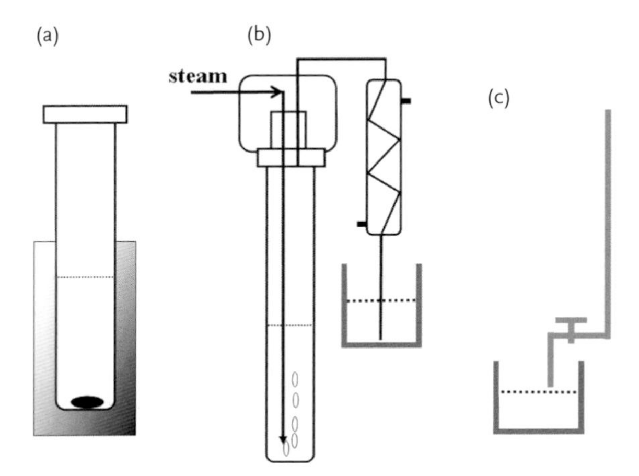

Figure 8.67 (a) digestion, (b) distillation, (c) titration.

$$(CHNO) + H_2SO_4 \rightarrow CO_2 + SO_2 + H_2O + NH_4^+ \tag{8.62}$$

2. Kjeldahl distillation

   After digestion the solution is allowed to cool and is then neutralized with concentrated sodium hydroxide solution. In a chemical equilibrium the solvated ammonium ions produce ammonia gas, which is quantitatively distilled by steam distillation.

$$NH_4^+ + OH^- \rightleftarrows NH_3(gas) + H_2O \tag{8.63}$$

   A common procedure to collect quantitatively the ammonia in the receiver involves the presence of boric acid dissolved in water.

3. Acid–base titration

   The concentration of the captured ammonium ions in the boric acid are determined by means of an acid-base titration commonly using standard solutions of sulfuric or hydrochloric acid.

$$NH_3 + HX + H_2O \rightarrow X^- + H_2O \tag{8.64}$$

Kjeldahl Procedures Parameter

Several interrelated conditions and parameters determine the accuracy and trueness of the nitrogen determination according to Kjeldahl. Among these are the sample size, the heating temperature of the acid digestion mixture, sulfuric acid/sulfate salts ratio added to elevate the acid boiling temperature, duration of digestion, catalyst addition, distillation and titration. Adjusting any one of these factors has an influence on the others.

Sample Size

The maximum amount of sample usable in a digestion depends on the behavior of the sample during oxidation and the volume of the reaction vessel. In the Kjeldahl reaction gaseous water and carbon dioxide are formed, which act as a propellant and can bring samples to foam. Care must be taken not to let the foam rise to more than half the height of the reaction vessel. The foam formation can be weakened by adding a spatula tip of stearic acid. Both the foam formation and the charring of the sample depend on the amount of sulfuric acid added.

Sulfuric Acid

In Table 8.10, correlations between the amount of required substance to be digested and sulfuric acid required and the optimal amount of Kjeldahl reaction mixture are shown.

Digestion

The procedure for the mineralization (= digestion) of organic materials according to Kjeldahl involves the complete oxidation of the sample with concentrated sulfuric acid at high temperatures. In this process sulfuric acid, sulfate salts and heavy metal catalysts are added to the sample. The catalyst lowers the activating energy

Table 8.10 Optimal amount of 98% sulfuric acid and associated quantity of Kjeldahl reaction mixture for controlled foaming and carbonization of a given sample volume.

Sample weight (g)	2	1	0.5	0.125
98% sulfuric acid (ml)	30	20	10	10
Kjeldahl Reaction mixture (g)	15	10	5	5

for the oxidation process and the addition of sulfate salts increases the boiling point of 98% sulfuric acid from 338 to 350–370 °C. This temperature range is maintained during the decomposition process.

In addition to the optimal parameter of a Kjeldahl digestion mentioned in Table 8.10, foaming and charring can be kept under control by gradually increasing the reaction temperature.

Catalysts

The speed and efficiency of the digestion is not only influenced by the temperature used but can also be improved by the addition of a suitable catalyst. Many systematic studies on the efficiency of various catalysts have been made and a conclusion from all of these investigations was that mercury, selenium and copper sulfate are the catalysts of choice. For certain applications titanium has also found some usage. Today copper and to some extent selenium are clearly the dominating catalysts.

Sulfuric Acid/Sulfate Salts Ratio

Kjeldahl digestion consists of a catalytically supported oxidation of organic sample materials in a mixture of boiling 98% sulfuric acid and sulfate salts. The function of the sulfate salts is to increase the boiling temperature of the sulfuric acid. In this way an initial boiling point of 350 °C is reached. During digestion, this increases to about 370 °C, as a part of the sulfuric acid is reduced to sulfur dioxide. In order to achieve a shorter digestion time, a small percentage, 0.5–2%, of catalytic substances are added to the sulfate salts. A typical catalyst mixture comprises some 95% sodium sulfate and/or potassium sulfate as well as the catalysts copper sulfate and titanium dioxide. Under these conditions, decomposition of the nitrogenous organic matrix takes place primarily by the formation of ammonium ions, while at temperatures above 390 °C as a side reaction elementary nitrogen is formed. As shown in Table 8.11, the relationship between sulfuric acid and the reaction mixture is responsible for the optimal reaction temperature. With the Kjeldahl digestion, therefore, attention should be paid to ensure that the reaction conditions described above are observed from the beginning to the end of the decomposition.

Steam Distillation and Titration

Organically bound nitrogen remains as ammonium ions in the acid medium and is distilled, after neutralization with concentrated sodium hydroxide, as ammonia

Table 8.11 Influence of the relationship between sulfuric acid and sulfate salt on the increase in the boiling temperature.

H_2SO_4 (ml)	20	20	20	20
K_2SO_4 (g)	0	5	10	15
Temperature (°C)	340	350	370	390

Table 8.12 Typical distillation parameters.

Water	50 ml
32% NaOH	90 ml
Distillation time	4 min

Table 8.13 Typical titration parameters.

Titration type	Boric acid titration
4% Boric acid	60 ml
Endpoint	pH 4.65

Table 8.14 Concentration of the titrant depending on the amount of nitrogen.

N (mg)	c(HCl) (mol l^{-1})	c(H_2SO_4) (mol l^{-1})
0.5	0.01	0.005
2–2.5	0.05	0.025
7–10	0.1	0.05
50–100	0.5	0.25

by steam distillation in a boric acid solution. The chemical reaction of ammonia condensed in boric acid is shown in Equation (8.65).

$$B(OH)_3 + NH_3 + H_2O \rightleftarrows NH_4^+ + B(OH)_4^- \tag{8.65}$$

The boric acid solution is titrated with standard hydrochloric acid or sulfuric acid solutions. The chemical reaction in the titration is described by Equation (8.66).

$$B(OH)_4^- + HX \rightarrow X^- + B(OH)_3 + H_2O \tag{8.66}$$

Typical distillation and titration parameters are shown in Tables 8.12 and 8.13.

Typically, concentrations of 0.01–0.5 N are used for the boric acid titration. In Table 8.14 suitable concentrations for titrants HCl or H_2SO_4 are indicated

Protein Determination

Kjeldahl nitrogen determinations are often used for the indirect determination of protein content in food. For comparative determinations of nitrogen content and

Table 8.15 Empirical protein factors.

Food	Protein factor
In general	6.25
Of animal origin	
Eggs and egg products	6.25
Meat and meat products	6.25
Fish and seafood	6.25
Milk, dairy products, cheese, whey	6.38
Cereals	
Wheat, barley and oats	5.83
Corn	6.25
Rice	5.95
Wheat	5.70
Fruit	
Fruit and fruit products	6.25
Vegetables	
Vegetables and vegetable products	6.25
Soya and Soya products	5.71
Nuts	
Tree nuts, coconuts	5.30
Peanuts	5.46
Almonds	5.18

protein, the concept has been around since 1900 to calculate protein content by multiplying the experimentally determined nitrogen content by an empirical protein factor f_{protein}. This made it possible to determine protein content in food using a simple and quick nitrogen analysis. Thereafter, one was able to renounce time-consuming direct determinations of protein content. The correlation between protein content %P and nitrogen content %N is given in Equation (8.67).

$$\%\text{protein} = \frac{\%\text{N} * 100\%}{16\%} = \%\text{N} * 6.25 = \%\text{N} * f_{\text{protein}} \quad (8.67)$$

In Table 8.15 some empirically ascertained protein factors f_{protein} have been compiled.

The protein levels detected in foods usually satisfy the regulatory requirements. Milk analysis is an exception in this regard. Milk contains, apart from proteins as a source of nitrogen, nitrogen that does not originate from proteins. This is primarily urea, which is described as non-protein nitrogen. In the regulatory specifications the determination of total protein content and that of non-protein nitrogen are treated separately. Milk producers are often compensated by milk processors based on the true protein content of the milk supplied. This true protein content can be expressed as the difference between the protein content determined by the Kjeldahl analysis and the non-protein nitrogen determined in the milk serum.

A recent problem in determining the protein content in milk occurred in China with the adulteration of milk by added melamine. The content of milk protein can be determined directly with a special method, known as the *Barnstein* method[1] [2]. Whether or not milk or a milk product has been adulterated with melamine can be discovered by comparing the directly determined protein content and the protein content determined according to Kjeldahl. If milk protein, non-protein nitrogen and melamine appear as an analyte, different analytical results can be found in adulterated milk for the milk protein determined with the Barnstein method and Kjeldahl's total protein content. Since the occurrence of cases of melamine in China, some specific methods for the direct and rapid determination of melamine levels in milk have been introduced.

1) *Barnstein* protein assay method (1900) The execution of the determination occurs in such a way that the material to be investigated is placed in hot water. Next specific quantities of copper sulfate solution and sodium hydroxide are added without filtering. If the material is now filtered, all protein substances should be found on the filter, whereas all non-protein nitrogen compounds are found in the filtrate. If a nitrogen determination according to Kjeldahl is now conducted, the result is that amount of nitrogen which exists as real protein. The difference between total nitrogen and protein nitrogen is customarily called amide nitrogen.

8.6.1.3 Material and Methods

Equipment

Apparatus for the determination of nitrogen and associated operating instructions should be available in the laboratory. It can be self-made equipment from laboratory glassware or digestion equipment and distillation units found on the market. The titrations can be performed manually with a buret or with a titrator. In highly automated laboratories, a sampler could be available with an automated digestion device and possibly with an automated distillation unit.

The report should describe the actual equipment used.

The experiments described here were performed with an Auto Kjeldahl Unit K-370 (Buchi, Switzerland).

Chemicals

98% Sulfuric acid	CAS- Number 7664-93-9	C	
Kjeldahl catalyst	–	O, N	
(47.7% K_2SO_4, 47.7% Na_2SO_4, 2.8% TiO_2, 1.8% $CuSO_4$)			
32% Sodium hydroxide	CAS- Number 1310-73-2	C	
Hydrochloric acid ($0.1\,mol\,l^{-1}$)	CAS- Number 7647-01-0	Xi	

Stearic acids	CAS- number 57-11-4	–	
Boric acid	CAS- Number 10043-35-3	T	
Mixed indicator according to Mortimer	–	O	
(Aqueous-ethanolic solution) Ammonium sulfate	CAS number 7783-20-2	–	
Ammonium dihydrogen phosphate (N% =12.18%)	CAS- number 7722-76-1	–	
Glycine (N%=18.66%)	CAS- Number 56-40-6	–	

Solutions

4% Boric acid (pH 4.65)

200 g of boric acid in 5 l with distilled water

Neutralizing Scrubber Solution

600 g of Na_2CO_3 and a pinch of bromothymol blue diluted with water to 3 l.

Standard Solutions

Ammonium Stock Solution

Dissolve 4.717 g of dried ammonium-sulfate in approximately 200 ml of distilled water and adjust to 1000 ml with distilled water after adding 2 ml of 98% sulfuric acid.

Ammonium Standard Solution

10.0 ml of the stock solution are diluted to 1000 ml (1 ml solution = 0.01 mg N).

Experiment

(A) Determination of protein content in milk by steam distillation

Do not use too small an amount of sulfuric acid. If the ratio of salt to acid is too large, the digestion temperature increases to a temperature range in which the degradation reaction leads not only to NH_3, but also to N_2. In this case a result lower than the actual value of nitrogen would be obtained. It should be noted that a portion of the sulfuric acid is consumed in the redox reaction and that at the end of the reaction there is still enough acid to keep the boiling temperature of the acid–salt mixture under 370 °C. Five blank samples are to be prepared and a fivefold determination for each type of milk is to be carried out.

Table 8.16 Typical distillation and titration parameters.

Distillation		**Titration**	
Water	70 ml	Boric acid 4%	60 ml
Caustic soda 32%	70 ml	Titration solution	HCl 0.1 $mol\,l^{-1}$
Distillation time	300 s	Type	Endpoint
Steam settings	100%	End-pH	4.65
Decomposition tube	300 ml		

Note: If the samples were already diluted with 50 ml of water, only add 20 ml of water for distillation.

Using an analytical balance, 5.0 g of milk sample with an accuracy of ± 0.1 mg is weighed in a digestion glass. Two Kjeldahl tablets[2)] of 5 g (10 g catalyst mixture) and 20 ml of 98% of sulfuric acid are added and the sample is mixed by gently swirling the sample glass.

The digestion unit should be connected to a suitable device for absorption of the ensuing acidic gases. When the liquid is not clear and blue-green, it is digested for another 30 min at 420 °C. Allow the samples to cool to room temperature. If the samples are not analyzed on the same day, dilute them with 50 ml of water to avoid crystallization. Only add water to cooled samples, as otherwise there is a strong reaction with concentrated acid and sample can be lost. Swirl the sample glass carefully to mix the decomposed sample with water.

Distillation and titration

The Auto Kjeldahl Unit K-370 was operated according to the parameters listed in Tables 8.11 or 8.12 and see also Table 8.16).

In this experiment, the sample sizes are determined by weighing, and thus a result will be expressed as protein content per 100 g of milk. If the sample volume for liquid milk is chosen by pipetting, a result can be obtained as protein content per 100 ml of milk.

Titration can be performed manually by reading the consumption of titrant on the buret or by a titrator.

The data for the sample size and the consumption of titrant are used to calculate the final results manually. In automated distillation devices with a built-in titrator there is software to analyze the measurement data.

Calculation

The results are calculated as a percentage of the nitrogen content. To calculate the protein content the nitrogen content is multiplied by a sample-specific protein factor. The following equations are used to calculate the results.

2) Recommended are tablets containing $CuSO_4$ because of the typical colors produced in the distillation step.

$$w(N) = \frac{(V_{sample} - V_{blank}) * z * c_{acid} * f * M(N)}{m_{sample} \cdot 1000} \tag{8.68}$$

$$\%(N)_{wt} = w(N) * 100\% \tag{8.69}$$

$$\%(P) = w(N) * f_{protein} * 100\% \tag{8.70}$$

w (N)	Weight percentage of nitrogen	(g N/g sample)
V_{sample}	Titrant volume for the sample	(ml)
V_{blank}	Mean of blank blank determinations	(ml)
z	Value for acid–base reaction	(1 for HCl, 2 for H_2SO_4)
c_{acid}	Concentration of titrant	($mol\,l^{-1}$)
f	Titer titrant	(often 1000)
M (N)	Atom mass nitrogen = 14.0069	($g\,mol^{-1}$)
m_{sample}	Sample mass	(g)
$f_{Protein}$	Specific protein factor for milk = 6.38	(dimensionless)
% $(N)_{wt}$	Weight percent of nitrogen	(g N/100 g sample)
% P	Percentage content of protein	(g P/100 g sample)

Once you have conducted 10 determinations of whole milk, partially skimmed milk and skimmed milk compare the values with the information on the package.

(B) The influence of various parameters on protein determination is to be investigated

Now vary the different parameters of the Kjeldahl method:

B.1 Digestion

Determine the effects of: sample volume, amount of catalyst, quantity of 98% sulfuric acid (parameters can be chosen).

B.2 Distillation

Determine the effects of: different temperature ramps and decomposition temperatures, amount 32% NaOH added and distillation time (parameters can be chosen).

B.3 Titration

Determine the effect of the concentration of the titrant (parameters can be chosen).

(C) The method validation using standard solutions

For the validation of the method (without digestion) ammonium sulfate or ammonium dihydrogen phosphate solutions are used.

To validate the whole method (including decomposition), glycine is used.

8.6.1.4
Questions

(1) Explain the difference between "normal" distillation and steam distillation.
(2) Why is boric acid used?
(3) What alternative is there to boric acid titration?
(4) Why cannot the Kjeldahl method be used to determine nitrate nitrogen?
(5) Are there any alternatives to the Kjeldahl method?
(6) Do you know other digestion methods?
(7) What are the advantages and disadvantages of the Kjeldahl method?

References

1 Egli, H. (2008) *Kjeldahl Guide, Büchi Labortechnik AG*, CH-9230 *Flawil*, ISBN 978-3-033-01688-0.

2 Barnstein, F. (1900) *Landwirtschaftliche Versuchsstationen*, **54**, 327.

8.6.2
Determination of Dissolved Oxygen (DO), Project: "Monitoring the Efficiency of the Biological Stage in a Sewage Treatment Plant"

8.6.2.1
Analytical Problem

The biological treatment of wastewater has proven to be a very effective process. Its degradation efficiency is due to the microorganisms used, which convert the components of the wastewater into food for their metabolism, first to provide energy, and secondly to produce new biomass, thus eliminating polluting compounds from the wastewater. This microbial decomposition of wastewater constituents requires energy, which in an aerobic environment is supplied by dissolved oxygen. The activated sludge must be oxygenated continuously and sufficiently. The oxygen content in the aeration tanks should not fall below the empirical value of about $5\,mg\,l^{-1}$.

In the activated sludge basin of a food company's wastewater treatment plant it has been increasingly observed that massive deposits and encrustations have been formed on the membranes used in the oxygen electrodes. This led to falsified test results. In addition, fatty deposits on the electrode surface led temporarily to a complete probe failure.

To meet the statutory requirements the Board has arranged that all the oxygen electrodes used in the ventilated zone should be checked periodically by means of comparative measurements (Winkler titration). As a first step a comparison of the two methods will be made and possible weaknesses of the two methods will be determined and eliminated.

Object of Investigation/Investigation System

Municipal and industrial sewage treatment plants

Samples

Activated sludge

Examination Procedure

The determination of the oxygen concentration using the Winkler (Lajos Winkler, 1863–1939) titration method is more than 110 years old. Today there are oxygen electrodes, which are easy to handle and which do not require the use of chemicals. The relatively new oxygen-sensitive electrodes facilitate continuous measurement and have broadened the knowledge of the presence, and particularly the amount of dissolved oxygen in all aquatic ecosystems.

These O_2-electrodes are superior to chemical methods with regard to measurement speed and their scope of application [1]. The 18th edition of *Standard Methods for the Examination of Water and Wastewater* [2] includes two methods for the determination of dissolved oxygen (DO) in wastewater, including the Winkler method (azide modification) and the electrometric method using membrane electrodes and a DO meter.

Both methods have certain limitations and neither can be used universally.

Analyte

Oxygen

Keywords

Partial pressure of oxygen, solubility of oxygen, Henry's law, salinity, electrochemical sensors, membrane electrodes

Bibliographical Research (Specific Topics)

Sawyer, C.N., McCarty, P.L., and Parkin, G.F. (2003) *Chemistry for Environmental Engineering*, 5th edn, McGraw Hill.

Stumm, W., and Morgan, J.J. (1996) *Aquatic Chemistry*, 3rd edn, Wiley Interscience.

ISO 5814:1990 *Water quality; determination of dissolved oxygen by the electrochemical probe method*, German version EN 25814:1992.

(1989) *Standard Methods*, 17th edn, American Public Health Association.

Project Process/Task

(A) Verification of Henry's absorption coefficient

Deionized water is bubbled with four different gas mixtures containing different percentages of oxygen. The measurements are carried out with an oxygen electrode. The calculated Henry coefficients and oxygen concentrations are compared with bibliographical data.

(B) Oxygen determination according to Winkler and electrochemical determination of oxygen with an oxygen electrode

The influence of temperature, matrix and salinity on the solubility of oxygen in solutions is investigated:

B.1 Determination of oxygen dissolved in distilled water at four temperatures (both methods).

B.2 Determination of dissolved oxygen in four solutions (solvent and salt solutions).

B.3 The sample throughput and sensitivity of the three methods is to be compared. Determination of dissolved oxygen in four solutions with different sodium chloride content (both methods).

These determinations, as implemented by the two methods, are compared.

B.4 The sample throughput and sensitivity of the three methods is to be compared.

(C) Project reporting

The results are to be submitted in the form of a report or presentation.

8.6.2.2 Introduction

Oxygen in Water

In virtually every liquid, more or less oxygen is dissolved. The dissolved oxygen concentration is one of the most important characteristic variables of natural waters. For example, at a temperature of 20 °C, at normal pressure and in a saturated state, water contains about $9\,mg\,l^{-1}$ of oxygen. The solubility of oxygen in water depends on the temperature, partial pressure and gas composition in connection with water and the dissolved compounds (such as salts, solvents, etc.). Each liquid takes up O_2 until the partial pressure in the liquid is in equilibrium with the atmosphere or gas phase.

The oxygen concentration determines, for example:

- Degradation processes in biological sewage treatment

  Aerobic bacteria and aquatic life such as fish must have DO to survive. Aerobic wastewater treatment processes use aerobic and facultative aerobic bacteria to break down the organic compounds found in wastewater into more stable products that will not harm the receiving waters. Wastewater treatment facilities such as lagoons or ponds, trickling filters and activated sludge plants depend on these aerobic bacteria to treat sewage.

- The living conditions for fish and small organisms in the aquatic environment

- Corrosion issues in water pipes (drinking water, boiler feed water)
- Shelf life of beverages (beer is poorly preserved in plastic bottles)

Oxygen Partial Pressure

All potentiometric and amperometric sensors for the determination of dissolved gases measure partial pressure and not analytical concentration. Partial pressure refers to the pressure exerted by a specific gas in a mixture of other gases. The total pressure exerted by a mixture of gases at constant V and T is equal to the sum of the partial pressure of each specific gas in air. The percentage of oxygen in dry air is generally 20.95%, resulting in a mole fraction of y_{O2}= 0.2095. The concentration of any individual component in a gas mixture is usually expressed as a mole fraction (y_x). The mole fraction of a component in a mixture is the fraction of moles of that component in the total moles of gas mixture.

Dalton's Law

At a constant temperature the partial oxygen pressure results from the equation

$$p_{O_2} = y_{O_2} \cdot p_L \tag{8.71}$$

$$p_{O_2} = 0.2095(p_L - p_W) \tag{8.72}$$

p_{O2} Oxygen partial pressure in air (bar)
y_{O2} Mole fraction of oxygen in air
p_L Atmospheric pressure[1] (bar)
p_W Vapor pressure[2] (bar)

Oxygen Saturation

If water is in equilibrium with the surrounding atmosphere, the water becomes saturated with oxygen. These values are dependent on pressure and temperature. The oxygen saturation index is also commonly known as the relative oxygen saturation. It indicates the proportion of the current oxygen saturation concentration in the saturation concentration under the prevailing temperature/pressure conditions. It is given as a percentage. If water, a 4 M KCl and a methanol/water solution are saturated with air, an almost identical p_{O2}-value results in all solutions and the oxygen electrode shows almost the same result in all solutions. The small differences are due to the different p_W-values in the various solutions. The concentrations, however, differ considerably.

Henry's Law

Partial pressure and gas solubility or gas concentration (x_{O_2}) are linked with each other by Henry's Law. Henry's Law is formulated quantitatively as follows:

$$p_{O_2} = x_{O_2} \cdot K_H \tag{8.73}$$

1) Height dependent: p (0 m) = 1.013 bar = 1 atm, p (500 m) = 0.954 bar.
2) Temperature dependent.

K_H Henry's constant (bar)
p_{O2} Oxygen partial pressure in air (bar)

$$x_{O_2} = \frac{O_2(\text{mole l}^{-1})}{O_2(\text{mole l}^{-1}) + H_2O(\text{mole l}^{-1})} \tag{8.74}$$

Henry's constant decreases sharply with increasing temperature. As a consequence, the solubility of gases generally decreases with increasing temperature. The value found in the literature for oxygen and water at 25 °C is 4.4×10^4 bar.

The vapor pressure of the dissolved oxygen is proportional to its quantity in the solution. The law applies to low pressures and low concentrations, that is, x_{O2} must be close to zero. One then speaks of "ideally diluted solutions".

Temperature Dependence

The Henry constant K_H, is, like every equilibrium constant, dependent on temperature. The effect of temperature on the oxygen solubility results from the van't Hoff equation:

$$\log K_1/K_2 = \Delta H/2.3R(1/T_1 - 1/T_2) \tag{8.75}$$

K_1, K_2 Equilibrium constants at temperatures T_1 and T_2
ΔH Reaction enthalpy (kJ mol^{-1})
R Ideal gas constant (8.3144 J K^{-1} mol^{-1})

Assumption: Reaction enthalpy constant in the temperature range T_1–T_2.

To measure the oxygen concentration, knowledge of the temperature is mandatory.

Iodometric Titration according to Winkler

$$\begin{array}{l} 4\,e^- + 4\,H^+ + O_2 \rightarrow 2\,H_2O \\ \underline{2\,Mn^{2+} + 4\,OH^- \rightarrow 2\,MnO_2(s) + 4\,H^+ + 4\,e^-} \\ 2\,Mn^{2+} + 4\,OH^- + O_2 \rightarrow 2MnO_2(s) + 2\,H_2O \end{array} \tag{8.76}$$

(brown precipitate)

After adding H_2SO_4, MnO_2 oxidizes I^-

$$\begin{array}{l} 4\,e^- + 2\,MnO_2(s) + 8\,H^+ \rightarrow 2\,Mn^{2+} + 4\,H_2O \\ \underline{4\,I^- \rightarrow 2\,I_2 + 4\,e^-} \\ MnO_2(s) + 4\,H^+ + 2\,I^- \rightarrow I_2 + Mn^{2+} + 2\,H_2O \end{array} \tag{8.77}$$

(yellow-brown solution)

The elementary I_2, is titrated with thiosulfate, $S_2O_3^{2-}$ and forms I^- and tetrathionate, $S_4O_6^{2-}$

$$\begin{array}{l} 2\,e^- + I_2 \rightarrow 2\,I^- \\ \underline{2\,S_2O_3^{2-} \rightarrow S_4O_6^{2-} + 2\,e^-} \\ 2\,S_2O_3^{2-} + I_2 \rightarrow S_4O_6^{2-} + 2\,I^- \end{array} \tag{8.78}$$

As an indicator, a small amount of starch solution is added. Once all the iodine is converted, the blue color disappears. The oxygen content can be calculated from the consumption of thiosulfate.

Nitrite Influence on the Determination

In the presence of I^- in an acidic NO^{2-} solution the following reaction takes place:

$$\begin{array}{l} 2\,e^- + 4\,H^+ + 2\,NO_2^- \rightarrow N_2O_2 + 2\,H_2O \\ 2\,I^- \rightarrow I_2 + 2\,e^- \\ \hline 2\,I^- + 2\,NO_2^- + 4\,H^+ \rightarrow I_2 + N_2O_2 + 2\,H_2O \end{array} \tag{8.79}$$

The I_2 produced is then titrated with $S_2O_3^{2-}$ causing a positive error. As oxygen enters the solution from the air, NO_2^- is regenerated to react again.

$$\begin{array}{l} 2\,e^- + 2\,H^+ + 1/2\,O_2 \rightarrow H_2O \\ 2\,H_2O + N_2O_2 \rightarrow 2\,NO_2^- + 4\,H^+ + 2\,e^- \\ \hline 1/2\,O_2 + N_2O_2 + H_2O \rightarrow 2\,NO_2^- + 2\,H^+ \end{array} \tag{8.80}$$

The addition of sodium azide, NaN_3, prevents this cyclical reaction

$$\begin{array}{l} 6\,e^- + 8\,H^+ + 2\,NO_2^- \rightarrow N_2 \uparrow + 4\,H_2O \\ 6\,N_3^- \rightarrow 9\,N2 \uparrow + 6\,e\text{-} \\ \hline 6\,N_3^- + 2\,NO_2^- + 8\,H^+ \rightarrow 10\,N_2 + 4\,H_2O \end{array} \tag{8.81}$$

If it is assumed that the nitrite concentration in the sample is low, the addition of NaN_3 can be omitted, as it is toxic.

Amperometric Oxygen Electrode

Membrane-covered electrochemical sensors form the basic principle of the electrochemical determination of oxygen concentration. The main components of the sensor are the oxygen-permeable membrane, the working electrode, the counter electrode, the electrolyte solution and possibly a reference electrode (Figure 8.68). Placing the two electrodes in solution and applying a potential between them is sufficient to reduce dissolved oxygen at the working electrode (typically ~−650 mV), one can determine the partial pressure of oxygen (pO_2) in an ideal solution because pO_2 is proportional to the limiting current. The importance of temperature for the oxygen measurement is already obvious from the temperature dependence of the different variables (Equations 8.72 and 8.75). In addition, the oxygen permeability of the membrane is also temperature-dependent. For this reason, in addition to the external temperature sensor (sample temperature), a further sensor is required, which is located in the sensor head. With these two temperature values, it is possible for the device to compensate for the temperature effect on the oxygen permeability of the membrane (IMT, isothermal membrane temperature compensation).

Flow at the Sensor

The incident flow to the sensor membrane must be continuous for oxygen measurements to be correct. Otherwise the diffusion of oxygen molecules into the

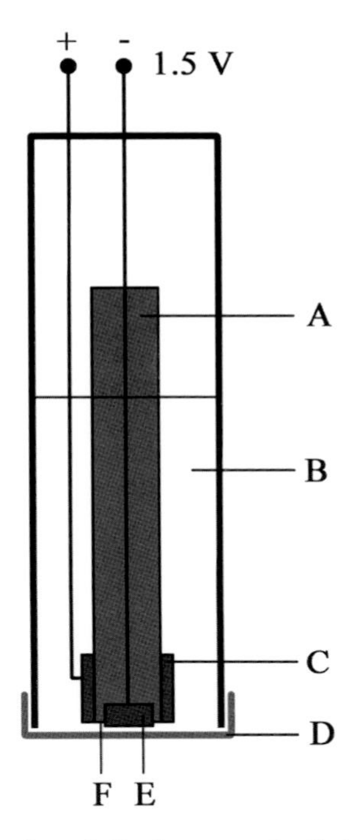

Figure 8.68 Structure of a Clark membrane-covered oxygen electrode. **A** insulation, **B** KCl solution, **C** ring-shaped Ag anode, oxygen-permeable membrane, **D** oxygen permeable membrane, **E** Pt cathode, **F** electrolyte film.

sensor head is too poor, simulating a too low oxygen concentration. This condition can be fulfilled either by stirring the sample, by moving the sensor, or by special stirring attachments.

Influence of Disturbing Gases

The membrane of the oxygen sensor is permeable not only for oxygen, but also for other gases. Nitrogen is inert and does not play a role as a contaminant gas. The high pH of the electrolyte solution protects the measurement from disturbing influences by ammonia. The problem, however, is carbon dioxide. While the buffering capacity of the electrolytic solution is sufficient for short-term stress, with prolonged stress carbon dioxide shifts the pH into the acidic range and leads to increased values.

Polarographic sensors can regenerate the buffering capacity better than galvanic sensors, since electrode reactions cause the formation of excess hydroxide ions. With high carbon contents (e.g., beer, champagne, lemonade) the buffer capacity of the electrolyte solution in the sensor is not sufficient and the pH is shifted into the acidic range leading to increased resulting values. The greatest risk of damage for oxygen sensors derives from hydrogen sulfide. Small quantities can be absorbed by the sensors but the sensors' lifetime is significantly shortened by permanent stress. Hydrogen sulfide is easily detected by its smell of rotten eggs, which can be noticed already at very low concentrations, without measurement effort.

8.6.2.3
Material and Methods

Equipment

The experimental procedure requires the following accessories:

- Rotameter or flowmeter
- Oxygen meter Oxi 330/340, inoLab Oxi (WTW)
- Oxygen sensor suitable for measuring
- OxiCal (calibration attachment)
- Oxi-Stirrer 300
- 300 ml Erlenmeyer flasks
- Full pipettes (2.5 ml)
- Magnetic stirrers
- 500 ml beaker
- Burets 25, 50 ml
- Measuring flasks 500 ml, 1000 ml
- Water bath/thermostat
- Nitrogen and air bottles

Chemicals

Manganese(II) chloride	CAS- Number13446-34-9	Xn
Sodium hydroxide	CAS- Number 1310-73-2	C
Potassium iodide	CAS- Number 7681-11-0	–
Sodium azide	CAS- Number 26628-22-8	T, N
Phosphoric acid	CAS- Number 7664-38-2	C
Zinc iodide	CAS- Number 10139-47-6	Xi
Starch	CAS- Number 9005-25-8	–
Sodium thiosulfate 0.01 mol l^{-1}	CAS- Number 7772-98-7	–

Reagents

Manganese(II) Chloride Solution

100 g manganese(II) chloride, $MnCl_2 \cdot 4\ H_20$, is dissolved in 150 ml of deionized water. Alkaline iodide-azide solution (precipitating agent)

70 g sodium hydroxide, NaOH, and 60 g of potassium iodide are dissolved in 150 ml of water. 2 g of sodium azide, NaN_3 are separately dissolved in 20 ml of deionized water and then both solutions are mixed.

Zinc Iodide–Starch Solution

Commercial zinc iodide–starch solution p.a. (e.g., Merck, Art. 5445) or add 4 g of soluble starch to a little double-distilled water and put into 100 ml of boiling water. Using a spatula, add approximately 0.5 g KI p.a. and dissolve.

Phosphoric Acid

600 ml phosphoric acid z. A., H_3P0_4, density $=1.71\,g\,ml^{-1}$, $(w = 85\%)$, is added to 400 ml of deionized water under constant stirring.

Sodium Thiosulfate Solution, $c(Na_2S_20_3) = 0.01\,mol\,l^{-1}$

1 ampoule Titrisol, Merck, Art 9909, is filled to 1000 ml with freshly boiled and cooled, deionized water in a volumetric flask. The solution is cooled and kept in the dark.

Experiment

(A) Verification of Henry's absorption coefficient

Deionized water is filled into a 300 ml Erlenmeyer flask. A magnetic stirrer ensures good mixing. Now a mixture of air and oxygen is bubbled into the water at constant flow rates. The gases are dosed together by means of two flow meters. The measurements are taken at room temperature. After reaching equilibrium, the oxygen content and saturation index are noted. The measurements are then conducted with three other air–nitrogen mixtures. From these measurements, the partial pressures, the mole fractions of oxygen and Henry's constants are calculated (Eq. 8.71 or 8.72). The values obtained are compared with bibliographical references.

Note:

$$x_{O_2} = \frac{O_2(\text{mol l}^{-1})}{O_2(\text{mol l}^{-1}) + 55.55\,\text{mol}_{H_2O}\,\text{l}^{-1}} \tag{8.82}$$

where

x_{o_2} is the mole fraction of oxygen.

The mole fraction of oxygen in air is y_{o_2}.

The water saturated with air ($y_{o_2} = 0.2095$) has an oxygen saturation of 100%. The mole fraction of oxygen in air y_{o_2} of a gaseous mixture is obtained by multiplying $y_{o_2} = 0.2095$ by the saturation (%).

(B) Oxygen determination according to Winkler and electrochemical determination of oxygen with an oxygen electrode

Work specification for the iodometric titration according to Winkler

Liquid flasks with ground glass stoppers are required, which can be filled completely with liquid without including air; 300 ml Erlenmeyer flasks are used. The volume must be accurately predetermined by weighing with water.

1. The liquid to be analyzed is filled into the Erlenmeyer flask which is stoppered, without any air bubbles enclosed.
2. Next, 2 ml manganese(II) chloride solution is added with a pipette. The pipette tip is kept well below the liquid level.
3. Add 2 ml of precipitating agent. Stopper: this expels 4 ml of liquid.
4. Mix liquid by inverting the Erlenmeyer several times. Allow sediment to settle and shake again.
5. Dissolve sediment by adding 5 ml of phosphoric acid. Close the flask, mix and let stand in the dark for 10 min.
6. Pour the contents into a 500 ml beaker. Titrate with 0.01 M sodium thiosulfate solution. The dark yellow color comes from the elementary iodine. In the course of the titration the yellow color becomes increasingly pale.
7. Towards the end of the titration add a few drops of starch solution and titrate again until the blue color disappears.

Calculation of the oxygen content

1 ml 0.01 M of sodium thiosulfate solution corresponds to 0.08 mg of oxygen.

The oxygen content in $\mathrm{mg\,l^{-1}}$ is calculated using the formula:

$$\text{conc.}_{O_2} = \frac{a.80}{b - 4\text{ ml}} \tag{8.83}$$

where a is the consumption of 0.01 M sodium sulfate solution (ml), and b is the sample volume (the sample volume is reduced by 4 ml by the addition of the main reagents) (ml).

B.1 Determination of the dissolved oxygen in deionized water at four temperatures.

Measure the exact temperature. The deionized water must be bubbled with air/mixture for at least 1 h to saturate it.

1. ice water (approx. 4 °C)
2. room temperature (depending on location)

3. thermostatted water at 35 °C
4. thermostatted water at 50 °C

B.2 Determination of the dissolved oxygen in three solutions containing different solvent concentrations.

1. 4 M KCL
2. 20% methanol: water (check resistance of the membrane)
3. 50% methanol: water (check resistance of the membrane)

B.3 Determination of the dissolved oxygen in four solutions with different salt contents.

1. $10\,g\,l^{-1}$ NaCl
2. $20\,g\,l^{-1}$ NaCl
3. $40\,g\,l^{-1}$ NaCl
4. $100\,g\,l^{-1}$ NaCl

B.4 The sample throughput and sensitivity of the three methods is to be compared.

8.6.2.4 Questions

(1) Describe the difference between activity (electrode) and concentration (Winkler titration).
(2) What influences the oxygen concentrations in solution more strongly – temperature or partial pressure?
(3) You need oxygen-free water. What possibilities do you know of for removing oxygen from water?
(4) What is the oxygen electrode (Clark-sensor) to be kept in and why?
(5) When is oxygen determination by Winkler preferable to the electrochemical measurement?
(6) When is measurement by electrodes to be given preference?
(7) Indicate the error sources of both methods.

References

1 DIN 38408-3 EN (2008) German standard methods for the examination of waste water and sludge – Gaseous components (group G) – Determination of ozone (G 3).

2 (1992) Standard Methods for the Examination of Water and Wastewater, APHA-AWWA-WEF, 18th edn, Method 4500-O.

8.7
Universal Separation Methods

8.7.1
Field Flow Fractionation (FFF), Project: "Characterization of Nanoparticles"

8.7.1.1
Analytical Problem

ETAP has developed a new generation of reactors for polymer production with improved devolatilization based on the continuous bulk polymerization. In the production of polymers, devolatilization is an important unit operation since polymerized products contain a certain amount of unreacted monomers. After the removal of the unreacted and inert materials the homogeneity of the product is analyzed to guarantee high polymer quality. In order to meet varied and stringent requirements on the molecular weight distribution, the Analytical Department of the company is now evaluating a relatively new analytical method: asymmetric flow field flow fractionation (FFF). Based on the results obtained the management will take a decision on whether to invest in this technology.

Object of Investigation/Investigation System
Nanoparticles

Samples
Polymers

Examination Procedure
To create structure–property correlations of polymer materials exact knowledge of the molecular weight distribution and the accompanying parameters is required. Information on chain branching is also vital. FFF technology represents a very universal separation method by virtue of its wide ranging applicability and flexibility. In addition, it offers excellent separation efficiency for very large and high-molecular weight particles, thus facilitating the separation of samples which could hitherto not be analyzed using the traditional methods. FFF is versatile and can be used in several fields of research. In the literature you will find applications from the fields of pharmaceuticals/biotechnology, chemistry/polymers, environmental sciences, food/agriculture/cosmetics, and nanotechnology. By virtue of its diverse advantages and the robust nature of the equipment, FFF technology will also claim a greater reputation among the analytical separation techniques in future.

Analytes
Polymer latex particles

Keywords
Polymer analysis, field flow fractionation (FFF), diffusion coefficient, cross-flow gradient, separation channel, (semi-permeable) membrane, electrophoretic mobility, thermal diffusion coefficient.

Bibliographical Research (Specific Topics)

Otte, T., Brüll, R., Macko, T., and Pasch, H. (2010) Optimisation of ambient and high temperature-asymmetric flow field flow fractionation with dual/multi-angle light scattering and infrared/refractive index detection. *Journal of Chromatography A*, **1217**, 722–730.

Messaud, F.A., Sanderson, R.D., Runyon, J.R., Otte, T., Pasch, H., and Williams, S.K.R. (2009) An overview on field flow fractionation techniques and their applications in the separation and characterization of polymers. *Progress in Polymer Science*, **34** (4), 351.

Schimpf, M., Caldwell, K., and Giddings, J.C. (2000) *Field Flow Fractionation Handbook*, Wiley-Interscience, ISBN 0-471-18430-6.

Wahlund, K.G., and Giddings, J.C. (1987) Properties of an asymmetric flow field flow fractionation channel having one permeable wall. *Analytical Chemistry*, **59**, 1332–1339.

Martin, M., and Hes, J. (1984) On-line coupling of thermal field flow fractionation with laser light scattering. *Chromatographia*, **15**, 426–432.

Project Process/Task

(A) Calibration of detectors using separation of bovine serum albumin (BSA) protein aggregates

The detectors are to be calibrated by separation of BSA protein aggregates.

(B) Separation of latex particles from cross-linked polystyrene as an example for the analysis of nanoparticles

Following calibration of the detectors, a latex mixture must be characterized.

(BB) Influence of the emulsifier content on the resulting diameter of the particles and the particle size distribution in the case of an emulsion polymerization

The ingredients used in manufacturing a polymer latex have a direct influence on almost all reaction and product parameters. This experiment is to determine the influence of the emulsifier content in the reaction mixture on the resulting size of the particles and the particle size distribution of polystyrene-polybutylacrylate-latex particles. Furthermore, the solid content of the latex samples is to be determined.

(C) Validation of the measurement method

The method is to be validated.

(D) Project reporting

The results are to be submitted in the form of a report (or presentation). Comparisons with bibliographical references are to be made.

8.7.1.2
Introduction

High-molecular weight polymers, biomolecules or particles in a size range of few nanometers up to several micrometers have been recently gaining in importance. Numerous applications are based on the special characteristics of these materials. For this reason it is imperative to know exactly the molecular weight or particle size and their distribution for later application as well as for the optimization of manufacturing processes or for use in quality control. In the case of macromolecular or biopolymers the chain structure is also a significant parameter, while in the case of particles, for example, the form plays a role. For this reason the often complex sample systems must be separated according to parameters such as particle size or molecular weight, and subsequently detected. In so doing a gentle and complete separation is vital, as otherwise information received on the sample system will be false even when using the best detectors [1–3]. The separation of very big or strongly adsorbing materials was, for example, hitherto only partially possible with chromatographic techniques such as, for example, GPC/SEC, HPLC or with methods such as ultracentrifugation, solubility fractionation or optical evaluation (microscopy). The column chromatographic separation methods are based on the separation of analytes in a mostly porous stationary phase. Due to the high flow velocities (shear forces) and the strong risk of adsorption by the large surface of the column packing, they are more suitable for low-molecular-weight applications. Polymers with strongly interacting particles, long-chain or highly branched polymers often cannot be (or only inadequately) characterized with these methods.

While off-line techniques provide better results, they are, for the most part, not universally applicable, are very expensive, or their separation processes are extremely time-consuming and painstaking. Today, characterization by means of field flow fractionation offers the possibility of a universal and high-resolution separation with subsequent detection of structures from the one-digit nanometer range [4] to micrometer particles [5] without lessening the flow velocity, filtration of the sample, or undesired adsorption in a stationary phase. The molecules or particles dissolved or dispersed in the solvent are separated in an empty separation channel according to size, composition or density. An external energy force, which works perpendicular to the direction of flow, is the basis for the separation of the sample particles. Depending on the type of equipment in use, a cross-flow gradient (asymmetric flow FFF–AF4), a temperature gradient (thermal FFF–TF3) or centrifugal field (centrifugal FFF–CF3) is applied. The cross-flow field serves to enhance the concentration of the analyte on the channel wall. The size-dependent self-diffusion of the particles then leads to an arrangement of structures of various sizes in flow layers according to the different speed of the parabolic flow profile in the channel, whereby size separation is achieved. In the case of thermal FFF a temperature gradient is used to separate the particles. As the resulting thermal diffusion also depends on the chemical nature of the analyte, this method can be used for separation according to size and chemical composition, whereby components with the same hydrodynamic radii but differing composition can also be separated. Centrifugal FFF uses a centrifugal field generated by rotating the channel. This method separates not only according to size, but also according to difference in density, whereby ideal possibilities of application in the area of particle samples arise, since the influence of the density facilitates very high separation efficiency, even for extremely small particles. All separation fields can be regulated as infinitely adjustable gradients in their intensity. Thus the separation efficiency can be individually adjusted to every separation problem and a "tailored" calibration curve can be obtained. It is then no longer necessary to limit the separation efficiency by, for example, specifying a certain column combination, as in SEC. In the FFF any detector combinations can be used. Particularly, multi-angle laser light scattering (MALLS) has proven valuable in combination with concentration sensitive detectors, such as, refractive index detectors (RI), IR or UV, as they make the radius and the molecular weight accessible, thus providing a plethora of information about the separated sample [6]. Other combinations, such as coupling of dynamic light scattering (DLS) [7], viscometers [8], or spectrometers (e.g., ICP-MS) [9], can be implemented without great effort.

Size Exclusion Chromatography (SEC)

Separation Principle

Size exclusion chromatography is a specific form of liquid chromatography, which is used particularly for separating macromolecules. The dissolved molecules are transported through a separation column with the help of an eluent. The solutes are then separated according to their size and leave the column at different elution times, and are subsequently detected. The separation columns in size exclusion chromatography are filled with different porous carrier materials, which are also called gels. The basis of separation in this method is not the affinity to the carrier material (no interaction with the surface of the stationary phase), but the different effective hydrodynamic radius of the dissolved molecules. Molecules that are smaller than the pore size of the carrier can enter the carrier pores and, therefore, have a longer path and longer transit time than larger molecules that cannot enter the pore. Molecules larger than the pore size elute together as the first peak in the chromatogram. This condition is called total exclusion. It is widely applicable to purification or desalting of proteins in complex samples, such as blood, which contain molecules of widely different sizes.

Limits of the Method

SEC is limited in its application to the relatively low molecular weight range of up to $10^6\,g\,mol^{-1}$ (depending on the polymer material). If larger molecular weights are applied a partial shear degradation of the macromolecules is observed. Theoretical computation shows that the strongest shear decomposition must take place in the column packing itself, since here there are several different velocity gradients in the various spatial directions. The resulting complex elongational flows lead to extreme shearing stress and thus to dissolution of the polymer chains, since the flow rates that are applied there do not permit sufficient stress relaxation. In addition, the maximum injection volume and the polymer concentration depend strongly on the column system used. Too high concentrations lead to misrepresentations in measurement results (band broadening) and similarly to increased shearing stress degradation due to the high viscosity of the solution. The separation efficiency depends mainly on the distribution of the pores of the SEC-column. Hence an increase in separation efficiency is generally only achieved with a column replacement. This dependence on the stationary phase reduces the possibility to optimize the separation. With extremely large molecules there is also a rapid loss of separation quality in the SEC. Superstructures such as aggregates or gels are partially separated beforehand by pre-filtering or column frits, which renders it impossible to analyze them. Thereby, quite frequently, a large portion of sample information gets lost. The use of SEC as a fast analytical method is extremely limited in the case of hydrophilic-based particles, since each change of eluent must be followed by a time-consuming conditioning of the column packing.

The History of FFF

Field flow fractionation was discovered in 1966 by Calvin Giddings (1930–1996), a scientist at the University of Utah in Salt Lake City, USA. Giddings founded the FFF Research Center at the University of Utah and developed the theory as well as the most varied FFF variations there with his team. Examples are thermal FFF, sedimentation FFF, flow FFF and split FFF (Figure 8.69). In the years that followed, Giddings was recommended twice for the Nobel Prize. Finally, in 1987 he published the first asymmetrical flow FFF together with K.G. Wahlund as another variation of flow FFF. In the field of detectors for FFF, M. Martin of the University of Paris was the first, in 1984, to publish work on the coupling of FFF with laser light scattering in. Thanks to the coupling of FFF as a method of separation with light scattering (LS) as a detection method, the door was open for the technology to be used henceforth for a number of applications. The FFF-LST technology has, in the meantime, gained the same significance for the characterization of proteins, polymers and particles as GC- and LC-MS has for characterizing small molecules.

The Separation Principle of FFF

FFF represents a family of separation techniques similar to chromatography. What all FFF methods have in common is the use of a specific, very flat separation channel instead of a column as used in chromatography (Figure 8.70). This separation channel does not have a stationary phase, but is rather only traversed by the solvent each time. By virtue of the small channel height (approx. 350 μm)

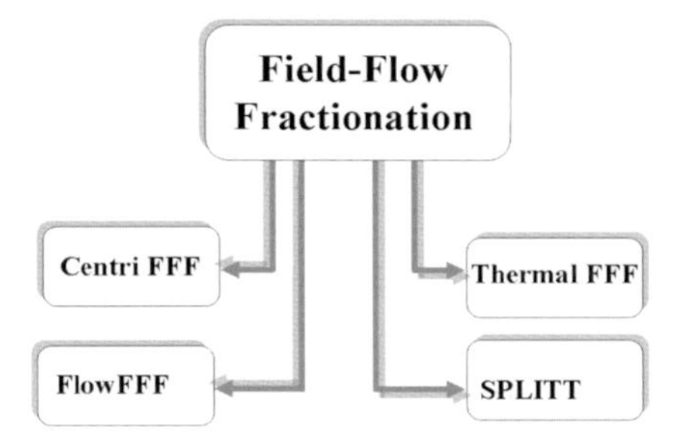

Figure 8.69 Different separation forces but same separation principle.

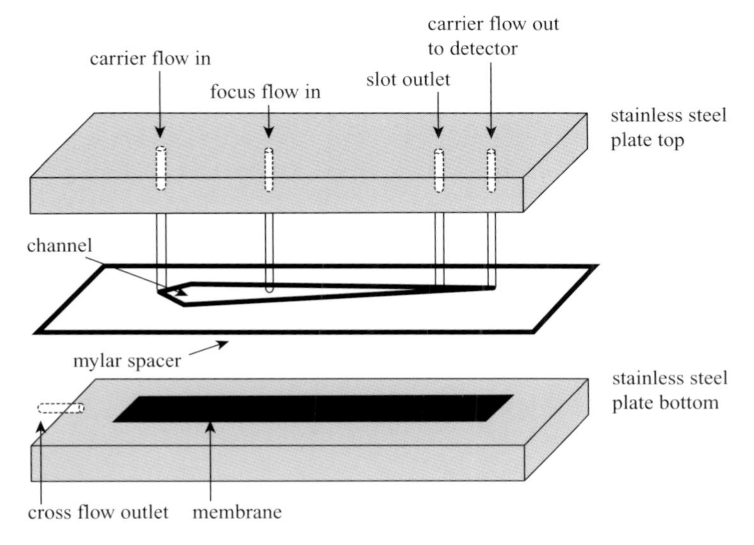

Figure 8.70 Asymmetric flow FFF unit.

a laminar flow with a parabolic flow profile is formed at the common flow rates of $1\,\mathrm{ml\,min^{-1}}$.

The actual separation of analytes is effected by a force field that is set up vertically to the parabolic flow. This force field can be another liquid flow, a temperature gradient, a centrifugal field or another physical force. The diverse FFF variants are thereby named in accordance with the force field used, for example, flow FFF [10], thermal FFF [11], centrifugal FFF [12]. Within the channel, the particles are guided by the separation force into the direction of the lower surface of the channel. As the particles do not pass over the channel floor they are scattered back to the center of the channel. Under the impact of the force field working from top to bottom and particle diffusion working in the opposite direction, from bottom to top, particles of differing sizes are separated. Depending on the diffusion coefficient, each sample species is present as a separate layer after the state of equilibrium has been achieved. Smaller particles with a stronger diffusion thus reach a higher level of the channel in areas of greater flow velocities and elute first. In contrast, large particles with a lower diffusion only reach areas with lower flow velocities and thus elute at a later point. The following is the relationship between the dimensionless retention parameter λ_R and the separation force with regard to all FFF techniques:

$$\lambda_R = \frac{kT}{Fw} = \frac{l}{w} \tag{8.84}$$

- *k* Boltzmann constant
- T Temperature
- w Channel height
- *l* Distance from the accumulation wall to the center of gravity of the analyte zone.
- F Field force that impacts the analyte

In chromatography the retention relationship *R* is frequently used to describe the separation:

$$R = \frac{t_0}{t_R} \tag{8.85}$$

where t_0 represents the dead time of the chromatographic system and t_R the retention time of the peak maximum of the separated components. In flow-FFF separation occurs normally within a layer near the wall with a negligible layer thickness. For this reason the gravitation center must similarly be very close to the wall and so the retention parameter becomes very small (< 0.02). Thus the following connection between the retention parameter and the retention relationship can be approximately assumed:

$$R = 6^{*}\lambda_R \tag{8.86}$$

It becomes clear that separation in flow FFF depends very strongly on the chosen cross-flow and even more strongly on the height of the channel, that is, on the chosen spacer.

Theoretically, molecular weights of up to approximately 1 kg mol^{-1} can be separated using flow FFF. In reality, however, this lower value is not achieved, since the pore size of the membranes used is limited, and also for the best cellulose membranes further molecular weight losses must be reckoned with (room temperature) for weights under 1–5 kg mol^{-1}. The so-called cut-off, however, has the advantage that lower molecular weight sample impurities are removed with the cross-flow so that they cannot negatively influence later detection.

Field Flow Fractionation

In FFF separation occurs in a thin band-like channel, through which eluent flows. Perpendicular to the main direction of flow a force field is established which interacts with the sample molecules. The field force presses the molecules in the direction of the accumulation wall. Due to the Brownian motion, however, the result is an independent diffusion of the particles. The part of the diffusion motion opposing the cross-flow causes a migration of the molecules towards the middle of the channel. Thus structures influenced differently by the cross-flow or which differ in their diffusion coefficient place themselves at different channel heights after reaching equilibrium. The separation channel is thus constructed so that there is a laminar layer flow. From this results a parabolic velocity profile, that is, the flow velocity continually decreases in the direction of the channel wall. According to their state of equilibrium, particles with varying characteristics (e.g., diffusion capability, chemical composition, electrical charge, weight) have varying flow layer velocities and will thus leave the channel at different elution times. The channel exit is, in most cases, connected to a detector, which subsequently detects the fractionated particles. The separation principle in the FFF channel is illustrated in Figure 8.71. In flow FFF the separation force is generated by a cross-flow. Thereby two types of applications can be distinguished, namely symmetrical and asymmetrical flow FFF. Regarding symmetrical flow FFF the upper and lower sides of the channel are composed of porous frits through which the cross-flow is guided. On the lower porous frit there is a membrane attached, which allows the passage of the solvent, but not of the sample particles. In contrast, in the case of asymmetrical flow FFF a channel with a plate top is normally used. As in the case of the symmetrical flow FFF, the lower channel limitation is a porous frit with a membrane attached thereto. As in the case of the other FFF variations, a longitudinal carrier flow is guided through the separation channel, which forms a laminar flow profile. Contrary to the other FFF methods, however, no external separation field is applied with asymmetrical flow FFF. The cross-flow is deviated as a partial flow of the main flow of the carrier liquid in the channel and led out through the

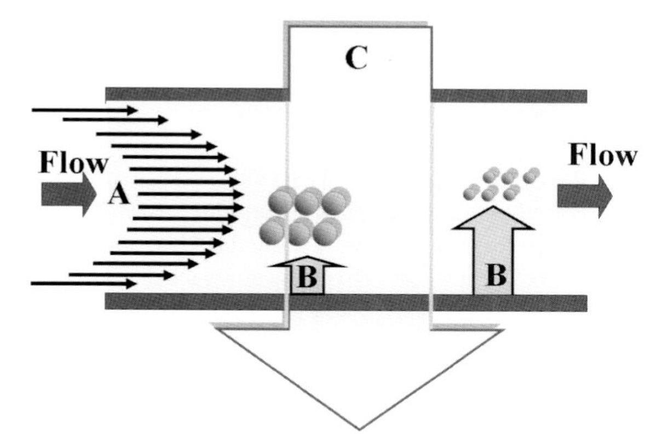

Figure 8.71 Separation forces in flow FFF. **A** parabolic flow profile, **B** diffusion, **C** force field.

lower membrane. As in the case of symmetrical flow FFF, the membrane prevents the sample particles from escaping from the channel with the cross-flow. Flow FFF can be used with aqueous and organic solvents. It is used in the fields of biotechnology and pharmaceuticals, chemistry and polymers, food and agriculture, as well as in the environment. The main applications are the characterization of proteins, antibodies, aggregates, viruses, chitosans, alginates, polyacrylamides, micelles, liposomes, humin colloids. This makes flow FFF the most universal FFF technology with the most wide-ranging scope of application. The following are some application examples:

Special Advantages of Flow FFF

1. Separation of proteins, polymers and particles
2. Large dynamic separation area
3. Use of aqueous and organic solvents
4. The possibility of focusing/concentration of the sample

Measuring Procedure

The injection phase is the initial phase. The solute is transported by the main flow from the sample loop into the AF4-channel. During this time the concentration flow and the cross-flow are active and their flow velocity is constant during injection. The velocity of the main flow and the cross-flow can thereby be freely chosen. The concentration flow is automatically adjusted each time, so that the desired detector flow is guaranteed. The main flow and the focus form counterflows, whereby the sample is focused as a narrow band at the point of focus during the injection (Figure 8.72). During the process the sample molecules are, at the same time, arranged by the already activated cross-flow into corresponding flow layers. The advantage of focusing is a drastic lessening of band broadening as well as the possibility to inject as long as desired, which facilitates the yield of an adequate quantity of very dilute solvents.

Following the injection phase, the focus flow is, in a short transition period, run down linearly to $0\,ml\,min^{-1}$. In so doing, the main flow must be increased by the same amount, in order to continue to guarantee a constant flow rate (Figure 8.72).

The elution phase (Figure 8.73) that follows consists of the continuous lowering of the cross-flow. The lowering can take place gradually in both a linear or exponential manner. Compared to a constant cross-flow guidance, this lowering has the advantage that very large species with a very weak independent diffusion are not retained for an excessively long time in the channel, and that thus the effective time for analysis can be clearly reduced. During the lowering, the main flow is reduced to the same extent as the cross-flow. During the entire separation process, the sample particles are thus able to switch to different flow layers depending on the strength of the temporary cross-flow. Thereby, separation can be impacted to a high degree by the cross-flow program chosen, since, in addition to the different diffusion coefficients of the particles the

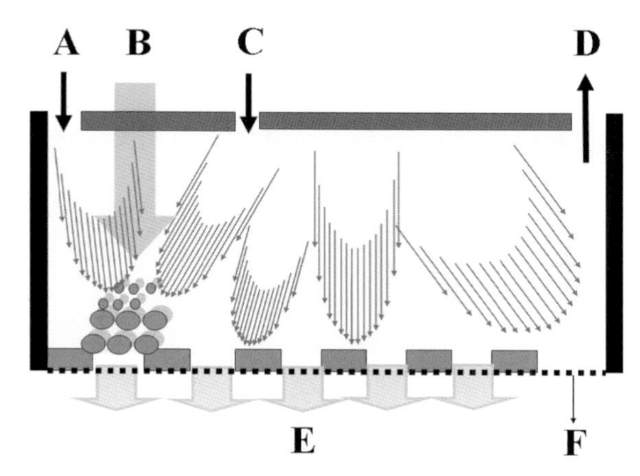

Figure 8.72 Injection-/Focussing phase. **A** inflow-main flow, **B** focus point, **C** inflow-focus flow, **D** exit to the detector, **E** cross-flow, **F** semi-permeable membrane.

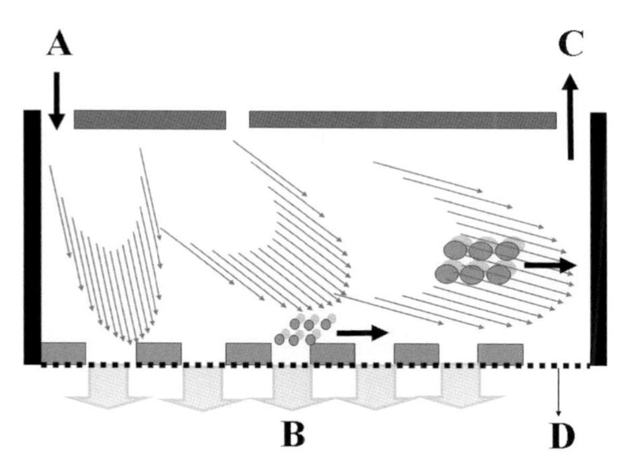

Figure 8.73 Elution phase. **A** inflow-main flow, **B** cross-flow, **C** exit to the detector, **D** semi-permeable membrane.

strength and the duration of the cross-flow now influences the arrangement of the particles.

Thermal FFF

Thermal FFF uses thermal diffusion to generate a retention. Thereby a temperature gradient is built up between the upper and lower limit of the FFF channel. The principle is illustrated in Figure 8.74. The channel is located between two metal blocks, whereby the upper block is exactly heated and the lower block is precisely cooled. The temperature difference thus arising induces a different strong thermal diffusion of the particles and polymers to be separated. The thermal

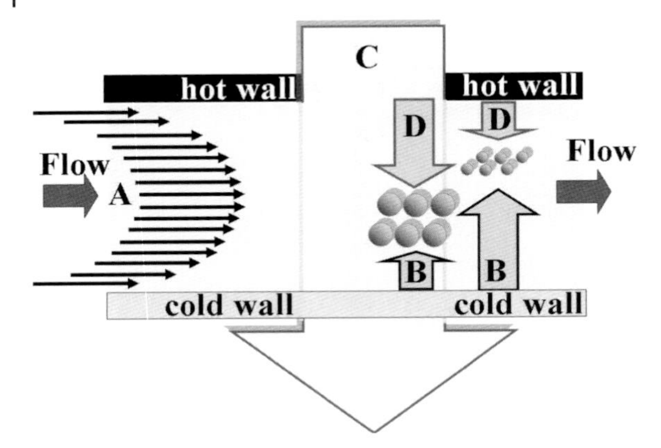

Figure 8.74 Functional principle of thermal FFF. A parabolic flow profile, B "normal" diffusion, C force field, D thermal diffusion.

diffusion counteracts the normal diffusion of the particles. The thermal diffusion and the normal diffusion, as well as the size of the temperature gradients, determine the layer thickness of the individual sample species in the channel. Components with a larger molecular mass move closer to the lower channel limitation in the area of lesser flow velocities. On the other hand, smaller components move farther away from the lower channel wall in faster flow lines and are thereby eluted on a more timely basis.

The thermal diffusion coefficient is independent of the molecular mass, contrary to the normal diffusion coefficient. However, it is influenced by the chemical composition of the analyte. From this the advantage ensues that the separation of the different parts of a sample in thermal FFF takes place after the molecular mass and chemical composition. Thermal FFF is used primarily with organic solvents. It is applied in the fields of chemistry/polymers and nanotechnology. The main applications are in the characterization of complex polymers and gel materials. The following are some application examples:

Special Advantages of Thermal FFF

1. High resolution for polymer/gel separations
2. Compatible with almost all common organic solvents
3. Separation after molecular mass and chemical composition

Centrifugal FFF

In centrifugal FFF a rotating ring channel is used, whereby the separating field is generated by the centrifugal force. The principle is illustrated in Figure 8.75. The channel is 250 μm high, so that a laminar flow builds up with a parabolic flow profile. The elution of the particles is determined by the number of revolutions of the ring channel, whereby the separation takes place according to the differences

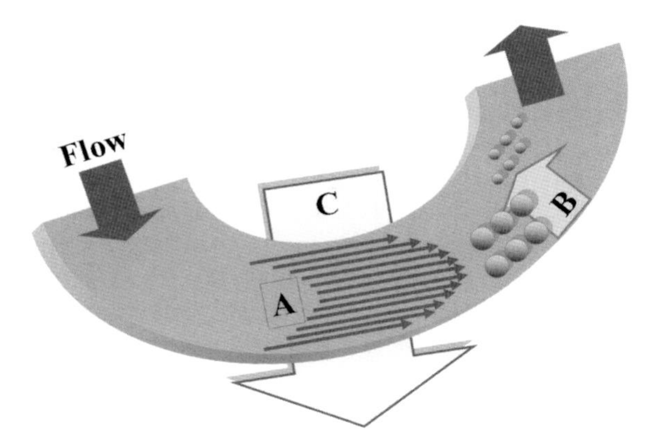

Figure 8.75 Separation principle of centrifugal FFF. A parabolic flow profile, B diffusion, C centrifugal field.

in the density and the component mass of the analyte. Smaller particles elute first and are then followed by the larger species. An analytical run starts at a high rotational speed to separate small components. Next the number of revolutions is reduced to elute larger components within an adequate time span. This enables polydisperse samples to be characterized. Typical areas of application are the separation of emulsions, colloids, particles, drug delivery-systems and cells.

Special Advantages of centrifugal sedimentation-FFF

1. High resolution for particle separation
2. Large dynamic separation area specially for particles
3. Separation according to particle size and thickness

SPLIT FFF

SPLITT FFF [13] is a preparative separation technique for particles in the micrometer area, which is in continuous operation. Separation is based on the different sedimentation rates of the particles in the channel under the influence of the earth's gravity (Figure 8.76). The construction of the SPLIT-channel enables a rapid separation in seconds or minutes since the sedimentation of the particles must only cover a relatively small distance, normally 500 μm. The hydrodynamic, stable, laminar liquid flow in the SPLIT channel forms the transport medium for the particles. Based on the continuous throughflow, SPLIT FFF is ideal for cleaning particle systems.

Special Advantages of SPLIT FFF

- Characterization of complex environmental samples, algae, sediment particles, chromatographic media,
- Preparative cleaning of a number of industrial particle products.

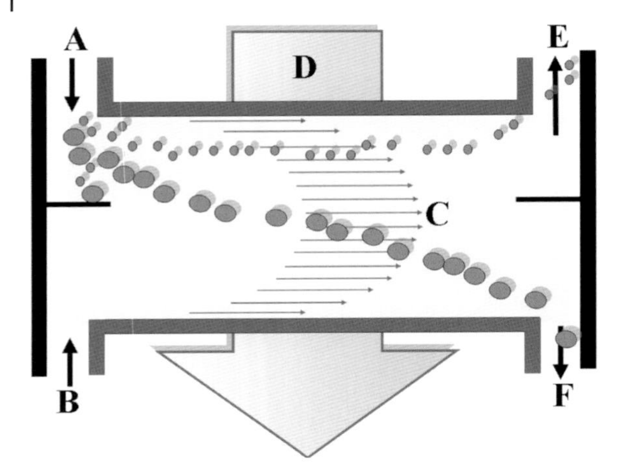

Figure 8.76 Principle of the SPLIT FFF. A sample flow IN, B carrier flow IN, C laminar flow, D gravitational field, E sample flow OUT, F sample flow OUT.

Summary

FFF technology is an efficient method of analysis for the rapid, gentle and high-resolution separation and size characterization of proteins, polymers and particles. FFF is particularly characterized by the following features:

- large dynamic mass-/size range analyzable (1 nm–100 μm, 10^3 Da–10^{12} Da)
- rapid, gentle separation without flow velocities in the channel
- no interaction with the stationary phase as with LC
- higher resolution than chromatographic separations >10^5 Da
- Flexible selection of separation conditions, for example, solvents, separation intensity
- possibility to fractionate the separated particles
- on-line/off-line-coupling with further analytical processes, for example, MS, EM, NMR.

Concentration Detection

Refractive Index (RI) Detector

The concentration detectors include all variations whose physical measurement signal changes proportionately to sample concentration. By virtue of this, the refractive index detector (RI) has been established, among others, for polymer analytics, since here the concentration dependence is not bound to special functional groups. The RI can be used universally for all substances which distinguish themselves in their refractive index from the eluent index. RI detectors are sensitive to pressure and temperature. A gradient elution with changes of solvent composition is not possible due to the changing refractive index of the solvent. Nowadays, detectors function in accordance with the principle of the differential refractometer.

Infrared (IR) Detector

Concentration detection using infrared is a particularly important aspect in the analysis of polyolefines. Polyolefines can only be dissolved and analyzed at temperatures above 130 °C and in solvents such as 1,2,4-trichlorobenzene. The refractive index increment is often infinitesimal in this high boiling solvent, so that concentration detection by means of RI can be very prone to error due to the poor signal/noise ratio. Here IR detection offers the great advantage of having very good signal intensity. Concentration detection via IR is merely limited by the selection of an IR-transmissive solvent without absorption at the required frequency. The background absorption is determined and corrected as a zero-value through a reference measurement with the pure solvent. Thus for the most part only unsaturated or chlorinated liquids are suitable as solvents, which either do not or hardly have saturated CH_2- and CH_3-groups.

The Evaporative Light Scattering Detector (ELSD)

The ELSD is very often used to detect non-volatile compounds. The individually eluting components are atomized through a nozzle with the help of an inert gas flow (often nitrogen or air pressure). The volatile solvent is then rapidly evaporated in a heated flight tube. The non-volatile components form fine solid particles. At the end of the flight tube, the particles reach a measurement cell, which is irradiated with light. The radiated light is scattered inelastically onto the particles. The thus scattered light is detected by a photodiode and in this way the detector signal is generated. A great advantage of ELSD is the fact that the mobile phase is completely evaporated and that its composition thus has no influence on the base line. For this reason, ELSDs are frequently used for detection in gradient chromatography. Due to its insensitivity towards the eluent and the high signal intensity, the ELSD has a one- to two-fold better signal/noise ratio than, for example, when using an RI detector. A disadvantage of the ELSD arises in the detection of polymers with low molecular weight. These are frequently relatively slightly volatile and can, under certain circumstances, evaporate together with the eluent and thus no longer be detectable. The use of ELSD to determine concentration is only possible to a very limited extent, since the detector signal depends on numerous influencing factors, which do not play a role in traditional concentration detectors. The main cause for this is the dependence of the scattered light signal on the particle form and size. Thus, there often exists, for example, a non-linear dependence of the detector signal on the molecular weight, concentration, chemical composition and molecular architecture of the components to be detected. In addition, the equipment parameters such as flow rate, evaporation temperature and gas flow rate also have an effect on signal intensity. For each unknown polymer–solvent system the dependences must, therefore, be checked prior to using ELSD and, if necessary, these dependences must be balanced by a calibration curve.

Scattered Light Detection [14,15]

Two types of light scattering measurements, namely static light scattering (SLS) and dynamic light scattering (DLS), have been used in the drug discovery process.

SLS measures the time average intensity of light scattered from the molecules in solution. The intensity of scattered light is proportional to both the concentration and the molecular weights of the molecules in solution. In practice, SLS measurements can be carried out either in a stand-alone mode or, more often, in conjunction with a size exclusion chromatography (SEC) system. SLS detectors, and especially multi-angle light scattering (MALS) detectors, are widely used in measuring the absolute molecular weight of proteins and other polymers.

Such measurements are independent of the shape of the molecules. SLS is used whenever any change in molecular weight caused, for example, by aggregation is likely to have a significant effect on the screening process. DLS, on the other hand, measures time-dependent fluctuations in the intensity of scattered light, which occur because particles are undergoing random Brownian motion in the solution. From these measurements the hydrodynamic radius and size distribution of the particles can be derived. The technique is extremely sensitive for measuring the amount of large particles in solution. For this reason DLS systems have recently become an important tool in high throughput compound screening.

8.7.1.3 Material and Methods

Equipment

- Asymmetric Flow FFF Postnova Analytics (Germany) Typ AF2000 MT

Detectors

- UV

  Scattered light detector (Precision Dynamics)

  Two-angle (90°/15°) static light scattering detector wavelength 808 nm

- RI

  (Depends on the available detectors and the samples to be detected. RI should always be the last detector, since fluctuations in back pressure can otherwise destroy cells).

Parameters

Separation channel	350 μm Mylar Spacer
Membrane	Regenerated cellulose RC 10, with 10 kDa cut off
Detector flow	0.5 ml min^{-1}
Dosage	Manual injection with 20 μl loop

Chemicals

NaN_3 CAS- Number 26628-22-8 T, N 

NaCl	CAS- Number 7647-14-5	–	
Styrene	CAS- Number 100-42-5	Xn	
SDS (sodium dodecyl sulfate)	CAS- Number 151-21-3	Xn	
APS (ammonium persulfate)	CAS- Number 7727-54-0	O, Xn	
NovaChem dispersing agents			
Albumin – Standard BSA (Sigma)			
Latex particles 22, 58 and 100 nm`			

Solutions

Buffer Solutions

- Salt solution: 0.1 M NaCl + 0.2 $g l^{-1}$ NaN_3 solution for BSA and Dextran and/or Pullulan separation
- 0.2 vol% NovaChem solution for Latex Particle Standards

Experiment

(A) Calibration of detectors using separation of bovine serum albumin (BSA) protein aggregates

The concentration of the BSA-solvent should be 1 $mg ml^{-1}$ and the injection volume 20 μl.

The BSA is dissolved and separated in a 0.1 molar salt solution. In so doing, the NaN_3 serves to suppress the biological growth in the solvent.

Prior to the actual separation, the RI detector is calibrated by means of direct injection without cross-flow. The sample is injected five times at a constant detector flow of 0.5 $ml min^{-1}$.

Each injection is started when the preceding peak has again reached the base line level. After that, each peak is individually assessed and the RI calibration constant is determined. An average is taken of the last four values for the RI calibration constant and the mean is used as a new RI-constant. In most cases, the first measurement must be rejected, since the sample loop is often not completely filled, so that air is injected.

The BSA is then separated with constant cross-flow after the hydrodynamic volume. Following separation, monomer, dimer and trimer Peaks can be recognized in the fractogram.

(B) Separation of latex particles from cross-linked polystyrene as an example of an analysis of nanoparticles

For the particle mixture, the salt buffer must be exchanged for NovaChem100 solution, since this special tenside mixture prevents the particles from aggregating and thus keeps the dispersion stable. Three different latex standards are mixed with 0.2 % NovaChem100 solution. The concentrations of the single standards are $5\,mg\,ml^{-1}$ for $d = 22\,nm$, $0.5\,mg\,ml^{-1}$ for $d = 58\,nm$ and $0.1\,mg\,ml^{-1}$ for $d = 100\,nm$. This mixture is subsequently separated with the AF4. This assessment again takes place by using normalization, which was carried out in Stage (A) with the BSA. Alternatively, there can be normalization to the smallest standard of the mixture. Only a slight concentration signal is received for the particles. The radius of gyration is determined from the light dispersion. This does not directly correspond to the hydrodynamic radius, but can be converted for certain geometries of particles (latex–spherical) into the hydrodynamic radius.

(BB) Influence of the emulsifier content on the resulting circumference of the particles and the particle size distribution in the case of an emulsion

Polymerization

In the case of emulsion polymerization, polymerization of poorly water-soluble monomers is conducted in an aqueous emulsion of the monomer. An example of such a system is the reaction of styrol to polystyrene. In this process, polymer particles are obtained, which are clearly smaller than the monomer droplets at the beginning of the reaction. In order to attain these, tensides are used, which build micelles from a certain concentration onwards. Monomer units from the stabilized monomer droplets wander about in these micelles. The initiation takes place in the aqueous medium, while the actual polymerization takes place in the micelles. During polymerization, the micelle grows due to the diffusing monomers or oligomers. The resulting particle size is dependent on the emulsifier and the micelle size formed by it. At the end of the reaction, the polymer latex is 10–500 nm in size. In the system used in the experiment, the particle size should be in the range 30–60 nm.

The advantages of this polymerization technique are:

- the use of aqueous solutions, no organic solvents are required
- the reaction heat can be well controlled by the aqueous phase.

An ensuing disadvantage of emulsifier polymerization is that a residue of emulsifier remains and other auxiliary materials in the polymer.

Execution

All polymerizations are conducted in a thermostatted glass reactor equipped with a mixer, a reflux condenser and two inlets for nitrogen and other ingredients. The polymerization mixtures consist of the monomer styrol, the solvent water, the emulsifier SDS (sodium dodecyl sulfate) and the initiator APS (ammonium persulfate). The solvent, the monomers and the emulsifier are placed into the reactor and heated to 50 °C under nitrogen.

Conditions

Monomer: $500\,g\,l^{-1}$ H_2O

SDS: $25\,g\,l^{-1}$ H_2O

ADS: $1.25\,g\,l^{-1}$ H_2O

Middle-sized particles are to be expected: radius 20 nm, circumference 40 nm.

Two reactors are in operation at the same time with different amounts of emulsion. The initiator is added quickly when the temperature is reached. At 10 min intervals, approximately 2 ml is taken as a sample, to determine the solid content. This process is repeated nine times. It is left to be mixed overnight when the particle sizes and the particle size distribution can be determined by means of field flow fractionation.

(C) Validation of the measurement method

Validate the method.

8.7.1.4 Questions

(1) What different FFF technologies are there and what distinguishes them from each other?
(2) Describe the set-up of an asymmetrical flow FFF system.
(3) How does the separation mechanism work in flow FFF?
(4) What advantages does FFF offer over chromatographic separation methods?
(5) What advantages does a cross-flow gradient offer over a constant cross-flow?
(6) List possible applications of FFF.

References

1 Parth, M., Aust, N., and Lederer, K. (2003) *International Journal of Polymer Analysis and Characterization*, **8**, 175.
2 Aust, N. (2003) *Journal of Biochemical and Biophysical Methods*, **6**, 323.
3 Zammit, M.D. (1998) *Polymer*, **39**, 5789.
4 Messaud, F.A., Sanderson, R.D., Runyon, J.R., and Otte, T. (2009) *Progress in Polymer Science*, **34**, 351.
5 Ratanathanawongs, S.K., and Giddings, J.C. (1992) *Analytical Chemistry*, **64**, 6.
6 Otte, T., Brüll, R., Macko, T., and Pasch, H. (2010) *Journal of Chromatography A*, **1217**, 722.
7 Caldwell, K.D., Li, J., Li, J.-T., and Dalgleish, D.G. (1992) *Journal of Chromatography A*, **604**, 63.

8 Mes, E.P.C., de Jonge, H., Klein, T., Welz, R., and Gillespie, D.T. (2007) *Journal of Chromatography A*, **1154**, 319.

9 Hasselöv, M., Lyven, B., Haraldson, C., and Sirinawin, W. (1999) *Analytical Chemistry*, **71**, 3497.

10 Otte, T. (2010) Characterisation of ultra-high molecular Polyolefines with Asymmetric Flow-Field Fractionation and Size Exclusion Chromatography, Dissertation, TU Darmstadt.

11 Thompson, G.H., Myers, M.N., and Giddings, J.C. (1969) Thermal field flow fractionation of polystyrene samples. *Analytical Chemistry*, **41**, 1219–1222.

12 Giddings, J.C., Yang, F.J.F., and Myers, M.N. (1974) Sedimentation field flow fractionation. *Analytical Chemistry*, **46**, 1917–1924.

13 Giddings, J.C. (1985) A system based on split-flow lateral-transport thin (SPLITT) separation cells for rapid and continuous particle fractionation. *Separation Science and Technology*, **20**, 749–768.

14 Martin, M. (1982) Polymer analysis by fractionation with on-line light scattering detectors. *Separation Science and Technology*, **19**, 685–707.

15 Yao, B., Collins, B., and Chen, M. (April/May 2005) Light scattering to detect compound aggregation in screening assays, DrugPlus International.

Appendix A
Selection of Recommended Sources by Subject Area

A.1
General Sources

A.1.1
Römpp

The proven Römpp Chemistry lexicon, created by some 180 technical writers, contains references and further reading on most key areas and, therefore, provides an ideal introduction to individual topics. In five volumes and five supplementary volumes, about 57 000 keywords in the areas of chemistry, food chemistry, environmental and process technology, biotechnology, genetic engineering, and natural materials are covered. Römpp Online is also available as a licensed version under http://www.roempp.com/prod/ online encyclopedia.

A.1.2
Wikipedia

The online encyclopedia Wikipedia (en.wikipedia.org/wiki/Main_Page, accessed August 2012) covers approximately 30 000 keywords in chemistry and its related fields.

A.1.3
CRC Handbook of Chemistry and Physics, David R. Lide (Ed.)

An indispensable handbook, with tables on physical properties of organic, inorganic and organometallic compounds, physical constants and mathematical formulas. Published bi-annually. Licensed online version at http://www.hbcpnetbase.com/ (accessed May 2012).

Practical Instrumental Analysis: Methods, Quality Assurance and Laboratory Management, First Edition.
Sergio Petrozzi.

A.1.4 Merck Index: Encyclopedia of Chemicals, Drugs and Biologicals

Reference to trade names of some 10 000 chemical, pharmaceutical and biological compounds and products, with information on trade names, toxicity, therapeutic applications and literature references. Licensed version also available online.

A.1.5 ChemSpider

Free search engine for substance information available at: http://www.chemspider.com/ (accessed May 2012).

A.2 Analytical Chemistry

A.2.1 Encyclopedia of Analytical Chemistry, RA Meyers (Ed.), John Wiley & Sons Ltd, Chichester (2000)

This 15-volume work discusses the theory, equipment required and applications of analytical methods.

A.2.2 Official Methods of Analysis, Association of the Official Analytical Chemists (1990)

Recommended, consistent and accurate methods for the analysis of agro-chemicals, contaminants, pharmaceutical and food products.

A.3 Inorganic and Organometallic Chemistry

A.3.1 Gmelin Handbook of Inorganic Chemistry

The Gmelin database was already mentioned in Section 4.2.7.4. The printed Gmelin handbook of inorganic chemistry is a valuable additional source for charts and historical information that could not be included in the database.

A.3.2
Dictionary of Inorganic (Metals and Organic) Compounds

Reference work organized by composition on the synthesis and characterization of inorganic and organometallic compounds. Also available online (licensed version): http://dioc.chemnetbase.com/ (accessed May 2012).

A.4
Chemical Engineering/Technical Chemistry/Process Engineering

A.4.1
Ullmann's Encyclopedia of Industrial Chemistry

This 40-volume encyclopedia contains multipage treatises on all the major products (manufacturing, service) of the chemical industry, as well as process engineering fundamentals. Also available online (licensing): http://www.mrw.interscience.wiley.com/ueic/ (accessed May 2012).

A.4.2
Kirk-Othmer Encyclopedia of Chemical Technology

27-volume encyclopedia of chemicals and chemical engineering similar to "Ullmann".

A.5
Chemicals: Directory of Suppliers

A.5.1
Databases Subject to Charge

Available Chemicals Directory, provided via Symyx DiscoveryGate or in companies on the Isentris platform.

CHEMCATS, access via SciFinder (Scholar or Web) of CAS

A.5.2
Free Access (Selection)

ChemExper (http://www.chemexper.com, accessed May 2012).
EMolecules (http://www.emolecules.com, accessed May 2012).
PubChem (http://www.pubchem.com, accessed May 2012).

A.6
Organic Chemistry

A.6.1
Science of Synthesis ("Houben-Weyl Methods of Molecular Transformations, Methods of Organic Chemistry")

Science of Synthesis contains proven methods of production of the most important classes of compounds. It contains descriptions of about 18 000 general manufacturing methods, indicating the original literature. On completion of the work, about 180 000 reactions and 800 000 structures will be covered. Structures and reactions can be searched for in the licensed online version.

A.7
Physico-chemical Data

A.7.1
CRC Handbook of Chemistry and Physics

Properties Sources Index EQI (http://www.eqi.ethz.ch/en/)
The Properties Sources Index EQI of the ETH Zurich contains databases, reference books and websites for approximately 1000 physico-chemical and other material properties (data definition, measurement). In German, English and French.

A.7.2
Landolt-Bornstein Numerical Data and Functional Relationships in Science and Technology

This approximately 360-volume reference work is available in only a few libraries, due to its high price, but in many cases it is the only source of accurate, high-quality physical data for molecules, radicals, mixtures, semiconductors, metals and alloys. Available as a licensed e-book at http://www.landolt-boernstein.com on the platform of Springer Verlag. Particular physical properties can best be found by searching the freely accessible subject index, or the previously mentioned EQI; connections can be researched through the Substance Index of the University of Hamburg (http://lb.chemie.uni-hamburg.de/, accessed May 2012).

NIST Chemistry Webbook, http://webbook.nist.gov/ (accessed May 2012).

Open source from the National Institute of Standards, includes thermodynamic data for approximately 7000 compounds and 8000 reactions, IR spectra of 16 000 compounds, mass spectra of 15 000 compounds, UV/Vis spectra of 1600 compounds, and GC data for 27 000 compounds.

A.7.3
TRC Thermodynamic Tables

Data collection of the Thermodynamic Research Center with thermodynamic data for organic compounds.

A.8
Polymers and Materials

A.8.1
DECHEMA Materials Table

Extensive data collection on the corrosion behavior of materials.

MatWeb (Material Data Sheets), http://www.matweb.com, accessed May 2012).

Database of over 69 000 properties and compositions of alloys, plastics, ceramics, and composites.

PolyInfo (NIMS Materials Database, Japan), http://mits.nims.go.jp/index_en.html, accessed May 2012).

Registration is required, but there is free access to this database of physical properties, chemical structures and literature citations on some 12 000 polymers and 6500 monomers.

A.8.2
Polymer Handbook (J. Brandrup, EH Immergut *et al.*)

Concise reference to nomenclature, polymerizing and physico-chemical properties of polymers. Also available online (licensed).

A.9
Spectra

A.9.1
Printed Data Collections

Aldrich and Sadtler created extensive data collections with ^{13}C and ^{1}H NMR spectra, as well as FTIR and IR spectra of organic and inorganic compounds. Access is easiest via the Formula tab. Data collections with mass spectra are available from Wiley, the Royal Society of Chemistry ("Eight Peak Index of Mass Spectra") and NIST. UV spectra collections have been produced by Wiley ("Organic Electronic Spectral Data") and TRC.

A.9.2 Online Products Requiring a License

- For around 5 million compounds, CrossFire Beilstein and Gmelin provide references to data of NMR, IR, MS and other spectroscopic methods.
- Chemical Abstracts, SciFinder (including Scholar and Web) contains about 500 000 measured ^{13}C-NMR, IR and mass spectra of Wiley and about 24 million calculated ^{1}H NMR spectra.
- ChemGate, Wiley, extensive database of IR, NMR and mass spectra, suitable for searching for fingerprints and for spectra prediction.
- KnowItAllU, Bio-Rad Laboratories, electronic counterparts to the Sadtler data collections mentioned above.

A.9.3 Free Access Online Products

NIST Chemistry Webbook, http://webbook.nist.gov/chemistry (accessed August 2012).

NMRShiftDB, http://www.nmrshiftdb.org (accessed August 2012), database of measured chemical shifts of organic compounds and for the prediction of NMR spectra (^{13}C, ^{1}H and other nuclei).

Spectral Database for Organic Compounds (http://riodb01.ibase.aist.go.jp/sdbs/cgi-bin/cre_index.cgi?lang=eng, accessed May 2012).

MS, ^{13}C and ^{1}H-NMR, IR, Raman and ESR spectra of the Japanese National Institute of Advanced Industrial Science and Technology (AIST), http://www.aist.go.jp/index_en.html (accessed August 2012), for approximately 30 000 compounds.

A.10 Toxicology and Safety

A useful overview of smaller databases on toxicology and safety is provided at http://www.infochembio.ethz.ch/cdbib.php (accessed May 2012).

TOXNET (http://toxnet.nlm.nih.gov/, accessed May 2012).

Comprehensive, freely accessible web platform of the National Library of Medicine comprising several databases on the toxicology of chemical compounds.

Appendix B
Statistical Tables

Table B.1 Student's (t) factors (DIN 1319-3).

Level of confidence				
***P* (%)**	**50%**	**90%**	**95%**	**99%**
α	**0.5**	**0.1**	**0.05**	**0.01**
Degrees of freedom (f)				
1	1.000	6.314	12.706	63.657
2	0.816	2.92	4.303	9.925
3	0.765	2.353	3.182	5.841
4	0.741	2.132	2.776	4.604
5	0.727	2.015	2.571	4.032
6	0.718	1.943	2.447	3.707
7	0.711	1.895	2.365	3.500
8	0.706	1.860	2.306	3.355
9	0.703	1.833	2.262	3.25
10	0.700	1.812	2.228	3.169
15	0.691	1.753	2.131	2.947
20	0.687	1.725	2.086	2.845
25	0.684	1.708	2.068	2.787
30	0.683	1.697	2.042	2.750
40	0.681	1.684	2.021	2.704
60	0.679	1.671	2.000	2.660
120	0.677	1.658	1.980	2.617
∞	0.674	1.645	1.960	2.576

n number of measurements.
For mean of values $f = n - 1$.
For data derived from linear regression $f = n - 2$.
Example:
t-factor ($n = 10$, $P = 95\%$) = 2.26.
t-factor ($n = 3$, $P = 99\%$) = 9.93.

Practical Instrumental Analysis: Methods, Quality Assurance and Laboratory Management, First Edition.
Sergio Petrozzi.

Table B.2 Grubbs comparative values (according to DIN 53804-1:2002).

Level of confidence	90%	95%	99%
n			
3	1.148	1.153	1.155
4	1.425	1.463	1.492
5	1.602	1.672	1.749
6	1.729	1.822	1.944
7	1.828	1.938	2.097
8	1.909	2.032	2.221
9	1.977	2.110	2.323
10	2.036	2.176	2.410
11	2.088	2.234	2.485
12	2.134	2.285	2.550
13	2.175	2.331	2.607
14	2.213	2.371	2.659
15	2.247	2.409	2.705
16	2.279	2.443	2.747
17	2.309	2.475	2.785
18	2.335	2.504	2.821
19	2.361	2.532	2.854
20	2.385	2.557	2.884

Table B.3 (Phi) $\Phi_{n,\alpha}$ factors for flash estimations (according to DIN 32 645).

Number of measurements (n)	$\Phi_{n,05}$	to$\Phi_{n,0,01}$	$\Phi_{n,0,005}$
4	2.6	5.1	6.5
5	2.3	4.1	5.0
6	2.2	3.6	4.4
7	2.1	3.4	4.0
8	2	3.2	3.7
9	2	3.1	3.5
10	1.9	3	3.4
11	1.9	2.9	3.3
12	1.9	2.9	3.2

Table B.4 Factors for confidence interval estimations (according to DIN 32 645).

Degrees of freedom (f)	K_U	K_O
2	0.52	6.38
3	0.57	3.73
4	0.60	2.87
5	0.62	2.45
6	0.64	2.20
7	0.66	2.04
8	0.68	1.92
9	0.69	1.83
10	0.70	1.75

K_U factor for lower limit.
K_O factor for upper limit.

Appendix C
Obligatory Declaration for Students

Practical Instrumental Analysis: Methods, Quality Assurance and Laboratory Management, First Edition.
Sergio Petrozzi.

Obligatory declaration for students

This form must be completed and signed by students before the start of the practical training.

Enter the locations of the following emergency facilities below and remember these locations:

Emergency showers ....

Fire blankets ....

Fire extinguishers ....

Water fire extinguishers ....

Emergency stop switch for gas ....

Phones ....

First aid kit ....

Eye showers ....

Emergency exits ....

Emergency meeting point ....

Safety instructions

The eyes may be injured by mechanical action, heat, chemicals or energy beams. Experience shows that especially the handling of chemical substances and also work with energy beams (lasers) may cause eye injury. Such injuries can be prevented in most cases by wearing protective glasses. Safety glasses have relatively large, shatterproof lenses and a shock-proof, heat-resistant frame. Of course safety-conscious behavior is also vital in dealing with chemicals or energy emissions. When working with large quantities of corrosive acids or alkalis, self-protective goggles and a face shield must be worn. Lenses should not be worn when working with chemicals because they enhance the action of chemicals between the lens and the eye and the removal of the lens is made very difficult by certain chemicals.

Protective clothing such as a lab coat is important in order to protect the whole body as well as clothing against irritating or harmful effects such as those of dust or chemicals; it must also be quickly removable in case of contamination or a fire. In addition long trousers and closed shoes are required for practical training (even in summer temperatures).

I have read the laboratory safety manual carefully; I understand it and agree to comply with the rules it contains.

Date: .... Signature..

Appendix D
The International System of Units (SI) – and the "New SI"

(Source: Bureau International des Poids et Mesures (BIPM); http://www.bipm.org/en/si/, accessed May 2012).

The 11th General Conference on Weights and Measures (*CGPM*), 1960, adopted the name *Système International d'Unités* (International System of Units, international), for the recommended practical system of units of measurement.

The 11th CGPM laid down rules for the prefixes, the derived units, and other matters. The *base units* are a choice of seven well-defined units which by convention are regarded as dimensionally independent: the metre, the kilogram, the second, the ampere, the kelvin, the mole, and the candela. *Derived units* are those formed by combining base units according to the algebraic relations linking the corresponding quantities. The names and symbols of some of the units thus formed can be replaced by special names and symbols which can themselves be used to form expressions and symbols of other derived units.

There are seven *base units* of the SI:

Metre	**m**
The metre is the length of the path travelled by light in vacuum during a time interval of 1/299 792 458 of a second.	
Kilogram	**kg**
The kilogram is the unit of mass; it is equal to the mass of the international prototype of the kilogram.	
Second	**s**
The second is the duration of 9 192 631 770 periods of the radiation corresponding to the transition between the two hyperfine levels of the ground state of the caesium 133 atom.	

Practical Instrumental Analysis: Methods, Quality Assurance and Laboratory Management, First Edition.
Sergio Petrozzi.

Ampere	**A**

The ampere is that constant current which, if maintained in two straight parallel conductors of infinite length, of negligible circular cross-section, and placed 1 m apart in vacuum, would produce between these conductors a force equal to 2×10^{-7} newton per metre of length.

Kelvin	**K**

The kelvin, unit of thermodynamic temperature, is the fraction 1/273.16 of the thermodynamic temperature of the triple point of water.

Mole	**mol**

1) The mole is the amount of substance of a system which contains as many elementary entities as there are atoms in 0.012 kilogram of carbon 12.
2) When the mole is used, the elementary entities must be specified and may be atoms, molecules, ions, electrons, other particles, or specified groups of such particles.

Candela	**cd**

The candela is the luminous intensity, in a given direction, of a source that emits monochromatic radiation of frequency 540×10^{12} hertz and that has a radiant intensity in that direction of 1/683 watt per steradian.

Coherent Derived Units The number of quantities in science is without limit, and it is not possible to provide a complete list of derived quantities and derived units. However, Table D.1 lists some examples of derived quantities, and the corresponding coherent derived units expressed directly in terms of base units.

Units with Special Names and Symbols; Units That Incorporate Special Names and Symbols For convenience, certain coherent derived units have been given special names and symbols. There are 22 such units. These special names and symbols may themselves be used in combination with the names and symbols for base units and for other derived units to express the units of other derived quantities. The special names and symbols are simply a compact form for the expression of combinations of base units that are used frequently, but in many cases they also serve to remind the reader of the quantity involved. The SI prefixes may be used with any of the special names and symbols, but when this is done the resulting unit will no longer be coherent. The values of several different quantities may be expressed using the same name and symbol for the SI unit. Thus for the quantity heat capacity as well as the quantity entropy, the SI unit is the joule per kelvin. Similarly for the base quantity electric current as well as the derived quantity magnetomotive force, the SI unit is the ampere. It is therefore important not to use the unit alone to specify the quantity. This applies not only to scientific and technical texts, but also, for example, to measuring instru-

Table D.1 Examples of coherent derived units in the SI expressed in terms of base units.

Derived quantity		SI coherent derived unit	
Name	**Symbol**	**Name**	**Symbol**
Area	A	Square metre	m^2
Volume	V	Cubic metre	m^3
Speed, velocity	υ	Metre per second	m/s
Acceleration	a	Metre per second squared	m/s^2
Wavenumber	σ, $\tilde{\upsilon}$	Reciprocal metre	m^{-1}
Density, mass density	ρ	Kilogram per cubic metre	kg/m^3
Surface density	ρA	Kilogram per square metre	kg/m^2
Specific volume	v	Cubic metre per kilogram	m^3/kg
Current density	j	Ampere per square metre	A/m^2
Magnetic field strength	H	Ampere per metre	A/m
Amount concentration[a], Concentration	c	Mole per cubic metre	mol/m^3
Mass concentration	ρ, γ	Kilogram per cubic metre	kg/m^3
Luminance	L_v	Candela per square metre	cd/m^2
Refractive index[b]	n	One	1
Relative permeability[b]	μr	One	1

a) In the field of clinical chemistry this quantity is also called substance concentration.
b) These are dimensionless quantities, or quantities of dimension one, and the symbol "1" for the unit (the number "one") is generally omitted in specifying the values of dimensionless quantities.

ments (i.e., an instrument read-out should indicate both the unit and the quantity measured).

A derived unit can often be expressed in different ways by combining base units with derived units having special names. Joule, for example, may formally be written newton metre, or kilogram metre squared per second squared. This, however, is an algebraic freedom to be governed by common sense physical considerations; in a given situation some forms may be more helpful than others. In practice, with certain quantities, preference is given to the use of certain special unit names, or combinations of unit names, to facilitate the distinction between different quantities having the same dimension. When using this freedom, one may recall the process by which the quantity is defined. For example, the quantity torque may be thought of as the cross product of force and distance, suggesting the unit newton metre, or it may be thought of as energy per angle, suggesting the unit joule per radian. The SI unit of frequency is given as the hertz, implying the unit cycles per second; the SI unit of angular velocity is given as the radian per second; and the SI unit of activity is designated the becquerel, implying the unit counts per second. Although it would be formally correct to write all three of these units as the reciprocal second, the use of the different names emphasises the different nature of the quantities concerned. Using the unit radian per second for angular velocity, and hertz for frequency, also emphasizes that the numerical value of the angular velocity in radian per second is 2π times the numerical value of the corresponding frequency in hertz.

In the field of ionizing radiation, the SI unit of activity is designated the becquerel rather than the reciprocal second, and the SI units of absorbed dose and dose equivalent are designated the gray and the sievert, respectively, rather than the joule per kilogram. The special names becquerel, gray, and sievert were specifically introduced because of the dangers to human health that might arise from mistakes involving the units reciprocal second and joule per kilogram, in case the latter units were incorrectly taken to identify the different quantities involved.

Units for Dimensionless Quantities, Also Called Quantities of Dimension One Certain quantities are defined as the ratio of two quantities of the same kind, and are thus dimensionless, or have a dimension that may be expressed by the number one. The coherent SI unit of all such dimensionless quantities, or quantities of dimension one, is the number one, since the unit must be the ratio of two identical SI units. The values of all such quantities are simply expressed as numbers, and the unit one is not explicitly shown. Examples of such quantities are refractive index, relative permeability, and friction factor. There are also some quantities that are defined as a more complex product of simpler quantities in such a way that the product is dimensionless. Examples include the "characteristic numbers" like the Reynolds number $Re = \rho \upsilon / \eta$, where ρ is mass density, η is dynamic viscosity, υ is speed, and L is length. For all these cases the unit may be considered as the number one, which is a dimensionless derived unit.

Another class of dimensionless quantities are numbers that represent a count, such as a number of molecules, degeneracy (number of energy levels), and partition function in statistical thermodynamics (number of thermally accessible states). All of these counting quantities are also described as being dimensionless, or of dimension one, and are taken to have the SI unit one, although the unit of counting quantities cannot be described as a derived unit expressed in terms of the base units of the SI. For such quantities, the unit one may instead be regarded as a further base unit.

Non-SI Units Accepted for Use with the SI, and Units Based on Fundamental Constants The International Committee for Weights and Measures (CIPM), 2004, has revised the classification of non-SI units from that in the previous (7th) edition of his Brochure.

Table D.2 gives non-SI units that are accepted for use with the International System by the CIPM, because they are widely used with the SI in matters of everyday life. Their use is expected to continue indefinitely, and each has an exact definition in terms of an SI unit. Table D.3 contains units that are used only in special circumstances. The units in Table D.3 are related to fundamental constants, and their values have to be determined experimentally.

Tables D.4 and D.5 (not shown here because they are for interest only in very specialised fields) contain units that have exactly defined values in terms of SI

Table D.2 Non-SI units accepted for use with the International System of Units.

Quantity	Name of unit	Symbol for unit	Value in SI units
Time	Minute	min	1 min = 60 s
	Hour[a)]	h	1 h = 60 min = 3600 s
	Day	d	1 d = 24 h = 86 400 s
Plane angle	Degree[b,c)]	°	$1° = (\pi/180)$ rad
	Minute	′	$1' = (1/60)° = (\pi/10800)$ rad
	Second[d)]	″	$1'' = (1/60)' = (\pi/648000)$ rad
Area	Hectare[e)]	ha	$1\,\text{ha} = 1\,\text{hm}^2 = 10^4\,\text{m}^2$
Volume	Litre[f)]	L, l	$1\,\text{L} = 1\,\text{l} = 1\,\text{dm}^3 = 10^3\,\text{cm}^3$
Mass	Tonne[g)]	t	$1\,\text{t} = 10^3\,\text{kg}$

a) The symbol for this unit is included in Resolution 7 of the 9th CGPM (1948; CR, 70).
b) ISO 31 recommends that the degree be divided decimally rather than using the minute and the second. For navigation and surveying, however, the minute has the advantage that one minute of latitude on the surface of the Earth corresponds (approximately) to one nautical mile.
c) The gon (or grad, where grad is an alternative name for the gon) is an alternative unit of plane angle to the degree, defined as $(\pi/200)$ rad. Thus there are 100 gon in a right angle. The potential value of the gon in navigation is that because the distance from the pole to the equator of the Earth is approximately 10 000 km, 1 km on the surface of the Earth subtends an angle of one centigon at the centre of the Earth. However the gon is rarely used.
d) For applications in astronomy, small angles are measured in arcseconds (i.e., seconds of plane angle), denoted as or ″, milliarcseconds, microarcseconds, and picoarcseconds, denoted mas, µas, and pas, respectively, where arcsecond is an alternative name for second of plane angle.
e) The unit hectare, and its symbol ha, were adopted by the CIPM in 1879 (PV, 1879, 41). The hectare is used to express land area.
f) The litre, and the symbol lower-case l, were adopted by the CIPM in 1879 (PV, 1879, 41). The alternative symbol, capital L, was adopted by the 16th CGPM (1979, Resolution 6; CR, 101 and Metrologia, 1980, 16, 56–57) in order to avoid the risk of confusion between the letter l (el) and the numeral 1 (one).
g) The tonne, and its symbol t, were adopted by the CIPM in 1879 (PV, 1879, 41). In English speaking countries this unit is usually called "metric ton".

units, and are used in particular circumstances to satisfy the needs of commercial, legal, or specialized scientific interests. ***It is likely that these units will continue to be used for many years***. Many of these units are also important for the interpretation of older scientific texts.

Each of the Tables D.2 and D.3 is discussed in turn below.

Table D.2 includes the traditional units of time and angle. It also contains the hectare, the litre, and the tonne, which are all in common everyday use throughout the world, and which differ from the corresponding coherent SI unit by an integer power of ten.

The SI prefixes are used with several of these units, but not with the units of time.

Table D.3 contains units whose values in SI units have to be determined experimentally, and thus have an associated uncertainty. Except for the astronomical unit, all other units in Table D.3 are related to fundamental physical constants. The first three units, the non-SI units electronvolt, symbol eV, dalton or unified atomic mass

Table D.3 Non-SI units whose values in SI units must be obtained experimentally.

Quantity	Name of unit	Symbol for unit	Value in SI units[a)]
Units accepted for use with the SI			
Energy	Electronvolt[b)]	eV	$1\,\text{eV} = 1.602\,176\,53\,(14) \times 10^{-19}\,\text{J}$
Mass	Dalton,[c)]	Da	$1\,\text{Da} = 1.660\,538\,86\,(28) \times 10^{-27}\,\text{kg}$
	Unified atomic	u	$1\,\text{u} = 1\,\text{Da}$
	Mass unit		
Length	Astronomical unit[d)] ua	$1\,\text{ua} = 1.495\,978\,706\,91\,(6) \times 10^{11}\,\text{m}$	
Natural units (n.u.)			
Speed	n.u. of speed (speed of light in vacuum)	c_0	$299\,792\,458\,\text{m/s}$ (exact)
Action	n.u. of action (reduced Planck constant)	$\hbar$	$1.054\,571\,68\,(18) \times 10^{-34}\,\text{J s}$
Mass	n.u. of mass (electron mass)	m_e	$9.109\,3826\,(16) \times 10^{-31}\,\text{kg}$
Time	n.u. of time	$\hbar/(m_e c_0^2)$	$1.288\,088\,6677\,(86) \times 10^{-21}\,\text{s}$
Atomic units (a.u.)			
Charge	a.u. of charge, (elementary charge)	e	$1.602\,176\,53\,(14) \times 10^{-19}\,\text{C}$
Mass	a.u. of mass, (electron mass)	m_e	$9.109\,3826\,(16) \times 10^{-31}\,\text{kg}$
Action	a.u. of action, (reduced Planck constant)	$\hbar$	$1.054\,571\,68\,(18) \times 10^{-34}\,\text{J s}$
Length	a.u. of length, bohr (Bohr radius)	a_0	$0.529\,177\,2108\,(18) \times 10^{-10}\,\text{m}$
Energy	a.u. of energy, hartree E_h (Hartree energy)	$4.359\,744\,17\,(75) \times 10^{-18}\,\text{J}$	
Time	a.u. of time	$\hbar/E_h$	$2.418\,884\,326\,505\,(16) \times 10^{-17}\,\text{s}$

a) The values in SI units of all units in this table, except the astronomical unit, are taken from the 2002 Committee on Data for Science and Technology (CODATA) set of recommended values of the fundamental physical constants, P.J. Mohr and B.N. Taylor, *Rev. Mod. Phys.*, **2005**, 77, 1–107. The standard uncertainty in the last two digits is given in parentheses.

b) The electronvolt is the kinetic energy acquired by an electron in passing through a potential difference of one volt in vacuum. The electronvolt is often combined with the SI prefixes.

c) The dalton (Da) and the unified atomic mass unit (u) are alternative names (and symbols) for the same unit, equal to 1/12 times the mass of a free carbon 12 atom, at rest and in its ground state. The dalton is often combined with SI prefixes, for example, to express the masses of large molecules in kilodaltons, kDa, or megadaltons, MDa, or to express the values of small mass differences of atoms or molecules in nanodaltons, nDa, or even picodaltons, pDa.

d) The astronomical unit is approximately equal to the mean Earth-Sun distance. It is the radius of an unperturbed circular Newtonian orbit about the Sun of a particle having infinitesimal mass, moving with a mean motion of 0.017 202 098 95 radians per day (known as the Gaussian constant). The value given for the astronomical unit is quoted from the International Earth Rotation and Reference System Service (IERS) Conventions 2003 (D.D. McCarthy and G. Petit eds., IERS Technical Note 32, Frankfurt am Main: Verlag des Bundesamts für Kartographie und Geodäsie, **2004**, 12). The value of the astronomical unit in metres comes from the JPL ephemerides DE403 (Standish E.M., Report of the International Astronomical Union (IAU) WGAS Sub-Group on Numerical Standards, Highlights of Astronomy, Appenzeller ed., Dordrecht: Kluwer Academic Publishers, **1995**, 180–184).

unit, symbol Da or u, respectively, and the astronomical unit, symbol ua, have been accepted for use with the SI by the CIPM. The units in Table D.3 play important roles in a number of specialized fields in which the results of measurements or calculations are most conveniently and usefully expressed in these units. For the electronvolt and the dalton the values depend on the elementary charge e and the Avogadro constant NA, respectively. There are many other units of this kind, because there are many fields in which it is most convenient to express the results of experimental observations or of theoretical calculations in terms of fundamental constants of nature. The two most important of such unit systems based on fundamental constants are the natural unit (n.u.) system used in high energy or particle physics, and the atomic unit (a.u.) system used in atomic physics and quantum chemistry. In the n.u. system, the base quantities for mechanics are speed, action, and mass, for which the base units are the speed of light in vacuum c_0, the Planck constant h divided by 2π, called the reduced Planck constant with symbol $\hbar$, and the mass of the electron me, respectively. In general these units are not given any special names or symbols but are simply called the n.u. of speed, symbol c_0, the n.u. of action, symbol ħ, and the n.u. of mass, symbol me. In this system, time is a derived quantity and the n.u. of time is a derived unit equal to the combination of base units $\hbar/m_e c_0^2$. Similarly, in the a.u. system, any four of the five quantities charge, mass, action, length, and energy are taken as base quantities. The corresponding base units are the elementary charge e, electron mass me, action ħ, Bohr radius (or bohr) a_0, and Hartree energy (or hartree) E_h, respectively. In this system, time is again a derived quantity and the a.u. of time a derived unit, equal to the combination of units $\hbar/E_h$. Note that $a_0 = \alpha/(4\pi R\infty)$, where α is the fine-structure constant and $R\infty$ is the Rydberg constant; and $E_h = e^2/(4\pi\varepsilon_0 a_0) = 2R\infty h c_0 = \alpha^2 m e c_0^2$, where ε_0 is the electric constant and has an exact value in the SI.

For information, these ten natural and atomic units and their values in SI units are also listed in Table D.3. Because the quantity systems on which these units are based differ so fundamentally from that on which the SI is based, they are not generally used with the SI, and the CIPM has not formally accepted them for use with the International System. To ensure understanding, the final result of a measurement or calculation expressed in natural or atomic units should also always be expressed in the corresponding SI unit. Natural units (n.u.) and atomic units (a.u.) are used only in their own special fields of particle and atomic physics, and quantum chemistry, respectively. Standard uncertainties in the least significant digits are shown in parenthesis after each numerical value.

Other Non-SI Units Not Recommended for Use There are many more non-SI units, which are too numerous to list here, which are either of historical interest, or are still used but only in specialized fields (for example, the barrel of oil) or in particular countries (the inch, foot, and yard). The CIPM can see no case for continuing to use these units in modern scientific and technical work. However, it is clearly a matter of importance to be able to recall the relation of these units to the corresponding SI units, and this will continue to be true for many years. The CIPM has therefore decided to compile a list of the conversion factors to the SI for such units and to make this available on the BIPM website at http://www.bipm.org/en/si/si_brochure/chapter4/conversion_factors.html (accessed May 2012).

Rules and Style Conventions for Expressing Values of Quantities Bureau International des Poids et Mesures, The International System of Units(SI), 8th edition 2006 Organisation Intergouvernementale de la Convention du Mètre

Value and Numerical Value of a Quantity, and the Use of Quantity Calculus The value of a quantity is expressed as the product of a number and a unit, and the number multiplying the unit is the numerical value of the quantity expressed in that unit. The numerical value of a quantity depends on the choice of unit. Thus the value of a particular quantity is independent of the choice of unit, although the numerical value will be different for different units. Symbols for quantities are generally single letters set in an italic font, although they may be qualified by further information in subscripts or superscripts or in brackets. Thus C is the recommended symbol for heat capacity, C_{m} for molar heat capacity, $C_{\mathrm{m},p}$ for molar heat capacity at constant pressure, and $C_{\mathrm{m},v}$ for molar heat capacity at constant volume.

The same value of a speed $v = dx/dt$ *of a particle might be given by either of the expressions*
$v = 25\,m/s = 90\,km/h$, *where 25 is the numerical value of the speed in the unit metres per second, and 90 is the numerical value of the speed in the unit kilometres per hour.*

Recommended names and symbols for quantities are listed in many standard references, such as the ISO Standard 31 *Quantities and Units,* the IUPAP SUNAMCO Red Book *Symbols, Units and Nomenclature in Physics,* and the IUPAC Green Book Quantities, *Units and Symbols in Physical Chemistry.*

Quantity Symbols and Unit Symbols Just as the quantity symbol should not imply any particular choice of unit, the unit symbol should not be used to provide specific information about the quantity, and should never be the sole source of information on the quantity. Units are never qualified by further information about the nature of the quantity; any extra information on the nature of the quantity should be attached to the quantity symbol and not to the unit symbol.

For example:
The maximum electric potential difference is Umax = 1000 V, but not $U = 1000\ V_{max}$.
The mass fraction of copper in the sample of silicon is $w(Cu) = 1.3 \times 10^{-6}$, *but not* $1.3 \times 10^{-6}\ w/w$.

Formatting the Value of a Quantity The numerical value always precedes the unit, and a space is always used to separate the unit from the number. Thus the value of the quantity is the product of the number and the unit, the space being regarded as a multiplication sign (just as a space between units implies multiplication). The only exceptions to this rule are for the unit symbols for degree, minute, and second for plane angle, °, ′, and ″, respectively, for which no space is left between the numerical value and the unit symbol.

$m = 12.3$ g where m is used as a symbol for the quantity mass, but $\varphi = 30° 22′ 8″$, where φ is used as a symbol for the quantity plane angle.

This rule means that the symbol ° for the degree Celsius is preceded by a space when one expresses values of Celsius temperature *t*.

$t = 30.2$ °C, but not $t = 30.2$°C, nor $t = 30.2$° C

Even when the value of a quantity is used as an adjective, a space is left between the numerical value and the unit symbol. Only when the name of the unit is spelled out would the ordinary rules of grammar apply, so that in English a hyphen would be used to separate the number from the unit.

a 10 kΩ resistor   a 35-millimetre film

In any one expression, only one unit is used. An exception to this rule is in expressing the values of time and of plane angles using non-SI units. However, for plane angles it is generally preferable to divide the degree decimally. Thus one would write 22.20° rather than 22° 12′, except in fields such as navigation, cartography, astronomy, and in the measurement of very small angles.

$l = 10.234$ m, but not $l = 10$ m 23.4 cm

Formatting Numbers, and the Decimal Marker The symbol used to separate the integral part of a number from its decimal part is called the decimal marker. Following the 22nd CGPM (2003, Resolution 10), the decimal marker "shall be either the point on the line or the comma on the line". The decimal marker chosen should be that which is customary in the context concerned. If the number is between +1 and −1, then the decimal marker is always preceded by a zero.

−0.234, but not −.234

Following the 9th CGPM (1948, Resolution 7) and the 22nd CGPM (2003, Resolution 10), for numbers with many digits the digits may be divided into groups of three by a thin space, in order to facilitate reading. Neither dots nor commas are inserted in the spaces between groups of three. However, when there are only four digits before or after the decimal marker, it is customary not to use a space to isolate a single digit. The practice of grouping digits in this way is a matter of choice; it is not always followed in certain specialized applications such as engineering drawings, financial statements, and scripts to be read by a computer.

For numbers in a table, the format used should not vary within one column.

43 279.168 29, but not 43,279.168,29 either 3279.1683 or 3 279.168 3

Multiplying or Dividing Quantity Symbols, the Values of Quantities, or Numbers When multiplying or dividing quantity symbols any of the following methods may be used: *ab, a b, a . b, a x b, a/b, b a,* $a\ b^{-1}$.

When multiplying the value of quantities either a multiplication sign, ×, or brackets should be used, not a half-high (centred) dot. When multiplying numbers only the multiplication sign, ×, should be used.

When dividing the values of quantities using a solidus, brackets are used to remove ambiguities.

Examples:
$F = ma$ for force equals mass times acceleration
$(53\,m/s) \times 10.2\,s$ or $(53\,m/s)(10.2\,s)$
25×60.5 but not $25.\ 60.5$
$(20\,m)/(5\,s) = 4\,m/s$
$(a/b)/c$, not $a/b/c$

Stating Values of Dimensionless Quantities or Quantities of Dimension One As discussed, the coherent SI unit for dimensionless quantities, also termed quantities of dimension one, is the number one, symbol 1. Values of such quantities are expressed simply as numbers. The unit symbol 1 or unit name "one" are not explicitly shown, nor are special symbols or names given to the unit one, apart from a few exceptions as follows. For the quantity plane angle, the unit one is given the special name radian, symbol rad, and for the quantity solid angle, the unit one is given the special name steradian, symbol sr. For the logarithmic ratio quantities, the special names neper, symbol Np, bel, symbol B, and decibel, symbol dB, are used.

Because SI prefix symbols can neither be attached to the symbol 1 nor to the name "one", powers of 10 are used to express the values of particularly large or small dimensionless quantities.

In mathematical expressions, the internationally recognized symbol % (percent) may be used with the SI to represent the number 0.01. Thus, it can be used to express the values of dimensionless quantities. When it is used, a space separates the number and the symbol %. In expressing the values of dimensionless quantities in this way, the symbol % should be used rather than the name "percent".

$x_B = 0.0025 = 0.25\%$, where x_B is the quantity symbol for amount fraction (mole fraction) of entity B.

The mirror reflects 95% of the incident photons.

In written text, however, the symbol % generally takes the meaning of "parts per hundred".

$\varphi = 3.6\%$, but not $\varphi = 3.6\%\ (V/V)$, where φ denotes volume fraction.

Phrases such as "percentage by mass", "percentage by volume", or "percentage by amount of substance" should not be used; the extra information on the quantity should instead be conveyed in the name and symbol for the quantity.

$x_B = 2.5 \times 10^{-3} = 2.5\,mmol/mol$

In expressing the values of dimensionless fractions (e.g., mass fraction, volume fraction, relative uncertainties), the use of a ratio of two units of the same kind is sometimes useful.

$u_r(U) = 0.3\,\mu V/V$, where $u_r(U)$ is the relative uncertainty of the measured voltage U.

The term "ppm", meaning 10^{-6} relative value, or 1 in 10^6, or parts per million, is also used. This is analogous to the meaning of percent as parts per hundred. The terms "parts per billion", and "parts per trillion", and their respective abbreviations "ppb", and "ppt", are also used, but their meanings are language dependent. For this reason the terms ppb and ppt are best avoided (In English-speaking countries, a billion is now generally taken to be 10^9 and a trillion to be 10^{12}; however, a billion may still sometimes be interpreted as 10^{12} and a trillion as 10^{18}. The abbreviation ppt is also sometimes read as parts per thousand, adding further confusion).

When any of the terms %, ppm, etc., are used it is important to state the dimensionless quantity whose value is being specified.

Appendix E
Evaluation Guide for Formal Reports

Group:

Schedule:

Project:

Theory **Remarks**

Literature processing _____/5
Theoretical Part/Questions _____/12
Suggestions _____/3

Experimental work

Experiments _____/3
Sample preparation/wet chemistry _____/10
Measurements (qualitative, quantitative) _____/17
Quantitative determinations (unknown samples) _____/10

Report

Title _____/2
Abstract/Summary _____/7
Introduction/Motivation _____/8
Description of practical work _____/7
Evaluations/Results _____/7
Validation and statistical evaluation of results _____/7
Discussion and Outlook _____/7
Bibliography _____/5
Sources _____/5
External form _____/5

Main focus of the evaluation

1) *Documentation of the experiment*
   It should be documented in such detail that another person could easily repeat the work.
2) *Traceability of raw data*

Practical Instrumental Analysis: Methods, Quality Assurance and Laboratory Management, First Edition.
Sergio Petrozzi.

3) *Evidence how results have been calculated*
Assessment of results using statistical methods

Overall evaluation

Theory: _____/Experimental work: _____/Report: _____/

TOTAL: _____/**120**

Mark

Final remarks

Signature/Date

Appendix F
Safety in the Analytical Laboratory

F.1
General Precautionary Measures

F.1.1
Measures for Personal Protection

Accidents can be avoided. Information will be provided in this section. In each case, the *"3 Step Principle"* will be followed:

Step 1: ELIMINATION OF DANGER
(Search for alternatives)
Step 2: CONTAINMENT OF DANGER
(Safety precautions)
Step 3: PROTECTION OF PEOPLE
(Behavioral guidelines)

F.1.2
Eye Protection

Mechanical effect, heat, chemicals or energy emissions can hurt the eyes. Experience has shown that the handling of chemical materials, above all, and in more recent times, work with energy beams (lasers) can lead to eye injuries. In most cases, such injuries can be prevented by the use of suitable *safety glasses.*

Safety glasses have relatively large, shatterproof lenses and an impact and heat resistant frame. When handling chemicals or lasers, it is important that safety conscious behavior, above all, is observed. These glasses protect your eyes better than the usual correction safety glasses since they provide side protection in addition to the safety lenses and frame already mentioned. If working with larger quantities of strongly corroding acids or caustic solutions, you should use closed eye goggles and carry a face shield.

You should *not wear contact lenses* when handling chemicals, since they can strengthen the effect of chemicals penetrating between the lens and the eyes and, in some cases, certain chemicals make the removal of the lenses difficult! When

Practical Instrumental Analysis: Methods, Quality Assurance and Laboratory Management, First Edition.
Sergio Petrozzi.

operating lasers, special laser eye protectors will be provided which provide protection for the specific range of wavelengths of the laser.

F.1.3
Skin Protection

Mechanical effects such as heat, cold, chemicals, microorganisms or energy beams can damage the skin. In a laboratory you can be exposed to many sources of danger. Skin protection, which consists at the very least of hand cream or gloves, must therefore be adapted to the respective sources of danger. Here, too, safety conscious behavior must first be observed. One of the most frequent mechanical injuries, cuts to the hands, can be prevented by the use of gloves while handling glass equipment and also by correct cutting techniques. When handling hazardous materials, thorough washing of the skin before and after the work in the laboratory is important.

F.1.4
Protective Clothing

Protective clothing, such as a laboratory coat, is important in order to protect the whole body and clothes against the disturbing or harmful effects of dirt and chemicals. In the case of contamination or fire, it should be rapidly and easily removable. In clean rooms, protective clothing is a necessity in order to prevent the contamination of the rooms by agents such as road dust, etc. Lab coats are therefore mandatory in clean rooms. In chemistry practical courses and workshops, the use of a laboratory coat is either mandatory or recommended.

F.1.5
Hearing Protection

Ear protectors are sometimes necessary, particularly in workshops or within the range of machine rooms. Remember that hearing damage is always irreversible. There are various earphones or plugs, which provide protection for the hearing. It is important that they are consistently worn, otherwise their protective effect is greatly reduced. For example, someone who works with a noise of 100 dB and wears no ear protection during 10% of his working time is still exposed to a (harmful) noise pollution of 90 dB. The Swiss National Accident Insurance Fund (SUVA) has classified a constant sound pressure level starting at 88 dB and lasting at least 8 hours per day as damaging to the hearing.

F.1.6
Respiratory Protection

The respiratory system and lungs can be damaged by dust, chemical gas or steam, and different types of gases. Inhalation of gas or steam can also be harmful to the central or peripheral nervous system or to other organs such as the liver or kidney. Here, too, safety conscious behavior is the cardinal rule for the prevention of accidents. For example, chemical reactions which can develop harmful steams or gases should be performed under the fume hood, with the ventilation switched on and the front shields closed whenever possible. Find out, especially before starting a reaction, which substance could develop; some dangerous gases such as carbon monoxide are odorless! With mechanical work you can, for example, moisten the workpieces, in order to prevent the creation of dust. If this is not possible, a ventilation source must be provided. The possibilities for respiratory protection are easy enough, from a simple dust respirator over half masks to full masks with the appropriate filter guards. If the oxygen supply is insufficient, this can be supplemented with the use of compressed air breathing masks.

F.2 First Aid

F.2.1 Rescue

As an initial course of action, the victim should be removed from the source of danger and immediate life-saving measures should be taken!

Consider self-preservation as the cardinal rule in each case!

- Electrical accidents:

  Turn off the power source. If this is not immediately possible, separate the victim from the electrified area by using a non-conductive material (such as a wooden slat) or the pull the victim away from the site by grabbing his or her clothes. You should insulate yourself by standing on an earthed base (such as a dry board or thick newspaper) making sure to touch nothing (wall, other person etc.)

- Accidents with gas or poisonous steam:

  Rescue only with the use of respiratory protection. Move victim immediately to fresh air.

  Continue further, if necessary, to resuscitation measures.

- Accidents with fire:

  Put out any flames on the victim and extinguish the fire (using an emergency shower, dousing blanket, or towels).

- Caustic burns:

  Immediately remove contaminated clothing. See also CAUSTIC BURNS.

Observe the Following:

- Examine breathing and pulse
- Do not allow the victim to become cold (cover with a blanket).
- Clear any obstacles blocking the air passage.
- Free the victim of restraining clothes.

F.2.2
Alerting Emergency Personnel

Report every accident immediately to a supervisor. If necessary, call the appropriate *emergency number* (possible from nearly all internal phones).

F.2.3
Treatment of Unconscious Victim

In order to avoid possible asphyxiation from foreign bodies or the tongue falling to the rear of the mouth, the *unconscious victim* must be immediately *positioned on his or her side* (see below). One should check whether *the breathing passage is free.*

Foreign bodies must be immediately removed!

F.2.4
Bleeding Wounds

Minor cuts:

- Allow to bleed, disinfect, and bandage

Major cuts:

- If possible, staunch bleeding (see below), bandage, see physician

Puncture wounds:

- Are *dangerous!* If possible, pull out object; seek medical treatment in all cases.

Finger injuries:

- Remove rings immediately. If the injury is serious, seek medical treatment

Staunching of bleeding:

- Suspend bleeding part as high as possible.
- *Apply pressure with a compress*: form a thick pad of cloth into a compress and apply pressure; tie down with a bandage or necktie.

If bleeding does not stop:

- Wrap a second layer of bandage over the first.
- *Apply additional direct pressure on the wound*: squeeze the artery that supplies blood to the limb or area against the bone with the heel of your hand. If necessary, the thigh artery can be squeezed by maximum bending at the hip.
- In case direct pressure on the artery is not possible, a finger or the fist can be pressed directly into the wound.

To ensure staunching of bleeding:

- Bandage light bleeding (possibly by applying brief finger pressure on the bandage).
- *Use pressure dressing for strong bleeding;* if insufficient, position higher, keep still. Seek medical treatment.

With severe bleeding there is a possibility of traumatic shock!
Consider including this link: http://www.howcast.com/videos/168459-How-To-Stop-Bleeding (accessed May 2012).

F.2.5 Shock

The following *symptoms* point to shock (not always all or at the same time):

- Sweat on the forehead, extreme unrest
- Pale, cold skin (esp. the extremities)
- Pulse becomes faster, then weaker and finally, hardly detectable

If the victim exhibits these symptoms, *the bleeding must first be staunched.* In addition loss of heat should be avoided, the pulse and breathing controlled, and the victim's sense of orientation observed. Seek medical treatment.

F.2.6 Eye Injuries

If *chemicals* come in contact with the eyes, they must be rinsed for at least 10 to 15 minutes with the help of an eye bath or gently flowing lukewarm water jet. A second person should keep the victim's eyes open. If the victim is wearing contact lenses, immediately remove them.

Eyes must be rinsed carefully with long flushing after coming into contact with the following substances:

- hydrofluoric acid
- alkaline compounds (soda, caustic potash solution, ammonia, amines, etc.)

Never try to neutralize substances that have penetrated the eyes with other chemicals!

Bring the victim to the ophthalmic clinic only *after* flushing.

If *foreign bodies* (e.g., glass fragments) penetrate into the eyes:

- The victim must be prevented from rubbing the eyes.
- If necessary, flush eyes as described above.
- Only loosely sitting foreign bodies may be removed, for example, with the corner of a clean cloth, never with tweezers.

Never try to remove imbedded foreign bodies!

- Transport the victim to the eye clinic in a *lying position.*

F.2.7 Burns

After extinguishing fire, immediately remove clothing that does not stick to the skin. Cool burned area of skin with cold water until pain is relieved (20 to 30 minutes). Then seek further medical treatment!

Skin only reddened:
- Repeat cooling of area if pain returns; if large area is affected, seek medical treatment.

Blistering:
- Do not use ointments. Do not perforate blisters. Seek medical treatment.

F.2.8 Caustic Burns

Immediately take off contaminated clothes; do not use them anymore. Rinse the affected area with large quantities of water. Remove organic substances with soap and water, never with solvents (alcohol, etc.). Otherwise, treat like a burn injury.

F.2.9 Poisoning

In case of poisoning, it is critical for correct medical treatment that a sample of the poison (substance, gas sample, vomit, etc.) is brought along!

Oral Poisoning
In principle, oral poisonings can be ruled out with the use of bulbs, pipette boys, etc.
Treatment of oral poisonings is differentiated according to its two substance classes:

1) *Standard poisons* (metal salts, organic substances, insecticide, etc.) and *water-insoluble solvents* (benzene, toluene, petroleum, ether, etc.). Induce vomiting as soon as possible, either by mechanical irritation of the throat wall or by the ingestion of lukewarm salt water

   (approx. 3 heaped teaspoons of salt per glass). Repeat until vomiting stops. Then seek medical treatment.

2) *Corrosive liquids* (acids, caustic solutions, etc.)

   Immediately drink as much water as possible, do not induce vomiting and seek medical assistance.

Important:

- *Do not give anything to drink to unconscious persons; they could suffocate!*
- *Never attempt to neutralize one ingested chemical with another!*

Percutaneous Poisoning
Immediately release the affected skin area and wash well with soap and water; seek medical help.

Gas Poisoning
After rescuing the victim from the danger zone, move him/her to the fresh air, calm the victim and observe their breathing, resuscitating if necessary (see F.2.3) and deliver to hospital by ambulance.

F.3
Working with Chemicals

This chapter helps to prepare you for the correct handling of chemicals and instructs you how to behave safely and recognize sources of danger and therefore avoid accidents (often caused by ignorance or wrong behavior).

- *Never* work *alone* in the laboratory.
- Due to the danger of fire and oral poisonings, *smoking is forbidden everywhere* in the chemistry laboratory.
- *Food and drinks* may not be stored or consumed in the laboratory. Food that is used for experimental purposes in the laboratory should be marked with a label.
- *Visitors* are only allowed into the laboratory with the express permission of the lab supervisor. Always equip them with *visitor safety glasses* for their own safety.

Personal Protective Measures

- Always wear *safety glasses* in the lab.
- If possible, avoid wearing contact lenses.
- Never work without a *laboratory coat.*
- When working with large quantities of corrosive liquids, always wear *closed eye goggles* and use a face shield.
- Use *gloves* when working with corrosive or toxic substances.

F.3.1
Chemicals

Mark each chemical bottle with a *label* that indicates at least the following information:

- Name of the substance
- Molecular formula
- Date of filling
- Name of the responsible person

Write only with water-resistant, black felt-tip pens or ballpoint pens, as other pens become illegible after a short time.

- The purchase of poisons for private purposes as well as any procurement of chemicals for a third party is forbidden.
- A *poison form* signed by the lab supervisor must be submitted for the purchase of certain special chemicals.
- Store only the smallest possible amount of chemicals in the laboratory.
- *Find out* carefully before you begin an experiment about the physical properties and toxic and (possibly dangerous) reaction properties of the chemicals involved.
- Chemical solutions should *never be pipetted with the mouth.* Always use pipetting aids (danger of oral poisoning!).
- For experiments, only use undamaged, *clean containers.* Never use containers that contain leftover chemicals.
- *Food containers* should *not* be used for the storage of chemicals.
- Use the smallest possible quantities for experiments.
- Place chemical containers, bottles, and devices *always as far back as possible* on the work surface (danger of falling off when knocked against).
- When heating materials in an open container, point the opening away from people.
- Use *boiling stones* when heating substances and make sure that the stones are added when the solution is *far* below the boiling point. Otherwise the liquid could suddenly cook over and spray all over (boiling retardation).
- Always immediately remove spilled chemicals and water puddles from the ground (danger of injuries).
- Use caution when *diluting concentrated acids* with water and when preparing solutions consisting of *solid alkali hydroxides* (e.g., KOH, NaOH). Always add acid or base to water while stirring, never in reverse (heat development!).

F.3.2
Solvents

Reduce the *stored supplies* to a minimum (fire risk). At the workbench and in the fume hood, use bottles that contain a maximum amount of 500 ml.

Solvents of the *ether group* (R1-O-R2, e.g., CH_3–O–CH_3) and many hydrocarbons (e.g., butene C_4H_8) form highly explosive peroxides after long standing. Before each handling, control the peroxide content with the Merckoquant peroxide test! If it is higher than 0.01% (=100 ppm, indicated as H_2O_2), the peroxides must be eliminated. Check periodically every 3 months. Do not open a bottle if you observe unusual viscosity or crystallization! Inform the lab supervisor.

When distilling peroxide-forming compounds, leave a small amount in the flask, never evaporate everything! Store distillates protected from light and if possible, with an inhibitor.

F.3.3 Handling of Glass and Glass Equipment

Be careful when working with glass and glass equipment, in particular, when for example, introducing glass tubes into rubber hoses. *Protect your hands* (use gloves). Puncture wounds due to broken-off glass tubes are dangerous!

Thin-walled glass containers, especially measuring and Erlenmeyer flasks should *not be evacuated*. Vacuum desiccators should be covered with parafilm. They should not be transported in an evacuated condition.

Thick-walled vacuum flasks should not be heated (suitable for evacuation).

When transporting *large glass flasks*, never hold by the neck or by the plug. Always *hold at the base* (danger of breaking). They may only be stored on lower shelves.

Never fill glass containers to over ninety percent of filling capacity. To avoid the occurrence of stuck glass plugs, it is advisable to use *screw-on caps* if possible.

Do not force open *jammed glass plugs*. Rather, loosen the plug by tapping it lightly with a wooden stick or slightly warming the outside (at the same time, if possible, cooling the inside).

Dispose of *glass waste*, after cleaning with the appropriate solvent, in their special container. Never dispose of them in the normal waste container (danger of injury to cleaning staff!).

F.3.4 Electrical Apparatus, Heating Sources

Electrical apparatus may not stand within the range of splashing water. Keep it always clean and free of corrosive substances.

Immediately report defective devices including *defective or corroded cables and plugs* to the lab supervisor for repairs. Do not try to repair damage yourself (insulating tape, etc.).

Use *open flames* as a heating source only if there are no other possible options.

Make sure before using the *Bunsen burner* that no flammable materials are in the proximity of the flame. In principle, work in organic laboratories should be done under the fume hood.

Do not use *water baths* to warm up containers with substances that strongly react with water (alkali metals, metal hydrides, organometallic compounds).

F.3.5
Fire Prevention

Find out before beginning work at a laboratory about the locations and the proper uses of the *emergency showers* and the *fire extinguishing equipment* as well as the *escape routes.*

Limit the quantity of *combustible liquids* kept at the laboratory to a minimum and keep them separate from strong oxidizing agents.

For the storage of *flammable liquids that need to be stored cool,* use refrigerators reserved only for this purpose and labeled.

F.3.6
Sources

In the evening, turn off the gas, water, electricity, compressed air, and vacuum, as well as the hood ventilation.

Evening experiments may only be performed with the permission of the supervisor, whereby certain hoses should be secured with clamps and a *night panel* posted.

F.3.7
Fume Hood

All work in which poisonous, flammable or otherwise dangerous or foul-smelling gas, steam, or aerosols could develop or escape should be performed under the fume hood.

The *use of the fume hood* is, nevertheless, *not a carte blanche* for the release of any amount of substances and chemicals. Toxic and corrosive gas and steam, as mentioned in the middle of the references written about devices, should be absorbed directly where they are generated. This prevents disturbance of the environment and costly damage to the ventilation system.

Close the front sash of the fume hood whenever possible.

Equipment should be placed *as far back as possible,* against the back wall of the hood.

Fume hoods equipped with switches to regulate ventilation should, for environmental reasons, only be set at the highest setting when in use.

Otherwise, precious heat would escape unnecessarily.

F.4
Chemical Reactions under Increased Pressure

Certain reactions must be performed under increased pressure in order to for example, raise the solubility of gas in liquid. The increased pressure entails certain additional risks that require special security precautions.

F.4.1 Chemicals

Extra caution is necessary with the following substances:

- Oxygen and nitric oxide (danger of explosion)
- Ethylene oxide (danger of explosion)
- Nitro compounds (danger of explosive decomposition)
- Hydrogen (danger of explosion)
- Acetylene, ammonia, hydrazine (form explosive compounds with copper)
- Halogen (corrosion of steel)
- Strong acids or alkaline media (corrosion)

Moreover, many gases are *toxic*. Find out before the start of a reaction exactly about the reaction properties and the toxicity of the chemicals.

F.4.2 Apparatus

For reactions under increased pressure, use only suitable *devices and containers meant* for this purpose.

Equipment may only be constructed at *places suitable* for this purpose (i.e., high-pressure laboratory).

When *introducing gas into liquids*, a device must be constructed that prevents the return of the liquids into the system or the collection vessel (use a safety flask).

Containers and devices which are placed under great pressure or heated in a closed condition must be equipped, as a rule, with a *manometer* (red mark at maximum pressure) and a *pressure valve* (safety valve and explosion shield). Also consider the gas that forms during the reaction.

If the device is to be heated or cooled, one should take into consideration the *thermal pressures* that develop.

Test that the device is *not leaky* before the reaction.

If experiments are performed with *flammable* gases and fluids that develop flammable vapors, the air present in the device must first be evacuated before the start of the reaction.

Make sure that at the start of the reaction that *no solvent containers* are present in the area.

If there is danger of exceeding the permissible running temperature or pressure, the reaction should be immediately interrupted (turn off heat, emergency cooling).

F.4.3 Working in Clean Rooms

Etching technology is used in micro-mechanics, micro- and quantum electronics and high frequency technology for the production of microscopic tools or electronic components. Because of their small size, they are not mechanically produced but rather etched from raw materials. For this purpose, strong acids and bases are utilized. In order to protect the work pieces from dust and other pollutants, the work is performed in clean rooms. The following are instructions that you must follow for security reasons and in order to ensure quality if you are to perform etching in the clean room.

F.4.3.1 **General Conduct**

Clean rooms may only be entered through the entryway and only in full *clean room gear* (lab coat, lab shoes or overshoes, hood). For visitors, disposable overalls are available for use.

Etching may be performed only after *instruction* by a responsible person. Use only machines which you have been trained to use.

Never work *alone* in the clean room.

Always wear *safety glasses*. When handling large amounts of acid or bases use a face shield and special gloves.

Handle the mask holder and other things that have come into contact with the workpieces only with gloves.

Avoid bringing unnecessary *papers* into the clean room. Never deposit paper in the working area of the flow box.

F.4.3.2 **Handling of Chemicals**

Beakers with *chemicals* must be covered and clearly labeled with at least the following information:

- Name of substance
- Molecular formula
- Date of filling
- Name of responsible persons

Acids and bases used for etching are *highly corrosive*. Before starting work, pay close attention to the first aid instructions for specific chemicals.

Solvents may only be heated on the heating plate under supervision.

Do not clean hot heating plates with solvents.

Use photo varnish only under yellow light.

Dispose of your chemicals after use, following the given guidelines (introduction) and clean your glassware.

F.4.3.3 **Devices**

Check *before turning on devices* that all necessary sources (cold water, compressed air, etc.) are available.

After the end of work, make sure that *devices* and sources *are turned off* properly (take into account cooling time, etc.).

Defective devices including *corroded or defective cables and plugs* must be immediately reported for repairs to a supervisor. Do not try to repair the damage yourself (insulating tape).

F.5
Disposal of Chemicals

The best disposal is to avoid the creation of waste (e.g., through bulk ordering, recycling).

F.5.1
Organic Chemicals

Common Organic Solvents
Chlorinated and unchlorinated are separated as a rule (1–2% chlorinated organic solution is tolerated).

Smaller amounts of chlorinated organic solutions are not worth separating. Aqueous solutions must be neutral.

Organic Solvents with Metal Compounds
As a rule, chlorinated and non-chlorinated are separated.

Disposal: collection container for organic solutions.

Liquid Ether
Disposal: Into normal solution waste, if they do not contain peroxide (peroxide test Merckoquant 10011).

Otherwise, hand in the original bottle with a warning indicating peroxide content.

Do not even try to open old ether bottles!

Used Oil

Small amounts: collection box for non-chlorinated organic solutions
Large amounts: separate collection box

Water–Oil Emulsion
Disposal: collection container

Solid Hydrocarbons, Alcohol, Ketone, Ester
($C_xH_yO_z$) Pack separately in cardboard, paper or plastic (polyethylene).

Disposal: normal garbage.

Explosive Materials That Explodes upon Impact, For Example, Friction
For example, organic perchlorate, certain azo-, nitrocompounds, peroxide (study the warning on the label.)

Gases
Return bottles before the expiry date. Release remaining pressure (ca. 2–10 bar)

Obtain gas only from firms that accept bottles with contents.

Other Chemicals
Pack separately in bottles with maximum 1-liter capacity.

In particular, mark mutagenic and carcinogenic materials (follow instructions on label)!

F.5.2
Inorganic Chemicals

Metal Salt solutions

All < 0.1 M: *no* heavy metal (esp. mercury, chrome, cadmium, lead) and *no* precious section2firstParametals, *no* complexing agents
– Introduce by way of the lab sink under cold running water.
Acid: *no* mercury, arsenic, cyanide, fluoride, sulfide or complexing agents
– Disposal: Collection container
Ammonia-containing: *no* mercury, arsenic or complexing agents
– Disposal: Collection container
Cyanide-containing: *no* mercury, arsenic or complexing agents
– Solutions must be alkaline. Disposal: Collection container
Mercury-containing: *no* complexing agents
– Disposal: Collection container
Arsenic-containing: *no* complexing agents, solutions must be alkaline.
– Disposal: Collection container
Selene-, antimony-, thallium-containing: *no* complexing agents
– Solutions must be alkaline. Disposal: In the arsenic container
With stable metal complexes for example, with EDTA-complexes: *no* mercury, arsenic or cyanide
– Disposal: Collection container
Kjeldahl solutions with mercury catalyst
– Disposal: Collection container

Acids

Chromosulfuric acid
– Disposal: Collection container
Hydrofluoric acid
Poison class 1; handle only with special gloves and face protection.
– Disposal: Collection container
Conc. hydrochloric acid
– Work only with gloves and eye protection. Disposal: Collection container
Conc. sulfuric acid
– Work only with gloves and eye protection. Disposal: Collection container
Other concentrated acids
– Collect separately. Disposal: Collection container
Nitrohydrochloric acid
– Dilute to 35%. Disposal: Collection container

Bases

Concentrated

Collect separately. Use eye shields and gloves when handling. Disposal: Collection container

Up to 0.1 M

– If possible neutralize. Introduce by way of the lab sink under cold running water.

Explosive, flammable or oxidizing material that develops explosive gases with water

Phosphorous
– Fill in glass bottles under water. Disposal station
Metals
– Fill in glass bottles or white tin cans under paraffin oil. Disposal station
Alkali amides
– Fill in glass bottles under toluene. Disposal station
Hydrogen peroxide
– Dilute strongly. Introduce by way of the lab sink under cold running water.

Oxidizing materials, for example, acetylide, azide, chlorate, fulminate, hydride, nitrate, nitrite, perchlorate, peroxide and perchloric acid.
– Pack appropriately. Waste disposal station

Explosive materials, for example, acetylide, azide, chlorate, fulminate, hydride, nitrate, nitrite, perchlorate, peroxide and perchloric acid
Pack appropriately. Waste disposal station

Others

Photographic Fixation Solution
– Disposal: Collection container
Photographic Developing Solution
– Disposal: Collection container
Film
– Disposal station

Gases
Return gas flasks before the storage expiration date! Release remaining pressure in bottle (ca. 2–10 bar)

Other Chemicals
Solid salt, materials that develop acidic bases, etc.
– Pack separately: Disposal station

Used Glass
Separation by color not necessary
For glass without solid remains, rinse once with water, remove lid; label need not be removed. Disposal: Collection container

Batteries
– Collect separately in recycling container

Plastic
Only collect PVC, polystyrene and PET separately
– Recycling container or household garbage

Aluminum, iron, steel
– Collect separately
Separate aluminum drinking cans from other aluminum (recyclable).
– Disposal: Recycling container or disposal station

Carton
Clean and free of plastics and other foreign material; staples are tolerated.
– Disposal: Recycling container

Paper
Clean and free of plastics and other foreign material, staples are tolerated.
– Disposal: Recycling container

Spilled chemicals
Many *slightly volatile compounds* (mostly solvents), upon contact with air, develop *explosive mixtures* that accumulate on the ground because they are heavier than air.

Avoid open flames and ventilate immediately! Unscrew safety valve (phone for help only when outside of the area containing explosive mixture!).

With mixtures that develop *highly toxic or caustic vapors* or *toxic dust*: block off the area, alert fire department.

EMERGENCY PROCEDURES

F.6 Gases

F.6.1 Compressed Gas Bottles with Small Leak

Place under the fume hood. Alert ....

F.6.2 Compressed Gas Bottle with Large Leak

Block off area if toxic, flammable, or oxidizing. Alert .... in every case

F.6.3 Explosive, Flammable or Oxidizing Materials That Develop Flammable Gas When Combined with Water

Avoid open flames. Unscrew the safety valve only when it is not in contact with flammable, oxidizing materials. Never turn off devices immediately (sparks!).

Phosphorus
Cover with wet sand and scoop into a container with water.

Alkali Metal

Block off area. Wait at a safe distance if a reaction with surrounding water follows. If not, cover alkali metal with sand and scoop into a container of paraffin oil.

Wear eye protection and gloves! Do not touch with bare hands under any circumstances!

Others

For example, certain acetylide, azo-compounds, azide, chlorate, fulminate, hydride, nitrite, nitro-compounds, perchlorate and perchloric acid

Block off area. EMERGENCY PROCEDURES

F.7 Liquids

F.7.1 Aqueous

Neutralize, wipe up

F.7.2 Organic

Soak up with suitable binding agent (for non-toxic liquids e.g., paper, wooden chips, otherwise vermiculite and oil binders).

Carbon disulfide: Hand in together with the binding material with a water overlay.
All others: Pack the binding material with absorbent compounds in plastic and hand in.

F.7.3 Mercury

In the first step the elementary mercury must be collected into a suitable container using a small brush or broom. The collected mercury can be brought in this form to the waste disposal center.

F.8 Working with Electricity

First some general remarks about electricity:

Normally, the human body is prevented from coming directly into contact with the flow of electricity due to the insulation of electrical devices. However, incorrect insulation or inappropriate handling of electrical installations can lead to an electrical circuit which can be closed by the human body. There are three frequent ways for current to flow.

1) Hands–Feet (resistance approx. 550 Ohm)
2) Hand–Feet (resistance approx. 750 Ohm)
3) Hand–Hand with isolated ground (resistance approx. 1000 Ohm)

The resistance of the body depends on the electrical path, the humidity of the skin and the articles of clothing, floors, and footwear, among other things.

Since electric current stimulates muscles it has flowed through to contract, it usually causes strong muscle cramping if the current flow is not interrupted. This becomes a problem, in particular, with the respiratory muscles that are stilled. With the heart muscle, whose function is based on electrochemical procedures, externally acting current can lead to heart ventricular fibrillations (irregular, cramp-like contractions of individual muscle fibers), whereby blood transport is practically interrupted and the victim is in a state of mortal danger. Even a current as low as 400 milliamperes can trigger such heart ventricular fibrillations. If the current flow manages to break through before the lapse of a heart period (approx. 0.8 seconds) high current can then be withstood without more extensive damage.

In the following table the possible effects of electric shock are represented according to current amperage and reaction time (threshold values):

- 0.05 mA tingle, perceptible with tongue
- 1 mA tingle, perceptible with finger
- 1 to 15 mA increased tingle, finally start of muscle cramps
- 15 to 20 mA "let go" range (hand cannot longer be removed from grasped conductor)
- 20 to 50 mA increasing muscle cramps, difficulty in breathing, respiratory arrest and in 3 to 4 minutes asphyxiation (if current flow is not interrupted)
- 50 mA ventricular fibrillation after a few seconds possible; death (flicker threshold)
- over 50 mA ventricular fibrillation in (fractions of) seconds; death
- over 3 A severe burns; death

In addition, the current flow has thermal effects. With small entry surfaces (e.g., small wires) currents of 220 V can already lead to burns, which can reach deep into the skin and develop into focal infections. Large burn wounds are even life threatening, due to the associated ample loss of liquid. Particularly during a longer action of direct current, there is also a danger of blood decomposition.

These examples show that even when handling relatively low voltages and currents, a certain degree of caution is necessary. The following instructions should, therefore, be followed when handling technical devices.

F.8.1 General Conduct

- Immediately report any *defects in the cable, plug or apparatus* for repairs to a supervisor. You should not try, under any circumstances, to repair a defect provisionally (insulating tape, etc.).

- Only *expert personnel* (operations electricians, house servicemen) may implement *installations* in the house electricity network.
- Electrical equipment may not be placed within range of *splashing water.*
- *Never bypass or change protective devices* (fuses, safety switches, etc.).
- Apparatus that cannot be grounded for instrumentation reasons must be secured by enough isolating transformers or circuit breakers.
- Use plastics or ceramics for *insulation.* Cardboard and wood are unsuitable for this purpose.
- *Disconnect* electrical devices *from the electrical source* before any manipulation.
- For *high voltage systems,* do not use banana plugs and extension cords with reciprocal plugs. Blank wires and crosspieces are dangerous as they impair the effectiveness of protective grounding!
- Make certain when laying out *extension cords* that nobody can stumble over them (if necessary, use tape to affix them to the floor). For permanent installations, the extension cord may not be longer than 5 meters. Extension cord *rolls* must be *completely unrolled* before operations, in particular if high performance devices are connected. (Rolled-up cables are fire risks because of increased resistance).
- When using *altered devices and plugs* (esp. untested ones), it should be clarified whether suitable security is available (in the ideal case, an FI-switch). Speak with responsible personnel to clarify any uncertainties.
- Always get expert help if you are in doubt about whether the equipment to be used is properly installed and functional.

F.9
Working with High Voltage

This section deals particularly with the use of high voltage for experimental purposes, for example, in the high voltage laboratory. It should not be forgotten, however, that many devices which must deliver high performance of any kind must also be operated with high voltage. This applies for example, to all lasers of class 4, to NMR devices, and to electron microscopes. The following applies to all these devices:

- Before each manipulation for which the housing must be opened, the equipment should be completely removed from the current supply.
- In case the fuse is to be unscrewed, a warning card should be attached to the fuse box so that no one inadvertently screws in the fuse too early! Areas where high voltage is in use and devices that are operated with high voltage must be marked by this danger warning:

F.9.1
General Facts

Find out about proper behavior in case of accidents with electricity before you begin setting up an experiment.

Find out before the execution of experiments with gases about their properties and potential dangers. Also bear in mind that during the experiment other (possibly toxic) compounds could develop.

– Use the red emergency switches only in case of an accident (they immediately disconnect the whole laboratory supply and set off an alarm for the technical staff, so that rapid assistance is ensured).

F.9.2
Experimental Setup

- High voltage outlet parts may be located *only within the enclosed experimental field*. They must be secured and constructed in such a way that no danger can exist outside of the experimental field.
- The entrance to the experimental field must be secured with fences and a blocking door.
- The supports of the grounding rod(s) and all stopper grids and doors must be within the safety area.
- *Before the first operation of the testing equipment*, the function of the safety circuit must be examined beforehand.
- When making *major changes* to the testing equipment, re-examine the security inside and outside of the experimental field.

F.9.3
Operation

- Do not use the safety switch for routine switching off of the testing equipment.
- Before entering the experimental field, make sure that:
- All sources of high voltage are switched off.
- The high voltage circuit has reliable contacts and stays grounded by means of the grounding rod.
- All condensers in the experimental field are unloaded and grounded, especially with direct and alternating current experiments.
- Check before switching on the high voltage testing equipment that:
- All persons have left the experimental field.
- The grounding bar has been removed and hung up in its proper place.
- If there is any uncertainty whatsoever, seek expert advice (e.g., specialized group: high voltage technology).

F.10
Handling of Compressed Gas Bottles and Gas

F.10.1
General Facts

The filling pressure of compressed gases normally amounts to 200 bar. With liquefied gases, the filling pressure corresponds to the steam pressure of the respective connection.

As the temperature rises, the pressure in the compressed gas bottles rises strongly.

F.10.2
Transport

Transport compressed gas bottles only with the valve protection cap in place.

Always use a cart when moving bottles, especially if the regulator is already installed and the valve protective cap cannot be put in place. Secure the gas bottle with a fireproof chain during transport.

Never attempt to roll, tow along, or drag a gas bottle over the ground. Avoid dropping and striking gas bottles against each other.

F.10.3
Storage

Compressed gas bottles may be stored, over long periods of time, only in dry, well-ventilated, fireproof areas.

The storage installations of combustible gases must be explosion-proof, and the area should be clear of any sources of heat or ignition. *The laboratory is not a suitable place for the storage of gas bottles!*

Fittings must be in place to prevent the gas bottles from falling over. It is advisable to separate the bottles according to their type (flammable, corrosive, not flammable).

Empty and full bottles must be stored separately. Empty bottles should be clearly marked (e.g., tape with label).

F.10.4
Valves and Fittings

Only knowledgeable personnel should handle the valves in any way. The gas bottles have distinct threaded connections according to the type of gas they contain.

For the corresponding gas, use only certified pressure regulators. This applies especially for compressed oxygen. Use only absolutely oil-free and fat-free fittings with flame-resistant seals.

F.10.5
At the Place of Use

If the cylinder is only needed for a short time at a specific place, it can be used in a lying position (except for liquid gas). Standing cylinders must at least be secured with a fire-resistant chain. If the cylinder will be used regularly or for a longer period of time, it should be secured with permanently mounted clips, clamps, chains, or similar.

Small gas bottles with a height amounting to not more than 2.5 times the diameter of its base can be kept in a standing position without additional stabilization.

No part of a pressurized gas cylinder may be set at temperatures over 40 °C for liquids and over 60 °C for compressed gases. The location must be selected accordingly (not in the proximity of heaters, hot liquefying baths, etc.). Remember that some gases react violently with certain materials.

When using *corrosive gases* (chlorine, hydrogen fluoride, etc.) the valve must be used frequently in order to avoid blockage. The pressure-regulator should not be left on the bottle unless it is used often. Close the valve after every use and rinse the pressure reduction valve with dry air or nitrogen.

(Source: PowerPoint Presentation: Safety Awards, 2007)
http://www.slideshare.net/jamestl2/top-10-unsafety-awards (accessed May 2012)

Caution is the parent of SAFETY
Proverb

F.11
Risk and Safety Phrases (R/S Phrases)

Refer to the original label for information concerning the danger of the substance (for example, see below). It indicates, on one hand, the hazard symbol and, on the other hand, it uses the so-called R and S phrases.

The R phrases indicate the risks, the S phrases indicate the safety recommendations.

Risk (R) Phrases

- R1: Explosive when dry
- R2: Risk of explosion by shock, friction, fire or other sources of ignition
- R3: Extreme risk of explosion by shock, friction, fire or other sources of ignition
- R4: Forms very sensitive explosive metallic compounds
- R5: Heating may cause an explosion
- R6: Explosive with or without contact with air
- R7: May cause fire
- R8: Contact with combustible material may cause fire
- R9: Explosive when mixed with combustible material
- R10: Flammable
- R11: Highly flammable
- R12: Extremely flammable
- R14: Reacts violently with water
- R15: Contact with water liberates extremely flammable gases
- R16: Explosive when mixed with oxidising substances
- R17: Spontaneously flammable in air
- R18: In use, may form flammable/explosive vapour-air mixture
- R19: May form explosive peroxides
- R20: Harmful by inhalation
- R21: Harmful in contact with skin
- R22: Harmful if swallowed
- R23: Toxic by inhalation
- R24: Toxic in contact with skin
- R25: Toxic if swallowed
- R26: Very toxic by inhalation
- R27: Very toxic in contact with skin
- R28: Very toxic if swallowed
- R29: Contact with water liberates toxic gas
- R30: Can become highly flammable in use
- R31: Contact with acids liberates toxic gas
- R32: Contact with acids liberates very toxic gas
- R33: Danger of cumulative effects
- R34: Causes burns
- R35: Causes severe burns
- R36: Irritating to eyes
- R37: Irritating to respiratory system
- R38: Irritating to skin
- R39: Danger of very serious irreversible effects
- R40: Limited evidence of a carcinogenic effect
- R41: Risk of serious damage to eyes

- R42: May cause senitisation by inhalation
- R43: May cause senitisation by skin contact
- R44: Risk of explosion if heated under confinement
- R45: May cause cancer
- R46: May cause heritable genetic damage
- R48: Danger of serious damage to health by prolonged exposure
- R49: May cause cancer by inhalation
- R50: Very toxic to aquatic organisms
- R51: Toxic to aquatic organisms
- R52: Harmful to aquatic organisms
- R53: May cause long-term adverse effects in the aquatic environment
- R54: Toxic to flora
- R55: Toxic to fauna
- R56: Toxic to soil organisms
- R57: Toxic to bees
- R58: May cause long-term adverse effects in the environment
- R59: Dangerous for the ozone layer
- R60: May impair fertility
- R61: May cause harm to the unborn child
- R62: Possible risk of impaired fertility
- R63: Possible risk of harm to the unborn child
- R64: May cause harm to breast-fed babies
- R65: Harmful: may cause lung damage if swallowed
- R66: Repeated exposure may cause skin dryness or cracking
- R67: Vapors may cause drowsiness and dizziness
- R68: Possible risk of irreversible effects

Combinations

- R14/15: Reacts violently with water, liberating extremely flammable gases
- R15/29: Contact with water liberates toxic, extremely flammable gases
- R20/21: Harmful by inhalation and in contact with skin
- R20/22: Harmful by inhalation and if swallowed
- R20/21/22: Harmful by inhalation, in contact with skin and if swallowed
- R21/22: Harmful in contact with skin and if swallowed
- R23/24: Toxic by inhalation and in contact with skin
- R23/25: Toxic by inhalation and if swallowed
- R23/24/25: Toxic by inhalation, in contact with skin and if swallowed
- R24/25: Toxic in contact with skin and if swallowed
- R26/27: Very toxic by inhalation and in contact with skin
- R26/28: Very toxic by inhalation and if swallowed
- R26/27/28: Very toxic by inhalation, in contact with skin and if swallowed
- R27/28: Very toxic in contact with skin and if swallowed
- R36/37: Irritating to eyes and respiratory system
- R36/38: Irritating to eyes and skin
- R36/37/38: Irritating to eyes, respiratory system and skin

- R37/38: Irritating to respiratory system and skin
- R39/23: Toxic: danger of very serious irreversible effects through inhalation
- R39/24: Toxic: danger of very serious irreversible effects in contact with skin
- R39/25: Toxic: danger of very serious irreversible effects if swallowed
- R39/23/24: Toxic: danger of very serious irreversible effects through inhalation and in contact with skin
- R39/23/25: Toxic: danger of very serious irreversible effects through inhalation and if swallowed
- R39/24/25: Toxic: danger of very serious irreversible effects in contact with skin and if swallowed
- R39/23/24/25: Toxic: danger of very serious irreversible effects through inhalation, in contact with skin and if swallowed
- R39/26: Very Toxic: danger of very serious irreversible effects through inhalation
- R39/27: Very Toxic: danger of very serious irreversible effects in contact with skin
- R39/28: Very Toxic: danger of very serious irreversible effects if swallowed
- R39/26/27: Very Toxic: danger of very serious irreversible effects through inhalation and in contact with skin
- R39/26/28: Very Toxic: danger of very serious irreversible effects through inhalation and if swallowed
- R39/27/28: Very Toxic: danger of very serious irreversible effects in contact with skin and if swallowed
- R39/26/27/28: Very Toxic: danger of very serious irreversible effects through inhalation, in contact with skin and if swallowed
- R42/43: May cause sensitization by inhalation and skin contact
- R48/20: Harmful: danger of serious damage to health by prolonged exposure through inhalation
- R48/21: Harmful: danger of serious damage to health by prolonged exposure in contact with skin
- R48/22: Harmful: danger of serious damage to health by prolonged exposure if swallowed
- R48/20/21: Harmful: danger of serious damage to health by prolonged exposure through inhalation and in contact with skin
- R48/20/22: Harmful: danger of serious damage to health by prolonged exposure through inhalation and if swallowed
- R48/21/22: Harmful: danger of serious damage to health by prolonged exposure in contact with skin and if swallowed
- R48/20/21/22: Harmful: danger of serious damage to health by prolonged exposure through inhalation, in contact with skin and if swallowed
- R48/23: Toxic: danger of serious damage to health by prolonged exposure through inhalation
- R48/24: Toxic: danger of serious damage to health by prolonged exposure in contact with skin

- R48/25: Toxic: danger of serious damage to health by prolonged exposure if swallowed
- R48/23/24: Toxic: danger of serious damage to health by prolonged exposure through inhalation and in contact with skin
- R48/23/25: Toxic: danger of serious damage to health by prolonged exposure through inhalation and if swallowed
- R48/24/25: Toxic: danger of serious damage to health by prolonged exposure in contact with skin and if swallowed
- R48/23/24/25: Toxic: danger of serious damage to health by prolonged exposure through inhalation, in contact with skin and if swallowed
- R50/53: Very toxic to aquatic organisms, may cause long-term adverse effects in the aquatic environment
- R51/53: Toxic to aquatic organisms, may cause long-term adverse effects in the aquatic environment
- R52/53: Harmful to aquatic organisms, may cause long-term adverse effects in the aquatic environment
- R68/20: Harmful: possible risk of irreversible effects through inhalation
- R68/21: Harmful: possible risk of irreversible effects in contact with skin
- R68/22: Harmful: possible risk of irreversible effects if swallowed
- R68/20/21: Harmful: possible risk of irreversible effects through inhalation and in contact with skin
- R68/20/22: Harmful: possible risk of irreversible effects through inhalation and if swallowed
- R68/21/22: Harmful: possible risk of irreversible effects in contact with skin and if swallowed
- R68/20/21/22: Harmful: possible risk of irreversible effects through inhalation, in contact with skin and if swallowed

R-phrases no longer in use

- R13: Extremely flammable liquefied gas
- R47: May cause birth defects

Safety-(S)-phrases

- (S1): Keep locked up
- (S2): Keep out of the reach of children
- S3: Keep in a cool place
- S4: Keep away from living quarters
- S5: Keep contents under . . . (*appropriate liquid to be specified by the manufacturer*)
- S6: Keep under . . . (*inert gas to be specified by the manufacturer*)
- S7: Keep container tightly closed
- S8: Keep container dry
- S9: Keep container in a well-ventilated place
- S10: Keep contents wet
- S11: *Not specified*

- S12: Do not keep the container sealed
- S13: Keep away from food, drink and animal foodstuffs
- S14: Keep away from . . . (*incompatible materials to be indicated by the manufacturer*)
- S15: Keep away from heat
- S16: Keep away from sources of ignition – No smoking
- S17: Keep away from combustible material
- S18: Handle and open container with care
- S20: When using do not eat or drink
- S21: When using do not smoke
- S22: Do not breathe dust
- S23: Do not breathe gas/fumes/vapour/spray (*appropriate wording to be specified by the manufacturer*)
- S24: Avoid contact with skin
- S25: Avoid contact with eyes
- S26: In case of contact with eyes, rinse immediately with plenty of water and seek medical advice
- S27: Take off immediately all contaminated clothing
- S28: After contact with skin, wash immediately with plenty of . . . (*to be specified by the manufacturer*)
- S29: Do not empty into drains
- S30: Never add water to this product
- S33: Take precautionary measures against static discharges
- S35: This material and its container must be disposed of in a safe way
- S36: Wear suitable protective clothing
- S37: Wear suitable gloves
- S38: In case of insufficient ventilation wear suitable respiratory equipment
- S39: Wear eye/face protection
- S40: To clean the floor and all objects contaminated by this material use . . . (*to be specified by the manufacturer*)
- S41: In case of fire and/or explosion do not breathe fumes
- S42: During fumigation/spraying wear suitable respiratory equipment (*appropriate wording to be specified by the manufacturer*)
- S43: In case of fire use . . . (*indicate in the space the precise type of fire-fighting equipment. If water increases the risk add –* ***Never use water***)
- S45: In case of accident or if you feel unwell seek medical advice immediately (show the label where possible)
- S46: If swallowed, seek medical advice immediately and show this container or label
- S47: Keep at temperature not exceeding . . . °C (*to be specified by the manufacturer*)
- S48: Keep wet with . . . (*appropriate material to be specified by the manufacturer*)
- S49: Keep only in the original container
- S50: Do not mix with . . . (*to be specified by the manufacturer*)

- S51: Use only in well-ventilated areas
- S52: Not recommended for interior use on large surface areas
- S53: Avoid exposure–obtain special instructions before use
- S56: Dispose of this material and its container at hazardous or special waste collection point
- S57: Use appropriate containment to avoid environmental contamination
- S59: Refer to manufacturer/supplier for information on recovery/recycling
- S60: This material and its container must be disposed of as hazardous waste
- S61: Avoid release to the environment. Refer to special instructions/safety data sheet
- S62: If swallowed, do not induce vomiting: seek medical advice immediately and show this container or label where possible
- S63: In case of accident by inhalation: remove casualty to fresh air and keep at rest
- S64: If swallowed, rinse mouth with water (only if the person is conscious)

Combinations

- S1/2: Keep locked up and out of the reach of children
- S3/7: Keep container tightly closed in a cool place
- S3/7/9: Keep container tightly closed in a cool, well-ventilated place
- S3/9/14: Keep in a cool, well-ventilated place away from . . . (*incompatible materials to be indicated by the manufacturer*)
- S3/9/14/49: Keep only in the original container in a cool, well-ventilated place away from . . . (*incompatible materials to be indicated by the manufacturer*)
- S3/9/49: Keep only in the original container in a cool, well-ventilated place
- S3/14 Keep in a cool place away from . . . (*incompatible materials to be indicated by the manufacturer*)
- S7/8: Keep container tightly closed and dry
- S7/9: Keep container tightly closed and in a well-ventilated place
- S7/47: Keep container tightly closed and at temperature not exceeding . . . °C (*to be specified by the manufacturer*)
- S8/10: Keep container wet, but keep the contents dry
- S20/21: When using do not eat, drink or smoke
- S24/25: Avoid any inhalation, contact with skin and eyes. Wear suitable protective clothing and gloves
- S27/28: After contact with skin, take off immediately all contaminated clothing, and wash immediately with plenty of . . . (*to be specified by the manufacturer*)
- S29/35: Do not empty into drains; dispose of this material and its container in a safe way
- S29/56: Do not empty into drains, dispose of this material and its container at hazardous or special waste collection point
- S36/37: Wear suitable protective clothing and gloves
- S36/37/39: Wear suitable protective clothing, gloves and eye/face protection
- S36/39: Wear suitable protective clothing and eye/face protection

- S37/39: Wear suitable gloves and eye/face protection
- S47/49: Keep only in the original container at temperature not exceeding . . . °C (*to be specified by the manufacturer*)

F.12
GHS (Globally Harmonized System of Classification and Labeling of Chemicals)

Since 2009 labeling of chemicals has been carried out in accordance with GHS. The old warning symbols have been replaced by new hazard pictograms, so that the validity of R-and S-sentences is limited to the transitional periods defined in GHS.

GHS, the system promoted by the United Nations (UN) for the classification and labeling of chemicals, is an abbreviation for "Globally Harmonized System of Classification and Labeling of Chemicals". Its classification criteria are designed to enable the hazards of chemicals to be communicated worldwide on labels and safety data sheets, using the same symbols and hazard and safety instructions. This should facilitate international trade and above all provide better protection for people in countries that do not have their own classification and labeling system. The GHS adopted at the UN level is a framework system. GHS is to be introduced in stages in the EU using the regulation on classification, labeling and packaging of substances and mixtures approved by the Parliament and Council, the so-called Classification, Labeling and Packaging (*CLP*); Regulation (EC) No 1272/2008). Substances had to be classified and labeled according to GHS by 01.12.2010, while the deadline for preparations (new mixtures) is 01/06/2015. GHS was first used in 2009.

F.12.1
Principles of the GHS

The different types of hazards which can be caused by substances, mixtures or products (items) are divided into hazard classes:

- Physical hazards: 16 hazard classes
- Health hazards: 10 hazard classes
- Environmental hazards: 2 hazard classes

Within the hazard classes, further classification into hazard categories is implemented, depending on the extent/severity of the effects (depending on the hazard class there are between 1 and 7 categories).

On the basis of their classification into hazard classes, products are labeled with corresponding pictograms, signal words, hazard warnings and safety instructions.

Further Information

- Globally Harmonized System of Classification and Labeling of Chemicals (GHS)

Third revised edition
http://live.unece.org/trans/danger/publi/ghs/ghs_rev03/03files_e.html (accessed May 2012).

- Website of the United Nations Economic Commission for Europe (UNECE) http://www.unece.org/trans/danger/publi/ghs/ghs_welcome_e.html (accessed May 2012).
- A Guide to The Globally Harmonized System of Classification and Labeling of Chemicals (GHS)
http://www.osha.gov/dsg/hazcom/ghs.html (accessed May 2012).

F.13
GHS Pictograms

Physical Hazards

http://www.unece.org/fileadmin/DAM/trans/danger/publi/ghs/GHS_presentations/English/phys_haz_e.pdf (accessed May 2012).

New pictogram	Hazard statement	Old pictogram
GHS 01 	Includes explosive substances and mixtures, explosive articles, and substances as well as mixtures and articles that are manufactured to produce a practical explosive or pyrotechnic effect. *H200, H201, H202, H203, H204, H240, H241* Explosive substances and mixtures – Solid or liquid substance or mixture which, by chemical reaction, is in itself capable of producing gas at such a temperature and pressure and at such a speed as to cause damage to the surroundings. Pyrotechnic substance/mixture – Substance/mixture designed to produce an effect by heat, light, sound, gas, or smoke or a combination of these as the result of non-detonative self-sustaining exothermic chemical reactions. Explosive article – Article containing one or more explosive substance or mixtures. Pyrotechnic article – Article containing one or more pyrotechnic substances or mixtures. Examples: picric acid, TNT	 **E**

New pictogram	Hazard statement	Old pictogram
GHS 02 	Extremely flammable gases, liquids, aerosols and solids; H220, H222, H223 H224, H225, H226, H228 – *Gases* ignitable when in mixture of = ≤ 13% in air or flammable range with air of ≥12% (regardless of the lower flammable limit). -*Aerosols*- this means aerosol dispensers–are any non-refillable receptacles made of metal, glass or plastics and containing a gas compressed, liquefied or dissolved under pressure, with or without a liquid, paste of powder, and fitted with a release device allowing the contents to be ejected as solid or liquid particles in suspension in a gas, as a foam, paste or powder or in a liquid state or in a gaseous state. – *Extremely flammable liquid and vapor*–Flashpoint <23 °C and initial boiling point = 35 °C. – *Highly flammable liquid and vapor*–Flashpoint <23 °C and initial boiling point >35 °C. – *Flammable liquid and vapor*–Flashpoint ≥ 23 °C and = 60 °C. Other groups: – Substances and mixtures which, in contact with water emit flammable gases, H260, H261 – Spontaneous and self-reactive liquids and solids; H250 – Self-heating substances and mixtures; H251, H252 – Organic peroxides, H241, H242 Examples: propane, butane, ether, acetaldehyde	 F  F+
GHS 03 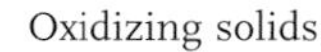 	Oxidizing solids, liquids and gases, *H270, H271, H272* – A *solid* which, while in itself is not necessarily combustible, may, generally by yielding oxygen, cause or contribute to the combustion of other material. Test results are assessed on the basis of (i) whether the mixture substance/cellulose ignites and burns; and (ii) the comparison of the mean burning time with those of the reference mixtures. – A *liquid* which, while in itself is not necessarily combustible, may, generally by yielding oxygen, cause, or contribute to, the combustion of other material. Test results are assessed on the basis of (i) whether the mixture substance/cellulose spontaneously ignites; and (ii) the comparison of the mean time taken for the pressure to rise from 690 kPa to 2070 KPa (gauge), with the mean times of the reference substances. – *Gases* that may, generally by providing oxygen, cause or contribute to the combustion of other material more than air does. Examples: oxygen, chlorine dioxide, hypochlorite	 O

(Continued)

New pictogram	Hazard statement	Old pictogram
GHS 04	Gases and gas mixtures – Compressed gases (under pressure), *H280* – Liquefied gases; *H280* – Dissolved gases; *H280* – Refrigerated liquefied gases; *H281* Examples: Compressed gas cylinders, LPG	–
GHS 05	Substances and mixtures which react chemically with metals, damaging materially or even destroying them (corrosion), H290	–

Health Hazards

http://www.unece.org/fileadmin/DAM/trans/danger/publi/ghs/GHS_presentations/English/health_env_e.pdf

New pictogram	Hazard statement	Old pictogram
GHS 05	Substances and mixtures that cause the following damage to health: – Burns (skin and irreversible tissue damage), *H314* – Serious damage to eyes, *H314, H318* Examples: hydrochloric acid, caustic soda, hydrofluoric acid	C
GHS 06	Chemicals that are fatal after being inhaled, swallowed or absorbed through the skin, already in small quantities, *Acute Toxicity – Oral* Category 1: LD_{50} = 5 mg/kg bodyweight Category 2: $LD_{50} \geq 5$ but <50 mg/kg bodyweight *Acute Toxicity – Dermal* Category 1: LD_{50} = 50 mg/kg bodyweight Category 2: $LD_{50} \geq 50$ but <200 mg/kg bodyweight *Acute Toxicity – Inhalation* Category 1: Inhalation (vapor) LD_{50} = 0.5 mg/l, or inhalation (dust/mist) LC_{50} = 0.05 mg/l.	T T+

New pictogram	Hazard statement	Old pictogram
	Category 2: Inhalation (gas) LC_{50} >100 but <500 ppm, or inhalation (vapor) LC_{50} >0.5 but <2.0 mg/l, or inhalation (dust/mist) LC_{50} >0.05 but = 0.5 mg/l. *H300, H301, H310, H311, H330, H331* Examples: Hydrofluoric acid, bromine, hydrogen cyanide	
GHS 07	Substances and mixtures which are less extremely hazardous to health with the following properties: – Acutely harmful if inhaled, swallowed or absorbed through the skin, *H302, H312, H332* – Irritant to skin or eyes, *H315, H319* – Causing allergic skin reactions (skin sensitization), *H317* – Irritation of the airways; *H335* – Numbing effect; *H336* Examples: hydrocarbons, limonene	Xi Xn
GHS 08	Substances and mixtures with various organ-specific toxic effects or long term health hazards. Properties: – Carcinogenic, mutagenic or toxic to reproduction (CMR), *H340, H341, H350, H351, H360, H361* – Specific, reversible or irreversible non-lethal effects on human health (organs) after single or prolonged exposure, *H370, H371, H372, H373* – Liquids which after ingestion cause severe lung damage (aspiration), *H304* – Substances that when inhaled may cause allergies or respiratory problems or sensitization of the airway; *H334* Examples: benzene, petroleum, isocyanates, methanol	Xn T or Xn

Environment

New pictogram	Hazard statement	Old pictogram
GHS 09	Substances and mixtures that cause acute and/or long-term adverse effects to aquatic organisms, for example, – Acute, *H400* – Chronic, *H410, H411*	N

(Continued)

New pictogram	Hazard statement	Old pictogram
	Very toxic to aquatic life 96 hr LC_{50} (fish) = 1 mg/l, and/or 48 hr EC_{50} (crustacea) = 1 mg/l, and/or 72 or 96 hr ErC_{50} (algae or other aquatic plants) = 1 mg/l. *Toxic to aquatic life* 96 hr LC_{50} (fish) = 10 mg/l, and/or 48 hr EC_{50} (crustacea) = 10 mg/l, and/or 72 or 96 hr EC_{50}(algae or other aquatic plants) = 10 mg/l. *Harmful to aquatic life* 96 hr LC_{50} (fish) = 100 mg/l, and/or 48 hr EC50 (crustacea) = 100 mg/l, and/or 72 or 96 hr EC_{50}(algae or other aquatic plants) = 100 mg/l. Examples: potassium hypochlorite solution (Javelle water), various insecticides, ammonia	

Hazard Statements

- *H2xx Physical hazards*
- *H3xx Health hazards*
- *H4xx Environmental hazards*
- *EUHxxx Country-specific hazards (European Union)*

Physical Hazards

- H200: Unstable explosive
- H201: Explosive; mass explosion hazard
- H202: Explosive; severe projection hazard
- H203: Explosive; fire, blast or projection hazard
- H204: Fire or projection hazard
- H205: May mass explode in fire
- H220: Extremely flammable gas
- H221: Flammable gas
- H222: Extremely flammable material
- H223: Flammable material
- H224: Extremely flammable liquid and vapor
- H225: Highly flammable liquid and vapor
- H226: Flammable liquid and vapor
- H227: Combustible liquid
- H228: Flammable solid
- H240: Heating may cause an explosion
- H241: Heating may cause a fire or explosion
- H242: Heating may cause a fire

- H250: Catches fire spontaneously if exposed to air
- H251: Self-heating; may catch fire
- H252: Self-heating in large quantities; may catch fire
- H260: In contact with water releases flammable gases which may ignite spontaneously
- H261: In contact with water releases flammable gas
- H270: May cause or intensify fire; oxidizer
- H271: May cause fire or explosion; strong oxidizer
- H272: May intensify fire; oxidizer
- H280: Contains gas under pressure; may explode if heated
- H281: Contains refrigerated gas; may cause cryogenic burns or injury
- H290: May be corrosive to metals

Health hazards

- H300: Fatal if swallowed
- H301: Toxic if swallowed
- H302: Harmful if swallowed
- H303: May be harmful if swallowed
- H304: May be fatal if swallowed and enters airways
- H305: May be harmful if swallowed and enters airways
- H310: Fatal in contact with skin
- H311: Toxic in contact with skin
- H312: Harmful in contact with skin
- H313: May be harmful in contact with skin
- H314: Causes severe skin burns and eye damage
- H315: Causes skin irritation
- H316: Causes mild skin irritation
- H317: May cause an allergic skin reaction
- H318: Causes serious eye damage
- H319: Causes serious eye irritation
- H320: Causes eye irritation
- H330: Fatal if inhaled
- H331: Toxic if inhaled
- H332: Harmful if inhaled
- H333: May be harmful if inhaled
- H334: May cause allergy or asthma symptoms or breathing difficulties if inhaled
- H335: May cause respiratory irritation
- H336: May cause drowsiness or dizziness
- H340: May cause genetic defects
- H341: Suspected of causing genetic defects
- H350: May cause cancer
- H351: Suspected of causing cancer
- H360: May damage fertility or the unborn child
- H361: Suspected of damaging fertility or the unborn child

- H362: May cause harm to breast-fed children
- H370: Causes damage to organs
- H371: May cause damage to organs
- H372: Causes damage to organs through prolonged or repeated exposure
- H373: May cause damage to organs through prolonged or repeated exposure

Environmental hazards

- H400: Very toxic to aquatic life
- H401: Toxic to aquatic life
- H402: Harmful to aquatic life
- H410: Very toxic to aquatic life with long lasting effects
- H411: Toxic to aquatic life with long lasting effects
- H412: Harmful to aquatic life with long lasting effects
- H413: May cause long lasting harmful effects to aquatic life

Country-specific hazard statements

European Union
The European Union has implemented the GHS through the CLP Regulation. Nevertheless, the older system based on the Dangerous Substances Directive will continue to be used in parallel until 2016. Some R phrases which do not have simple equivalents under the GHS have been retained under the CLP Regulation:

Physical properties

- EUH001: Explosive when dry
- EUH006: Explosive with or without contact with air
- EUH014: Reacts violently with water
- EUH018: In use may form flammable/explosive vapor-air mixture
- EUH019: May form explosive peroxides
- EUH044: Risk of explosion if heated under confinement

Health properties

- EUH029: Contact with water liberates toxic gas
- EUH031: Contact with acids liberates toxic gas
- EUH032: Contact with acids liberates very toxic gas
- EUH066: Repeated exposure may cause skin dryness or cracking
- EUH070: Toxic by eye contact
- EUH071: Corrosive to the respiratory tract

Environmental properties

- EUH059: Hazardous to the ozone layer

Other EU hazard statements
Some other hazard statements intended for use in very specific circumstances have also been retained under the CLP Regulation. Note that in this case the numbering of the EU specific hazard statements can coincide with GHS hazard statements if the "EU" prefix is not included.

- EUH201: Contains lead. Should not be used on surfaces liable to be chewed or sucked by children.
- EUH201A: Warning! Contains lead.
- EUH202: Cyanoacrylate. Danger. Bonds skin and eyes in seconds. Keep out of the reach of children.
- EUH203: Contains chromium (VI). May produce an allergic reaction.
- EUH204: Contains isocyanates. May produce an allergic reaction.
- EUH205: Contains epoxy constituents. May produce an allergic reaction.
- EUH206: Warning! Do not use together with other products. May release dangerous gases (chlorine).
- EUH207: Warning! Contains cadmium. Dangerous fumes are formed during use. See information supplied by the manufacturer. Comply with the safety instructions.
- EUH208: Contains <name of sensitizing substance>. May produce an allergic reaction.
- EUH209: Can become highly flammable in use.
- EUH209A: Can become flammable in use.
- EUH210: Safety data sheet available on request.
- EUH401: To avoid risks to human health and the environment, comply with the instructions for use.

F.13.1 Precautionary Statements

- P1xx General precautionary statements
- P2xx Prevention precautionary statements
- P3xx Response precautionary statements
- P4xx Storage precautionary statements
- P5xx Disposal precautionary statements

General precautionary statements

- P101: If medical advice is needed, have product container or label at hand
- P102: Keep out of reach of children
- P103: Read label before use

Prevention precautionary statements

- P201: Obtain special instructions before use
- P202: Do not handle until all safety precautions have been read and understood
- P210: Keep away from heat/sparks/open flames/hot surfaces – No smoking
- P211: Do not spray on an open flame or other ignition source

- P220: Keep/Store away from clothing/.../combustible materials
- P221: Take any precaution to avoid mixing with combustibles
- P222: Do not allow contact with air
- P223: Keep away from any possible contact with water, because of violent reaction and possible flash fire
- P230: Keep wetted with . . .
- P231: Handle under inert gas
- P232: Protect from moisture
- P233: Keep container tightly closed
- P234: Keep only in original container
- P235: Keep cool
- P240: Ground/bond container and receiving equipment
- P241: Use explosion-proof electrical/ventilating/light/.../equipment
- P242: Use only non-sparking tools
- P243: Take precautionary measures against static discharge
- P244: Keep reduction valves free from grease and oil
- P250: Do not subject to grinding/shock/.../friction
- P251: Pressurized container – Do not pierce or burn, even after use
- P260: Do not breathe dust/fume/gas/mist/vapors/spray
- P261: Avoid breathing dust/fume/gas/mist/vapors/spray
- P262: Do not get in eyes, on skin, or on clothing
- P263: Avoid contact during pregnancy/while nursing
- P264: Wash . . . thoroughly after handling
- P270: Do not eat, drink or smoke when using this product
- P271: Use only outdoors or in a well-ventilated area
- P272: Contaminated work clothing should not be allowed out of the workplace
- P273: Avoid release to the environment
- P280: Wear protective gloves/protective clothing/eye protection/face protection
- P281: Use personal protective equipment as required
- P282: Wear cold insulating gloves/face shield/eye protection
- P283: Wear fire/flame resistant/retardant clothing
- P284: Wear respiratory protection
- P285: In case of inadequate ventilation wear respiratory protection
- P231 + 232: Handle under inert gas. Protect from moisture
- P235 + 410: Keep cool. Protect from sunlight

Response precautionary statements

- P301: IF SWALLOWED:
- P302: IF ON SKIN:
- P303: IF ON SKIN (or hair):
- P304: IF INHALED:
- P305: IF IN EYES:
- P306: IF ON CLOTHING:

- P307: If exposed:
- P308: If exposed or concerned:
- P309: If exposed or you feel unwell:
- P310: Immediately call a POISON CENTER or doctor/physician
- P311: Call a POISON CENTER or doctor/physician
- P312: Call a POISON CENTER or doctor/physician if you feel unwell
- P313: Get medical advice/attention
- P314: Get medical advice/attention if you feel unwell
- P315: Get immediate medical advice/attention
- P320: Specific treatment is urgent (see . . . on this label)
- P321: Specific treatment (see . . . on this label)
- P322: Specific measures (see . . . on this label)
- P330: Rinse mouth
- P331: Do NOT induce vomiting
- P332: If skin irritation occurs:
- P333: If skin irritation or a rash occurs:
- P334: Immerse in cool water/wrap in wet bandages
- P335: Brush off loose particles from skin
- P336: Thaw frosted parts with lukewarm water. Do not rub affected areas
- P337: If eye irritation persists:
- P338: Remove contact lenses if present and easy to do. continue rinsing
- P340: Remove victim to fresh air and keep at rest in a position comfortable for breathing
- P341: If breathing is difficult, remove victim to fresh air and keep at rest in a position comfortable for breathing
- P342: If experiencing respiratory symptoms:
- P350: Gently wash with soap and water
- P351: Rinse continuously with water for several minutes
- P352: Wash with soap and water
- P353: Rinse skin with water/shower
- P360: Rinse immediately contaminated clothing and skin with plenty of water before removing clothes
- P361: Remove/Take off immediately all contaminated clothing
- P362: Take off contaminated clothing and wash before reuse
- P363: Wash contaminated clothing before reuse
- P370: In case of fire:
- P371: In case of major fire and large quantities:
- P372: Explosion risk in case of fire
- P373: DO NOT fight fire when fire reaches explosives
- P374: Fight fire with normal precautions from a reasonable distance
- P375: Fight fire remotely due to the risk of explosion
- P376: Stop leak if safe to do so
- P377: Leaking gas fire–do not extinguish unless leak can be stopped safely
- P378: Use . . . . for extinction
- P380: Evacuate area

- P381: Eliminate all ignition sources if safe to do so
- P391: Collect spillage
- P301 + 310: IF SWALLOWED: Immediately call a POISON CENTER or doctor/physician
- P301 + 312: IF SWALLOWED: Call a POISON CENTER or doctor/physician if you feel unwell
- P301 + 330 + 331: IF SWALLOWED: Rinse mouth. Do NOT induce vomiting
- P302 + 334: IF ON SKIN: Immerse in cool water/wrap in wet bandages
- P302 + 350: IF ON SKIN: Gently wash with soap and water
- P302 + 352: IF ON SKIN: Wash with soap and water
- P303 + 361 + 353: IF ON SKIN (or hair): Remove/Take off immediately all contaminated clothing. Rinse skin with water/shower
- P304 + 312: IF INHALED: Call a POISON CENTER or doctor/physician if you feel unwell
- P304 + 340: IF INHALED: Remove victim to fresh air and keep at rest in a position comfortable for breathing
- P304 + 341: IF INHALED: If breathing is difficult, remove victim to fresh air and keep at rest in a position comfortable for breathing
- P305 + 351 + 338: IF IN EYES: Rinse continuously with water for several minutes. Remove contact lenses if present and easy to do–continue rinsing
- P306 + 360: IF ON CLOTHING: Rinse immediately contaminated clothing and skin with plenty of water before removing clothes
- P307 + 311: IF exposed: Call a POISON CENTER or doctor/physician
- P308 + 313: IF exposed or concerned: Get medical advice/attention
- P309 + 311: IF exposed or you feel unwell: Call a POISON CENTER or doctor/physician
- P332 + 313: If skin irritation occurs: Get medical advice/attention
- P333 + 313: If skin irritation or a rash occurs: Get medical advice/attention
- P335 + 334: Brush off loose particles from skin. Immerse in cool water/wrap in wet bandages
- P337 + 313: Get medical advice/attention
- P342 + 311: Call a POISON CENTER or doctor/physician
- P370 + 376: In case of fire: Stop leak if safe to do so
- P370 + 378: In case of fire: Use . . . for extinction
- P370 + 380: In case of fire: Evacuate area
- P370 + 380 + 375: In case of fire: Evacuate area. Fight fire remotely due to the risk of explosion
- P371 + 380 + 375: In case of major fire and large quantities: Evacuate area. Fight fire remotely due to the risk of explosion

Storage precautionary statements

- P401: Store . . .
- P402: Store in a dry place
- P403: Store in a well-ventilated place

- P404: Store in a closed container
- P405: Store locked up
- P406: Store in a corrosive resistant/ . . . container with a resistant inner liner
- P407: Maintain air gap between stacks/pallets
- P410: Protect from sunlight
- P411: Store at temperatures not exceeding . . . °C/ . . . °F
- P412: Do not expose to temperatures exceeding 50 °C/122 °F
- P420: Store away from other materials
- P422: Store contents under . . . .
- P402 + 404: Store in a dry place. Store in a closed container
- P403 + 233: Store in a well-ventilated place. Keep container tightly closed
- P403 + 235: Store in a well-ventilated place. Keep cool
- P410 + 403: Protect from sunlight. Store in a well-ventilated place
- P410 + 412: Protect from sunlight. Do not expose to temperatures exceeding 50 °C/122 °F
- P411 + 235: Store at temperatures not exceeding . . . °C/ . . . °F. Keep cool

Disposal precautionary statements

- P501: Dispose of contents/container to . . .

Index

Practical Instrumental Analysis: Methods, Quality Assurance and Laboratory Management, First Edition.
Sergio Petrozzi.

m

n